GIS: A Visual Approach

Second Edition

Bruce E. Davis

ONWORD PRESS
THOMSON LEARNING

Australia Canada Mexico Singapore United Kingdom United States

GIS: A Visual Approach

by Bruce E. Davis

Publisher:
Alar Elken

Executive Editor:
Sandy Clark

Acquisitions Editor:
James Gish

Managing Editor:
Carol Leyba

Development Editor:
Daril Bentley

Editorial Assistant:
Jaimie Wetzel

Executive Marketing Manager:
Maura Theriault

Executive Production Manager:
Mary Ellen Black

Production Manager:
Larry Main

Manufacturing Coordinator:
Betsy Hough

Technology Project Manager:
David Porush

Cover Design:
Cammi Noah

Printed in Canada
10 9 8 7 6 5 4 3 2 1

For more information, contact
OnWord Press
An imprint of Thomson Learning
Box 15-015
Albany, New York 12212-15015

Or find us on the World Wide Web at
http://www.onwordpress.com

Library of Congress
Cataloging-in-Publication Data
Davis, Bruce Ellsworth.
GIS : a visual approach / Bruce E. Davis.— 2nd ed.
p. cm.
ISBN 0-7668-2764-X (alk. paper)
1. Geographical information systems.
I. Title.
G70.212 .D38 2001
910'.285—dc21
00-52290

About the Author

Bruce Davis is a geographer, with a B.A. from the University of California, Santa Barbara, M.S. from the University of Southern Mississippi, and Ph.D. from the University of California, Los Angeles. He has taught in various schools throughout the United States, both as an eclectic geographer and a GIS specialist. With over twenty-five years experience in remote sensing and GIS, he has served as director of several centers, as a GIS consultant, NASA visiting scholar, applied practitioner, a Fulbright Scholar to the University of the South Pacific, and educator. Dr. Davis is the former Director of the GIS Unit at USP in Fiji. He is currently chairman of the Department of Geography at Eastern Kentucky University.

Acknowledgments

Any worthwhile labor involves numerous folks, regardless of the single name up front. My sincerest gratitude is offered first to editor Daril Bentley, whose kind yet tenacious push kept me more on track and less tardy than was my natural inclination. His patience and guiding hand are much appreciated. To my colleagues, who have patiently forgiven my absences and distractions, I owe special gratitude. Thank you Dave, Alice, Rick, Janie, Dennis, Keshav, Glenn, Will, and especially dear Cecelia. To ESRI, Inc., for their kind permission in sharing data, operations, assistance, and information. For my children, to whom is owed much more than I can ever give, you have all that I can offer. And to Tina, partner for life and coauthor of all things positive, I give myself. To all of you, I am most appreciative.

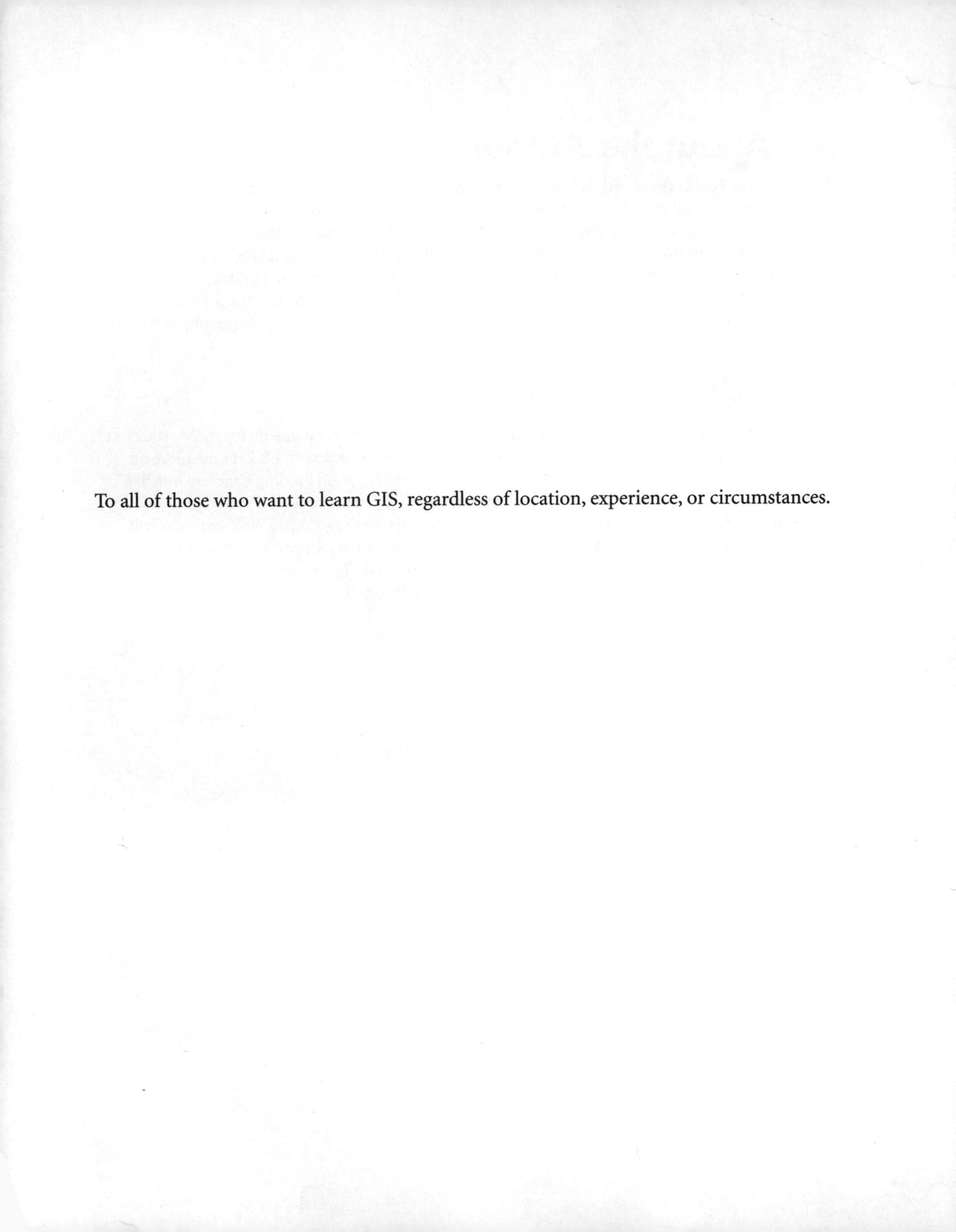

To all of those who want to learn GIS, regardless of location, experience, or circumstances.

CONTENTS

INTRODUCTION

GIS IS A HIGHLY DYNAMIC FIELD, GROWING AT THE VERY RAPID PACE of technological change and increasing number of applications. As a multibillion-dollar industry, its place as a major information technology is without debate. GIS is one of the few modern tools to live up to, and exceed, its early promise. It has spread into a global infrastructure, reaching to the emerging nations and remote corners of the planet. But in this extensive and swift growth, the beginning student (academic or professional) is often left behind in favor of those with advanced backgrounds. There are too many students (inside and out of schools) who want to learn GIS, despite their lack of graduate-level academic credentials. This text is intended to provide an introduction to GIS for everyone who wants to learn.

While in the South Pacific I found that most students are able to learn GIS, regardless of technological experience or level of English. The first edition of this book (1996) presented a format composed of many illustrations and uncomplicated, brief text. I am still satisfied with that approach and find it effective in courses in both the emerging and developed worlds. Introductory GIS does not have to be difficult for anyone.

Given the progress in technology and GIS in the past five years, it is time for a second edition, but one that builds on the first, not merely replacing it with updated technical capabilities. This edition expands the text to offer better explanation and greater breadth. Like the first edition, it is written in understandable English and is centered on an illustration for each major point being presented. For the most part, there is an illustration on the left and its text on the right of two facing pages, a "visual set approach" pioneered in the first edition. Several considerations were used in developing this approach to learning GIS:

- GIS is a visual methodology and technology, and it is best learned by illustration of its concepts and operations.
- Many English-speaking students, and most ESL (English as a Second Language) students, appreciate relatively uncomplicated words and phrases when learning a new technology.
- A complete review of GIS is not possible in an introductory text. Too many concepts and too much technical detail tend to confuse readers and therefore defeat the primary purpose. The aim of this book is to provide a foundation for basic understanding and for growth in the field. I struggled with the amount of information presented and hope that an acceptable balance between too little and too much has been achieved.
- Chapters are not exclusive; they are intended to be cumulative, building a sequence of principles and operations. It is recommended to follow the chapter outline when possible.
- In GIS, there are usually several ways to accomplish tasks. The illustrations and examples used here are meant to be simple, "generic" representations of GIS concepts and operations in general. Hands-on experience will reveal other approaches and techniques.
- Unfortunately, despite today's computer technology advances, the GIS user must still understand computer systems as well as geographic applications. Some attention is given to systems considerations where possible, but there is no substitute for experience.

An introductory book cannot replace hands-on experience with GIS. The reader is strongly encouraged (required if possible) to learn GIS at the computer along with this text. There are numerous programs available, from the very simple that demonstrate basic operations, to the highly advanced capable of sophisticated applications. ESRI's ArcView GIS software is used for illustration and operations here, and two wonderful companions for hands-on learning are ESRI's *Getting to Know ArcView GIS* and OnWord Press's *INSIDE ArcView GIS.* Numerous other GIS vendors also offer software suitable for learning GIS; check the many Internet web sites for further information.

Thank you for using *GIS: An Illustrated Approach, Second Edition.* I hope it opens doors for you and expands your horizons.

CHAPTER 1

GIS AND THE INFORMATION AGE

"You on the cutting edge of technology have already made yesterday's impossibilities the commonplace realities of today."

—Ronald Reagan, U.S. President, 1985

"Unless we can find some way to keep up with, and our sights on, tomorrow, we cannot expect to be in touch with today."

—Dean Rusk, U.S. Secretary of State, 1963

Introduction

IN ORDER TO UNDERSTAND GIS AND HOW IT WORKS, we must understand its significance and value as a major information system. As societies enter the Information Age, technology becomes more influential and more important. GIS plays a substantial role in this process. This chapter first presents the nature and meaning of information in society and then previews GIS by providing a basic description of it and discussing its infrastructure and roles.

The word *paradigm* is used in this book. It is a conceptual word that combines philosophy, how we think, the influences on our thinking, how things work, and models of ideas. In this sense, a paradigm is our mindset about something. Here, it usually refers to our changing ideas about technology. As technology changes, so do our perceptions, and therefore our paradigms change ("shift," as is commonly used in this context). GIS affects not only the actual procedures of computer operations and the nature of geographic applications but our thinking and our professional methodologies.

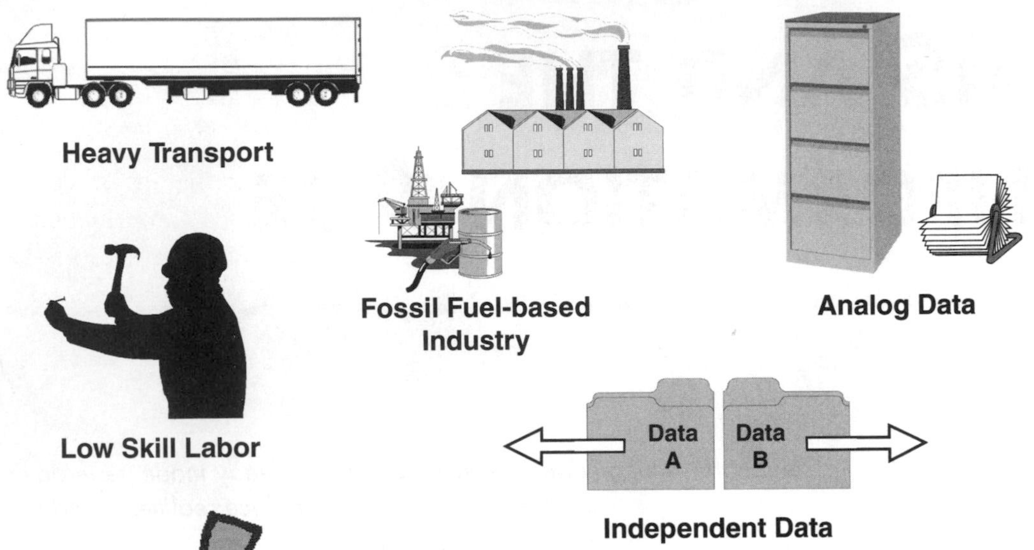

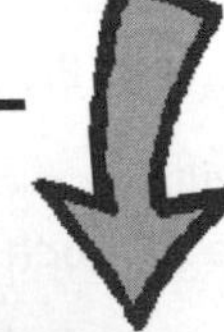

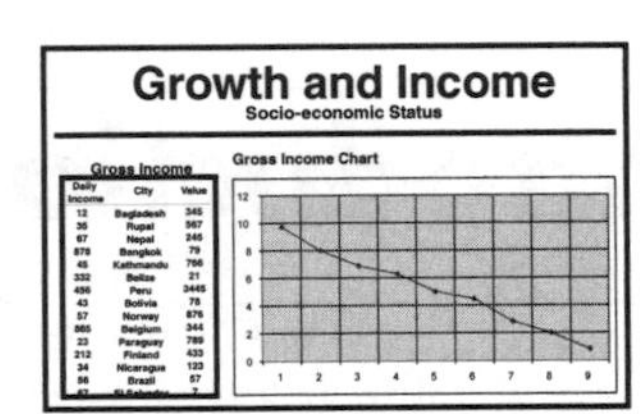

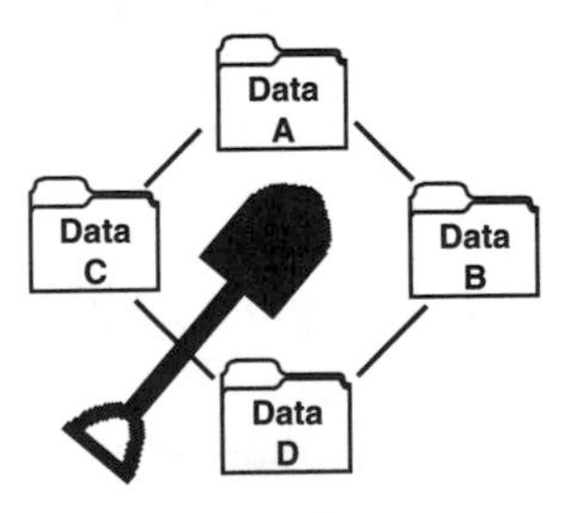

Fig. 1-1: The Industrial Age and the Information Age.

The Information Age

"Today's focus in geography [or any other discipline] on data handling and data manipulation reflects the national [and international] emphasis on meeting the needs of the post-industrial information age."

—Saul Cohen, President, Association of American Geographers, 1990

This quote says that the world is changing and that our methods of meeting the needs of those changes are also changing. The world is entering the "Post-Industrial Information Age"—a time when information is becoming a major product of, and foundation for, economic and social progress. Increasing emphasis on information systems is apparent and necessary. GIS, as a major information system technology, is an important component in this evolution.

Figure 1-1 points out the transition Cohen and others have considered. Many nations are changing from economies based on heavy industry, with petroleum-based energy and heavy transportation, to economies with more emphasis on data, information, and service. Large, centralized factories and semiskilled workers have been a way of national economic life and culture. This is the old style of economic survival, and the old model for development.

In the past, data was analog, mostly written or typed, stored on paper in various forms, and accessible by manual searching and handling. Information was largely disconnected, "independent," and existed for a single purpose, seldom integrated with other data outside its original and immediate need. Information systems then were little more than storage units, perhaps with some type of index, but without much analytical or integration function.

Modern times are rapidly passing from the old ways into a period in which sophisticated technology is changing the traditional power base and culture. Today, much of the world is experiencing social, scientific, and educational revolutions that are basically technologically driven. Computers and other electronic equipment (such as satellites) are the machinery of new economies and cultures.

Data and information are the "currencies" of the Information Age, and skilled, technology-trained workers are becoming a major labor force. The Internet and World Wide Web, effectively born only a few years ago, are becoming a major communications medium (often referred to as the "Information Highway"), and have already established a true global communications infrastructure (the network of basic facilities, services, and other components needed to make a large system work).

The "Integrated Data and Data Mining" image in figure 1-1 deals with the new techniques and new paradigms in finding and using the wealth of available global data for special purposes. It is just one example of a totally new concept in data technology that

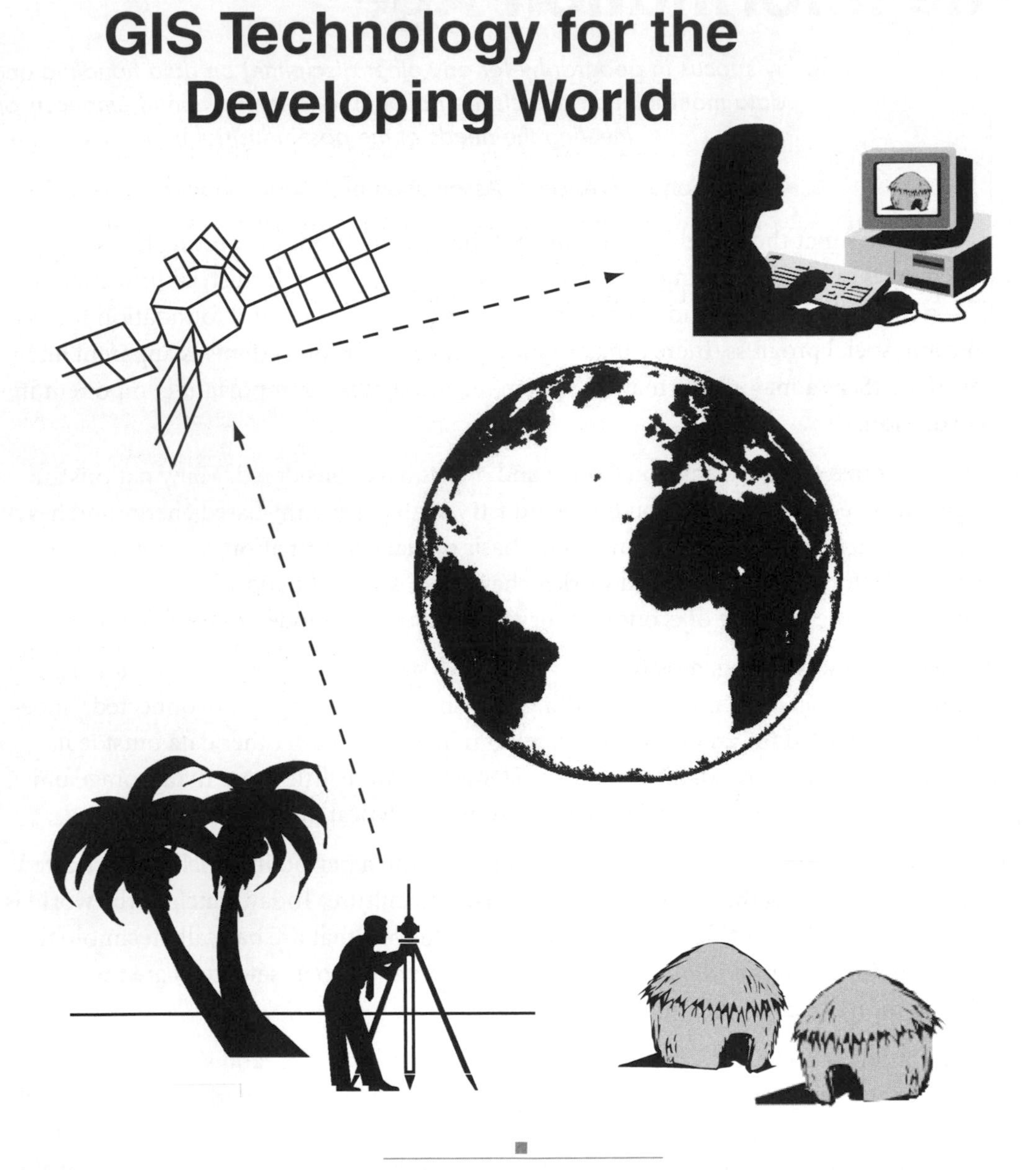

Fig. 1-2: The Information Age is a global revolution.

will be a substantial part of the Information Age. GIS is also an important part of this new age and will undoubtedly become even more important in the near future.

As figure 1-2 indicates, the Information Age is a global revolution, reaching the developing world as well as technologically advanced nations. The developing world ("Third World") is faced with the very heavy burden of moving quickly into the Information Age if it is to keep up and avoid being overwhelmed by this global evolution. Even though the Industrial Age has yet to be fully realized in many nations, the Information Age will not wait patiently. The pressures to catch up are high for much of the world, and the demands for doing so are troublesome and expensive, yet exciting and progressive.

According to a report of the 1992 Rio de Janeiro, Brazil, Earth Conference, "The gap in the availability, quality, coherence, standardization, and accessibility of data between the developed and the developing world has been increasing, seriously impairing the capacities of countries to make informed decisions concerning environment and development." Clearly, there is a need for better information and information systems in all parts of the world.

GIS is a central component in the world's environmental information structure, and it will continue to play a primary role. The gap between the Haves and the Have-Nots of the information world is increasing. There is a huge difference between the information capabilities and wealth of the developed Western world and those of the developing world. The gap should be closing, not becoming wider. Developing nations need good, reliable information and information systems for survival and progress. GIS may be perceived as a potential "appropriate technology" for helping to achieve these goals.

Technologic Change: Stages of Acceptance

Reluctance to Adopt

Galileo's ideas

Horseless carriage

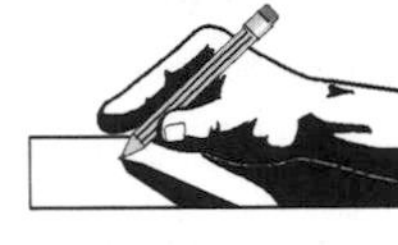

Calculator versus manual calculation

Cautious Acceptance

Computer views

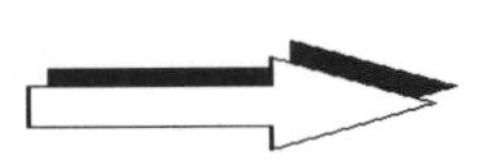

New ways of doing old things

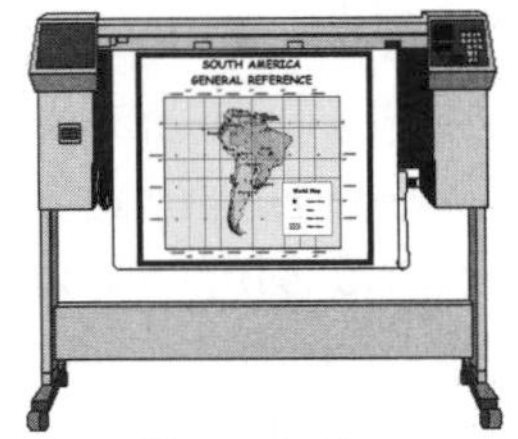

Map plotter

Full Use

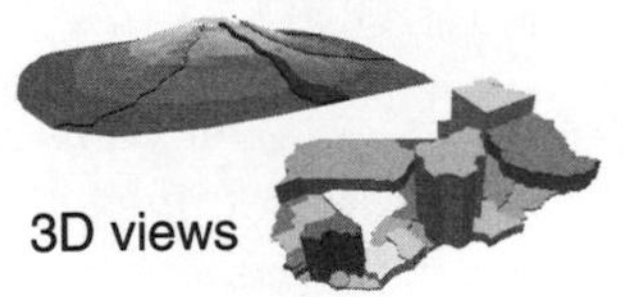

3D views

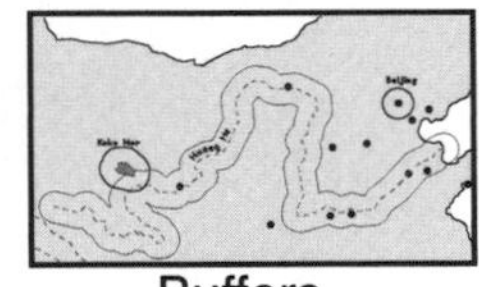

Buffers

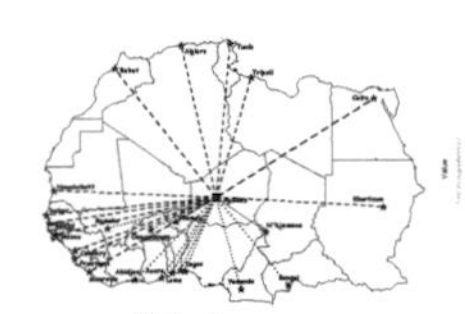

Distance analysis

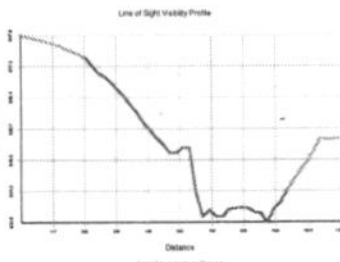

Profiles

Innovative Use

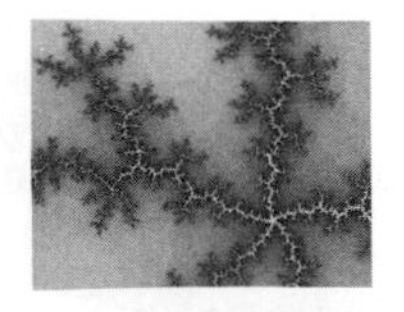

Fractal analysis

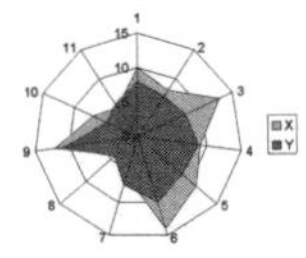

Advanced analysis

Virtual reality

Holographic education

Where to next?

Fig. 1-3: New technologies move through four basic stages of acceptance.

Stages of Acceptance

New technologies can be exciting, yet problematic and challenging. Almost any new technology, whether a simple tool or an advanced electronic machine, usually goes through four stages of acceptance. These stages, depicted in figure 1-3, can be experienced by an individual, an institution (e.g., a school), or even a government.

1. *Reluctance to adopt:* At first, we often reject new technology, because existing (old) ways are satisfactory. We are comfortable with the current state of technology and ways of doing things (basically a conservative approach to the world). Also, resistance can come from fear of the unknown, ignorance of what is involved in the use of the technology, fear of learning something new, or fear of the changes it may bring. Thus, we tend to stay with traditional methods. Many people and institutions react to and accept change slowly. We need to be convinced that change will be productive. Although Galileo brought new insights of the cosmos derived from his telescopic observations (a new technology), the new ideas were not accepted immediately. The first automobiles were resisted by many communities out of fear of the perceived dangers, pollution, and the potential impact on existing transportation and its economics. Approval of calculators in education was an issue for a while because of concern over perceived ill effects on students.

2. *Cautious acceptance:* After overcoming initial fears of the new technology (perhaps because others are using it), there is cautious acceptance. The first uses are not particularly advanced, often being simple reproduction of traditional products—new ways of doing old things. In GIS, this was making maps by computer (computer-aided drawing), followed by integrating database with the maps for more sophisticated analysis and presentation. Such first tentative steps led to further progress.

3. *Full use:* After gaining some experience and comfort, we begin making full use of a technology and recognize that capabilities can be expanded. We see new things that can be done and new ways of accomplishing tasks. New modes of analysis and different ways of presentation, for example, begin opening up new opportunities and new applications. Illustrated are just a few GIS capabilities that will be explored in coming chapters.

4. *Innovations:* The most exciting part of new technology is the potential for real innovation; that is, completely new methods that help construct new concepts in our field. Various products previously unimaginable are developed from the new technology. Illustrated are several advances that may be forthcoming, such as fractal analysis of landscapes and advanced graphing. Consider the potential of virtual reality for many GIS applications. Holographic displays will aid science and education. From such giant steps comes new paradigms—new philosophies, new ways of thinking, and new ways of looking at the world. Where do we go from these wonderful possibilities? The imagination seems to be the only limitation.

Price, Cost, and Value of Information

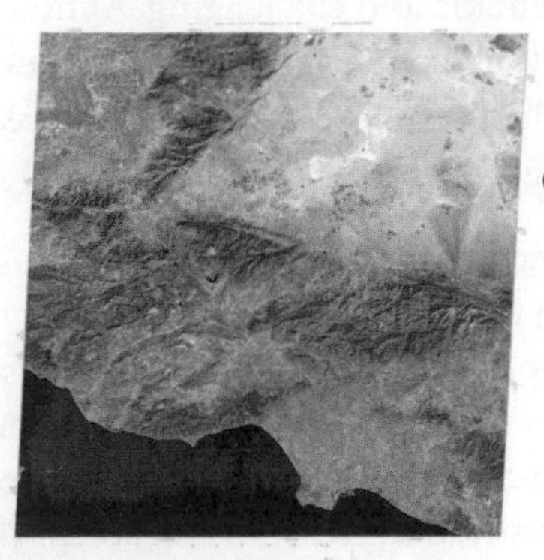

Image 1: Price of Data/Information

Southern California
Satellite View
$4,000 data, but
34,000 km^2 coverage

Image 2: Cost of Information

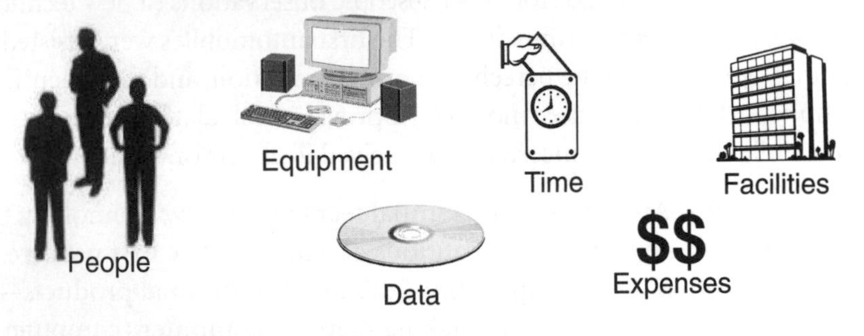

Image 3: Cost of No Information

Image 4: GIS, Modern Information Technology

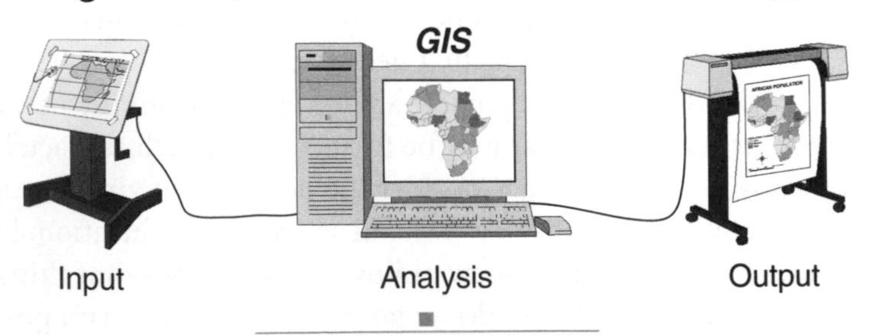

Fig. 1-4: The price and cost of information versus the cost of no information.

Value of Data and Information

As the world moves into the Information Age, a new major "currency" is meaningful data. A critical question of our time concerns the value and use (as well as misuse) of data and its ultimate form as "information." What is the value of information? How important is it? What can it be used for? What are the benefits? GIS may be perceived as a technology, but these concerns are critical social and economic questions deriving from its use. Figure 1-4 illustrates both technical components and concepts of information as a value paradigm. The price of data and information is the monetary charge, the fee that must be paid.

Sometimes data is free or low priced, but often the fee seems very high. A satellite image can have a U.S. $4,000 price, more than many agencies or institutions can afford. However, because space imagery covers a very large area, the price per unit area can be relatively low and cost effective. Image 1 shows a Landsat image of Southern California, USA, covering about 34,000 square kilometers. Consider the price of performing field data collection for the entire region; the $4,000 now seems to be a bargain.

The *cost* of data and information involves much more than monetary price. It is the sum total of investments that make up the overall GIS infrastructure, which includes other types of resources, such as human effort to convert data into information, the equipment needed to process data, facilities, time, and other expenses (wages, utilities, supplies). Image 2 shows some of these support components.

Perhaps the more important question concerns the cost of *no* information. We may pay dearly for lacking this crucial resource. For example, critical "informed decisions" about environmental management may be impossible without useful information. Lives can be affected and economies ruined without the necessary information. Those who have too little or no information will always be behind the more informed, always trying to catch up, but usually falling further behind (image 3). (Note the connection between the words *informed* and *information.*)

What are the value, price, and cost of an information system, the major tool of the Information Age? The price and cost can be very high, but an information system is an excellent investment, considering its value. GIS is a major system in the new age (image 4). It may be expensive in the beginning, but its effective use can be highly cost efficient.

Data Visualization

Image 1: Which is preferred for the initial view?

Map

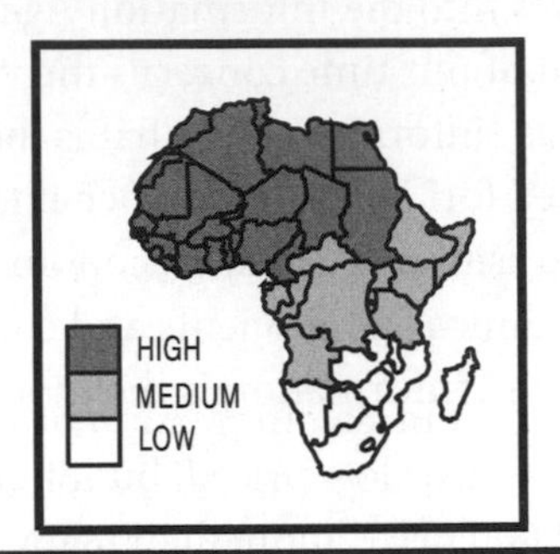

Database

NATION	DATA QUANTITY	DATA QUALITY	NATION	DATA QUANTITY	DATA QUALITY	NATION	DATA QUANTITY	DATA QUALITY
Algeria	45	High	Gambia	44	High	Rwanda	20	Medium
Angola	20	Medium	Ghana	44	High	Sao Tome-Principe	21	Medium
Benin	40	High	Guinea	40	High	Senegal	31	High
Burkina Faso	42	High	Guinea-Bissau	34	High	Sierre Leone	32	High
Burundi	21	Medium	Kenya	27	Medium	Somalia	22	Medium
Cameroon	18	Medium	Lesotho	10	Low	South Africa	11	Low
Cape Verde	17	Medium	Liberia	40	High	Sudan	45	High
Central African Rep.	25	Medium	Libya	40	High	Swaziland	12	Low
Chad	45	High	Madagascar	15	Low	Tanzania	17	Medium
Congo	20	Medium	Malawi	12	Low	Togo	40	High
Cote D'Ivoire	40	High	Mauritania	44	High	Tunisia	45	High
Djibouti	27	Medium	Morocco	48	High	Uganda	41	High
Egypt	45	High	Mozambique	7	Low	Zaire	21	Medium
Equatorial Guinea	29	Medium	Namibia	9	Low	Zambia	11	Low
Ethiopia	18	Medium	Niger	40	High	Zimbabwe	13	Low
Gabon	45	High	Nigeria	40	High			

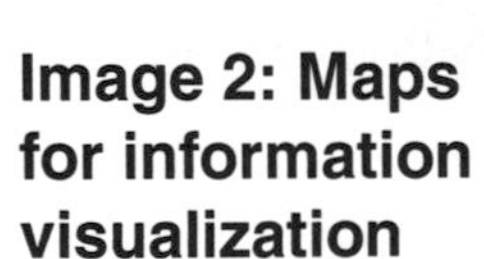

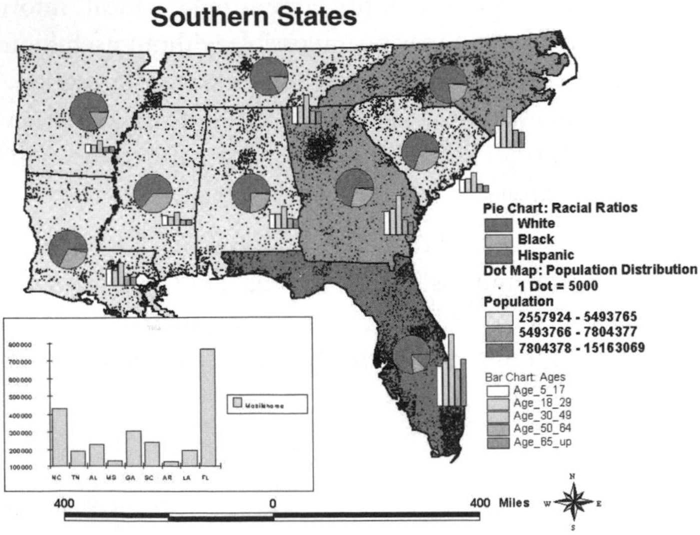

Fig. 1-5: Data visualization.

Visualization of Data

New technologies are supporting advances in the increasingly important techniques of data visualization, the graphical representation of complicated data for easier understanding. Image 1 of figure 1-5 presents two sets of data, a map of Africa and its database. Both say the same thing, though one is presented in detailed numeric tabular form, whereas the other is a visual map. Which one would you prefer to help make the data more understandable, especially at first?

The map gives the best initial impression of what the information means. The table is best for detailed analysis. Most people prefer the visual approach first, followed by details as needed. Humans are visual animals who rely on sight as the major sensory means of critical information gathering and in forming initial impressions.

Visualization is the presentation of data in graphic form. Tables and lists of numbers are usually difficult to understand without careful study, but visualization is a convenient and effective means of communicating complex information. Computer technology enhances visualization by offering sophisticated techniques of changing data into pleasing and understandable displays. Greater importance is being placed on data visualization today, and GIS is a leading technology in this movement. Maps are standard products that are produced easily, supported by analytical charts and other visual devices.

Image 2 demonstrates the many possibilities of standard GIS map presentation (though it is a bit crowded at this small scale). It includes the U.S. southern states, population distribution using dot mapping, pie charts for each state's racial populations, a bar chart for each state giving age distributions, and a bar chart at the bottom giving the number of mobile homes for each state. GIS uses data visualization effectively and efficiently.

What Is GIS?

Geography The Real World

Information Data and Information

NATION	DATA QUANTITY	DATA QUALITY	NATION	DATA QUANTITY	DATA QUALITY	NATION	DATA QUANTITY	DATA QUALITY
Algeria	45	High	Gambia	44	High	Rwanda	20	Medium
Angola	20	Medium	Ghana	44	High	Sao Tome-Principe	21	Medium
Benin	40	High	Guinea	40	High	Senegal	31	High
Burkina Faso	42	High	Guinea-Bissau	34	High	Sierre Leone	32	High
Burundi	21	Medium	Kenya	27	Medium	Somalia	22	Medium
Cameroon	18	Medium	Lesotho	10	Low	South Africa	11	Low
Cape Verde	17	Medium	Liberia	40	High	Sudan	45	High
Central African Rep.	25	Medium	Libya	40	High	Swaziland	12	Low
Chad	45	High	Madagascar	15	Low	Tanzania	17	Medium
Congo	20	Medium	Malawi	12	Low	Togo	40	High
Cote D'Ivoire	40	High	Mauritania	44	High	Tunisia	45	High
Djibouti	27	Medium	Morocco	48	High	Uganda	41	High
Egypt	45	High	Mozambique	7	Low	Zaire	21	Medium
Equatorial Guinea	29	Medium	Namibia	9	Low	Zambia	11	Low
Ethiopia	18	Medium	Niger	40	High	Zimbabwe	13	Low
Gabon	45	High	Nigeria	40	High			

Systems Technology Support System

GIS: A computer-based technology and methodology for collecting, managing, analyzing, modeling, and presenting geographic data for a wide range of applications.

Fig. 1-6: The fundamental components and nature of a GIS.

What Is GIS?

GIS refers to three integrated elements (depicted in figure 1-6).

- *Geography:* The real world; spatial realities.
- *Information:* Data and information; their meaning and use.
- *Systems:* Computer technology and its support infrastructure.

GIS therefore refers to three aspects of the modern world and offers new methods of using them. As will be discussed, GIS is much more than a computer system—it is also a methodology (a system of techniques and principles) in science and applications, a new profession, and a new industry and business. GIS has become a new Information Age paradigm.

GIS Overview

There are numerous definitions of GIS, none of which actually describe or explain it sufficiently. Sometimes we simply say "advanced computer mapping," but a better definition is "A computer-based technology and methodology for collecting, managing, analyzing, modeling, and presenting geographic data for a wide range of applications." Geographic data is detailed in Chapter 2, but it basically refers to map information; that is, anything that can be located on Earth (or elsewhere), and any description or accompanying information.

GIS has evolved to be the foundation for a new discipline, sometimes termed GISci for Geographic Information Science. It is also called Geomatics and is the major part of Geotechniques (or Geotechnologies). This new discipline uses the overall GIS paradigm—the fundamental concepts, principles, ideas, methods, data, and approaches to applying GIS to many types of tasks and applications. Some universities offer degrees in GIS, although most still incorporate it within Geography or a similar department.

GIS Overview Notes

The following are points to keep in mind regarding GIS.

- GIS is pronounced by each letter, as a three-syllable word that sounds like "*gee-eye-ess*" (not the single syllable "jis").
- There is no difference in meaning between geographical and geographic; it is correct to use either term.
- The plural of GIS may be "GIS," because the acronym stands for both Geographic Information System (singular) and Geographic Information Systems (plural). However, the use of "GISs" is common, as in this book. "GISes" is not acceptable.
- The term *GIS systems* is valid when referring to the computer systems that support GIS (even though it is a bit redundant).

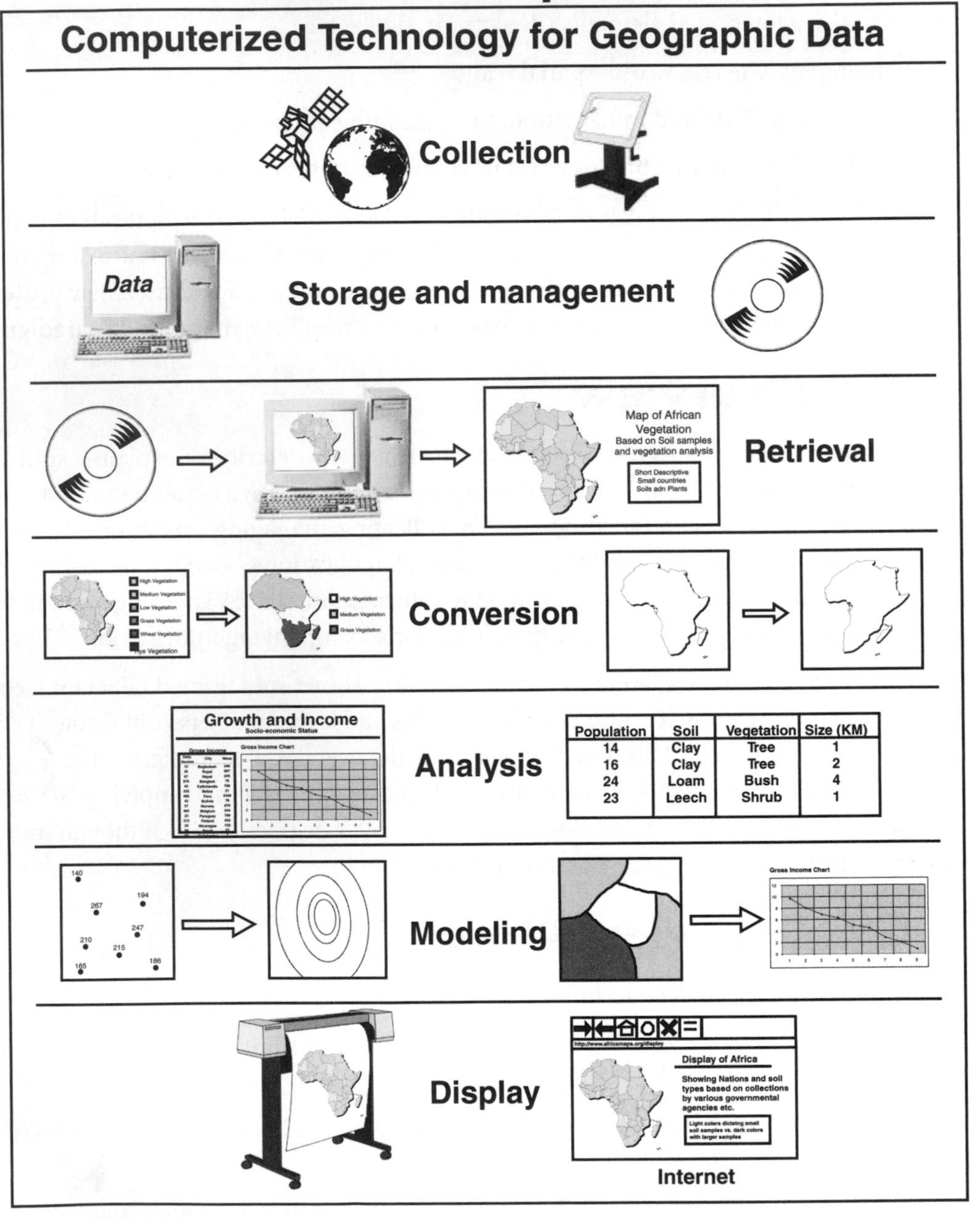

Fig. 1-7: The functions of GIS.

GIS Description

GIS has numerous functions. Its primary roles are presented in the following, dealing with geographic data from input to output. These concepts, depicted in figure 1-7, are discussed in detail in the chapters that follow.

- *Collection*: Gathering data from many sources, including global communications media. The illustration shows a remote sensing satellite, which is a major data source today. The digitizing table is for converting paper maps into digital computer map data by electronic tracing. Other means of collecting data include field work, the Internet, CD-ROMs, and text.
- *Storage and management:* Efficient digital storage is necessary. Administering and keeping track of data, including integration of various types of data sets into a common database or sets of data. Database management is replacing the physical map storage structure. Computer hard disks, CD-ROMs, the Internet, and other media are used.
- *Retrieval*: Easy and efficient selection and viewing of data in a variety of ways, including computer monitor display, printed maps, and the Internet.
- *Conversion:* Changing data from one form to another or one map format to match another. For example, converting from one geographic projection to another, rescaling, reclassifying, and other computer "tricks" to make the data more useful.
- *Analysis:* Analyzing data to produce insight and new information. Using various techniques of data investigation, statistical procedures, and other methodologies.
- *Modeling:* Simplifying the data to understand how things work or to explain what the data means; a generalization of the data or a simple explanation of reality. A contour map from elevation data is a model of an area based on a few points of data or a graph to present another version of map information.
- *Display:* Presenting data in various ways (such as maps, graphs, and reports) for easy understanding.

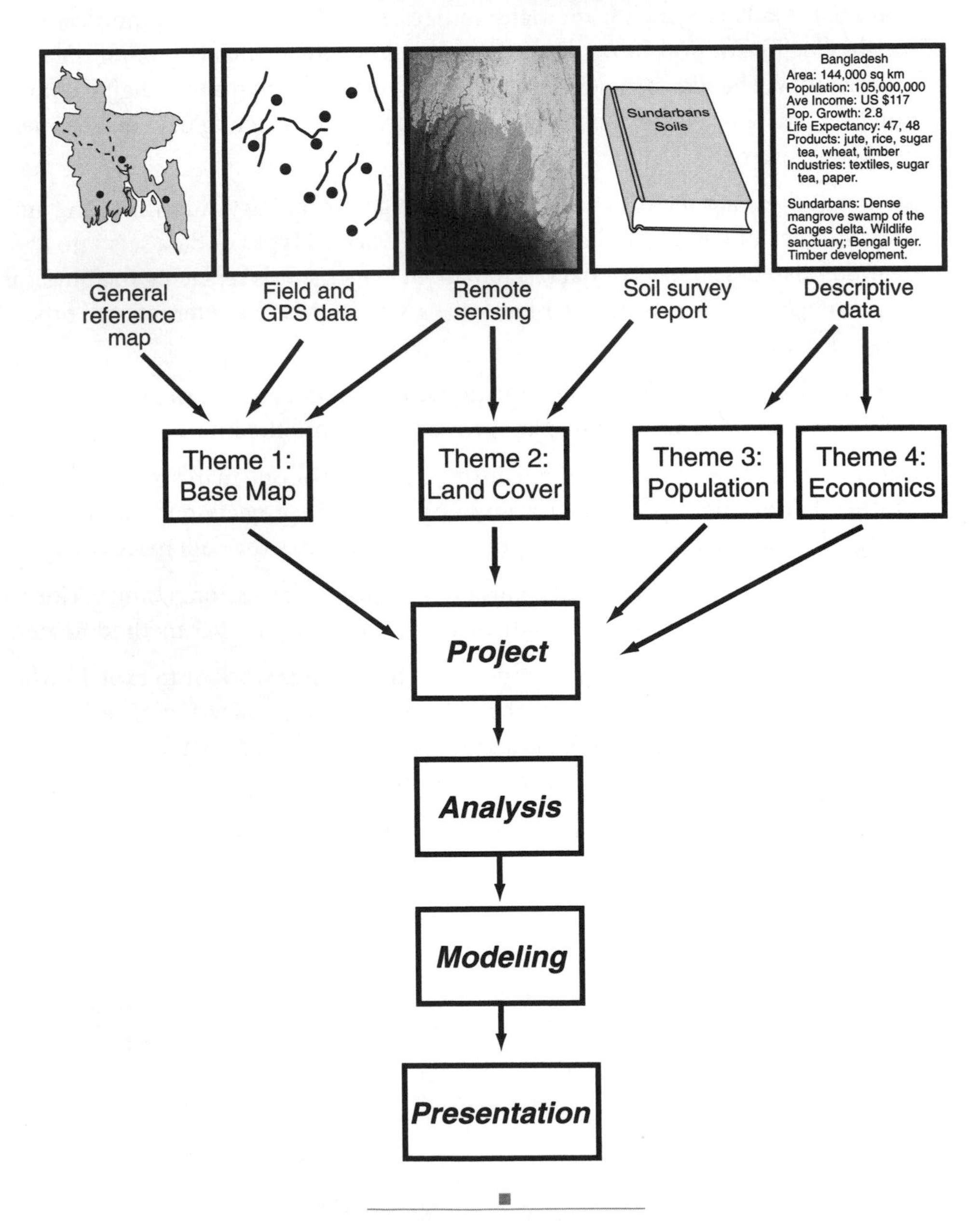

Fig. 1-8: Components of a GIS project.

GIS Projects

A GIS project is the foundation of GIS work. It consists of at least one central database (though numerous databases are often used) and graphics presenting features and themes, as shown in figure 1-8. The databases and graphics are functionally linked. A theme is a layer of data made of one or more features (geographic objects) of a similar topic. For example, an ecosystem theme can include many areas with various vegetation types. A project usually contains numerous themes, such as soils, demographics, field surveys, and wildlife.

A GIS project organizes many types of data into a common structure that supports its purpose and objectives. GIS data comes from many sources and in a variety of formats. Therefore, a project has a great flexibility in the types of data it can use. The integration of databases with graphics gives a special strength to GIS that few other informational technologies possess. GIS project data can include diverse types of media, illustrated with an application on the endangered Sundarbans, a large mangrove area of coastal Bangladesh.

- *Maps:* The most common type of geographic data is from existing maps. In the near future, digital data will probably be the dominant input. Maps of Bangladesh can be obtained from many sources. Projects usually need a base map that serves as a basic data foundation.
- *GPS (Global Positioning System):* A special satellite system that provides highly accurate location and elevation data from anywhere in the world. GPS is a valuable part of field data.
- *Imagery:* Remote sensing, such as satellite or aircraft digital imagery, provides a major source of GIS data. This can also include scanned pictures, such as air photos or field photography. Illustrated is a satellite view of the Sundarbans.
- *Reports (text data):* Reports and text documents dealing with spatial subjects, such as a soil survey or research from another project.
- *Tabular data:* Lists of numeric data, such as descriptive, census, or economic data.

After the project has been assembled (which sometimes can require 75 percent or more of the project time), other GIS work begins, such as analysis, modeling, and presentation. These steps are discussed in detail in the chapters that follow.

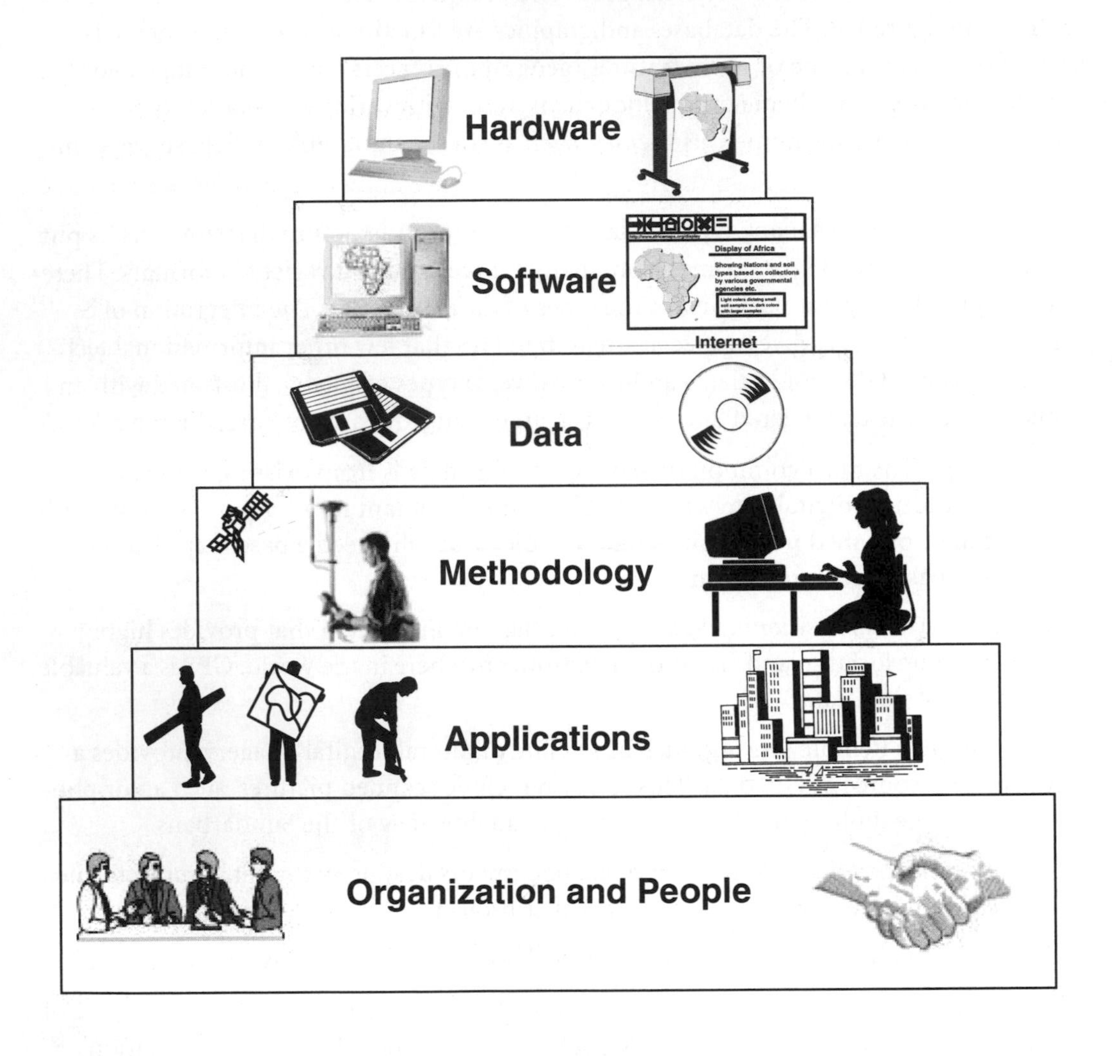

Fig. 1-9: Pyramid of GIS components in importance from bottom to top.

GIS Infrastructure

There are (at least) six primary components of GIS, as depicted in figure 1-9. The basic infrastructure is presented, appropriately, as a pyramid, in order of importance from the bottom up.

- *Organization and people:* The most important part of a GIS structure, the base of the pyramid, is the organization and people. Without them, the GIS and its infrastructure have little purpose and will not work. If GIS is to fit into the organization as a significant useful tool, it must have dedicated people and facilities assigned to it.
- *Applications:* Applications are the uses, questions, or "customers" of GIS; the purpose for its production (for example, environmental analysis, city planning, or development). This may be a research project or an "enterprise" GIS that supports the various missions of an organization (government department, or business).
- *Methodology:* The various procedures, techniques, and ways of using GIS and GIS data in applications. The questions asked will determine the methods used to answer them. Much of this book concerns methodology, but chapters 7 through 10 focus on specific analytical techniques. Methodology can also determine the data to be used.
- *Data:* Data is the heart of GIS operations. Major emphasis in GIS operations is on the data, from input through analysis, and ultimately to the presentation.

  NOTE: *At times, a project may have to use data that is already obtained. Therefore, the nature of the data can determine the methodology to be used. The Methodology and Data stages can be interchangeable, depending on circumstances.*
- *Software:* The computer programs needed to run GIS. There are many GIS programs available, from the simple low-cost product to expensive and very powerful ones. Support software is also used for statistical analysis, word processing, graphing, Internet operations, and many specialized purposes.
- *Hardware:* The machinery on which GIS operates—computers, printers, plotters, digitizers, and other types of equipment. Like software, there are many options in hardware brands, types, functions, and costs.

The pyramid was constructed with people at the base, the foundation of GIS. Primary importance must be given to the people who are to run the project, and to the organizational structure in which they work. All elements are necessary, of course, but hardware, at the top, should be the last consideration in developing a GIS operation (all the other components will help to determine the hardware needs).

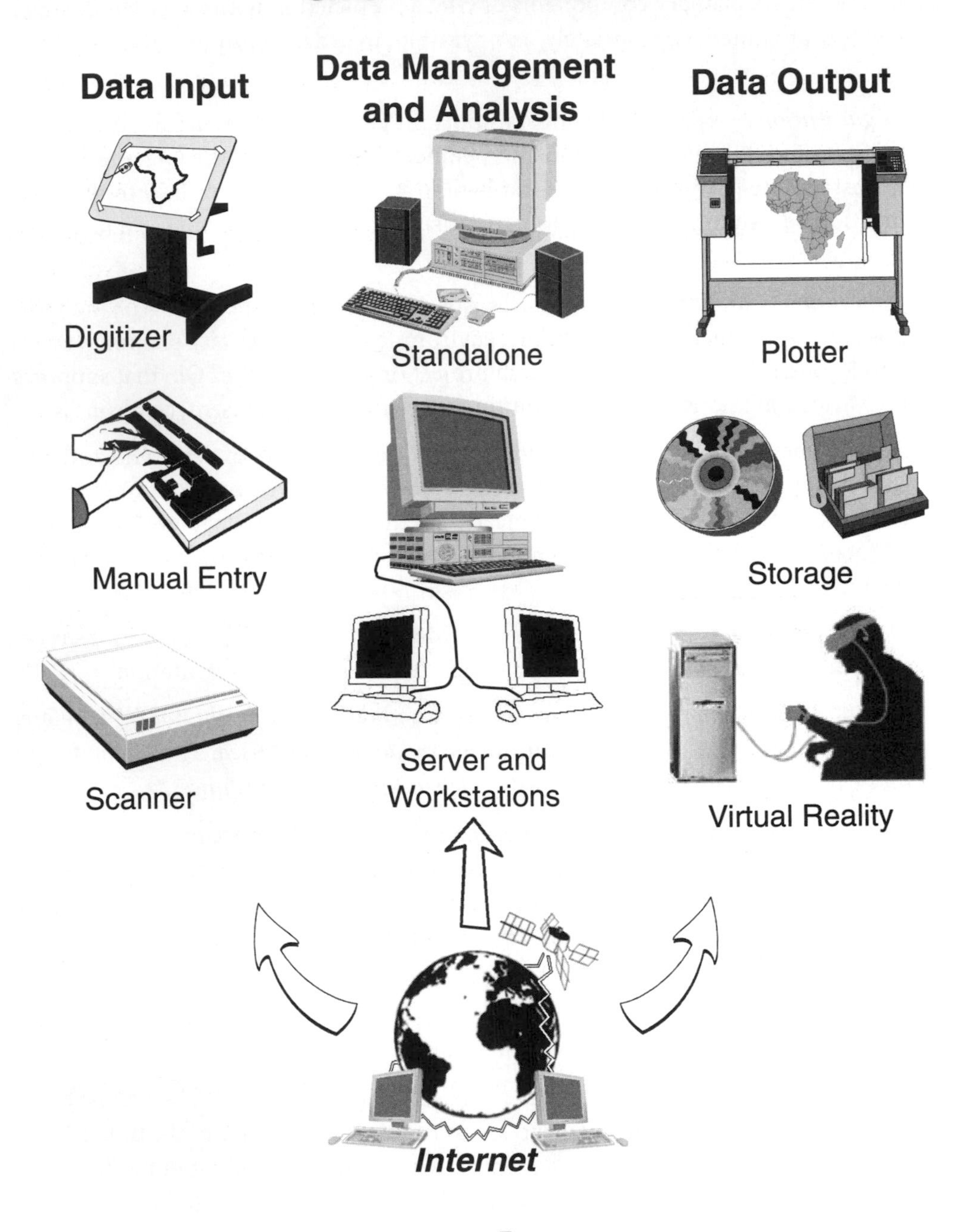

Fig. 1-10: Components of a GIS hardware infrastructure.

GIS Hardware Components

GIS hardware is not the most important component in the GIS pyramid, but it is the most visible. Hardware can be relatively simple and inexpensive, or it can be complex and expensive. The minimum configuration for a project requires a way to input data, support for management and analysis, and tools for output. A standard hardware infrastructure includes the components described in the following and depicted in figure 1-10. Other hardware could be added, such as scanners, special printers, CD-ROM writers, and others.

- *Data input:* Data is entered into GIS in a number of ways. The most common is to copy maps by manual tracing on digitizer tables. Text and numeric data can be entered from the keyboard as well. Other input devices include CD-ROM and tape readers, scanners, GPS, and networks such as the Internet. Data entry is discussed in detail in Chapter 5.
- *Data management and analysis:* The main task of a GIS is to manage and analyze data. The choice of computer equipment (called the "platform") can vary from low-cost standalone microcomputers to high-end workstations and network servers. There are many brands and types of computers, and there are many reasons, including the intended projects, for selecting a specific platform. Some GIS software can operate on old, low-end machines, but the professional user tries to keep up with the latest technology. Moreover, the Internet is a new type of GIS platform, serving as a data source, as an interactive presentation medium, and in the near future as a "Web-based GIS" (meaning that the primary software and data will reside on the Internet, to be rented by the user for performing project work). Chapter 13 presents technology advancement and GIS in the future.
- *Output:* GIS can provide a variety of products, including simple display on the computer monitor, sophisticated maps, and cutting-edge technological presentation, such as virtual reality. Plotters are common mapping devices, producing large, high-quality maps. Digital data can be output directly to tapes, CD-ROM or DVD disk, or a network for storage, and then input into another GIS. The Internet is rapidly becoming a major medium for GIS data and display. Advanced technologies are being combined with it to create terrific new versions of data presentation.

Interestingly, note that the Internet is a contributor to all three parts of a GIS hardware infrastructure. Even though it is not technically classed as hardware, the "Net" serves as a major GIS component and its importance and value are sure to increase.

Old and New GIS

Medieval Cartographer

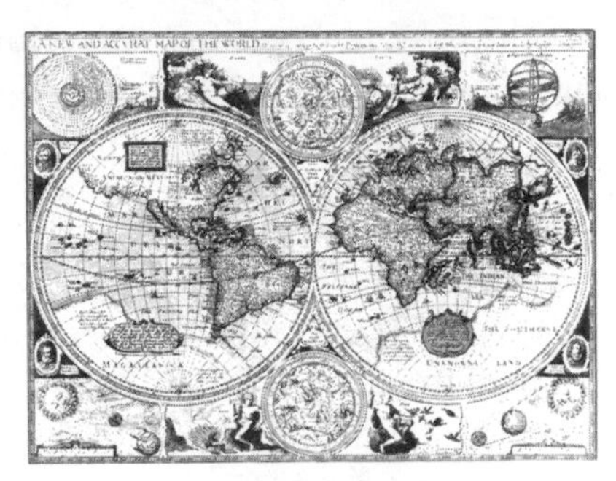

Ornate artistry

B.C. GIS Technician (Before Computers)

Artistic and functional

Modern GIS Professionals

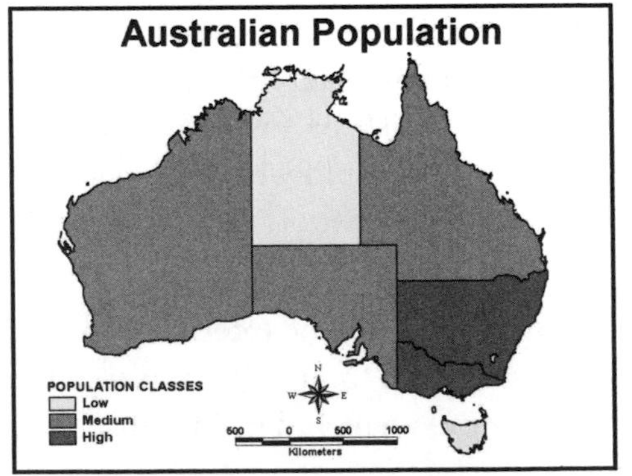

Simple and utilitarian

Fig. 1-11: Outdated GIS technology and new GIS technology.

Old and New GIS

Geographic analysis has been around longer than maps, probably back thousands of years, when early humans planned hunting and movement. Perhaps the first GISs were scratches in the dirt or sand. Actually, a map in any form is a simple but effective geographic information system because it has most of the necessary information to accomplish many spatial data tasks.

Traditional GIS (before it was called GIS) included paper maps and manual drawing tools. Highly skilled cartographers spent many hours drawing maps, and map users labored over the analytical tasks. Simple measurements required time, and overlays (combined maps) were lengthy and special accomplishments.

This was the "BC" era—Before Computers. The G (geography) was the same as today, the I (information) was traditional (and has become increasingly sophisticated), but the S (system) was rather elementary. Today, traditional "pen and ink" cartography by skilled draftspersons is being replaced by GIS. Hand-rendered cartography, other than specialized art, is becoming outdated. Figure 1-11 shows two old-style cartographers with their traditional maps, one from the very ornate (artistically elaborate) period of several hundred years ago and the other more recent, making less elaborate maps, though both still used manual methods. At bottom are the modern cartographers; people who use advanced technology, reach around the world for data, and collaborate to produce sophisticated data visualization products that include rather simple but functional thematic maps and a host of support graphics, databases, and analysis, all delivered rapidly and conveniently.

Modern GIS arrived when computers became powerful, more easy to operate, more affordable, and generally available to many users (the applications people). GIS as we know it today involves computer technology to accomplish the traditional tasks of yesterday and new tasks of today. Geographic data and applications have evolved tremendously because of it. Technology has created many changes, including geography (the world and the way people view and interact with it), data (new types, such as digital and satellite imagery), and methods of analysis. Certainly the modern cartographers in the illustration view the world much differently than their earlier counterparts. In effect, modern GIS is a new paradigm.

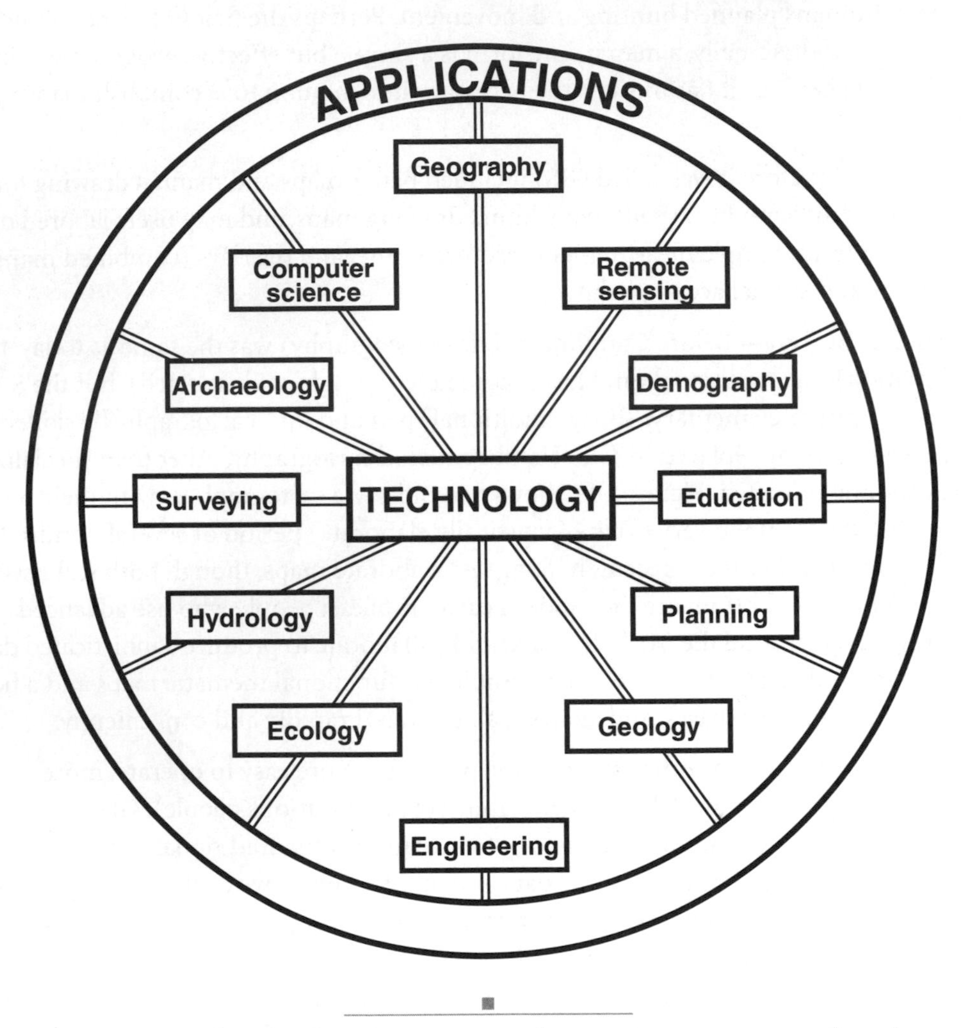

Fig. 1-12: GIS technology has many applications in many areas of work.

GIS Is Multidisciplinary

As depicted in figure 1-12, GIS is multidisciplinary—it uses data and techniques from many professions and academic disciplines and its applications are found in a diversity of fields. It is not just a tool for academic research but functions as a true integrated technology. Phone companies, banks, advertising firms, many types of commercial enterprises, emergency services, and many other public and private activities have adopted GIS as a major information and decision-support technology. Its rapid growth will continue.

GIS data and applications typically use world data of some type (e.g., environmental), which by nature is not limited to a single theme. The world is holistic and its many parts, physical and cultural, are connected and integrated in many ways. Few processes or activities in nature or culture are isolated and independent of all others. GIS is an ideal technology to use with multidisciplinary topics.

Today, GIS borrows from numerous disciplines and applications. The illustration shows a conceptual view of GIS as a wheel, with the applications as the rim (the ultimate purpose of the wheel), technology as the hub, and various fields represented as spokes. Many more spokes could be used, but the basic idea is that GIS is a multidisciplinary, multi-application, multi-purpose, multi-technique, and multi-dimensional integrated technology and methodology. (That's a lot of descriptive adjectives, but they are all valid.)

5. Business

6. Paradigm

Fig. 1-13: GIS plays numerous distinct, but often interrelated, roles.

GIS Roles

As noted, GIS is much more than a computer system. It functions in a variety of roles in society. GIS is all of the following, depicted in figure 1-13.

- *Technology:* GIS is a system of modern hardware, software, data, peripherals (associated equipment), and people.
- *Methodology:* GIS is a set of techniques, procedures, and principles for managing and analyzing geographic data. It has led to new ways of accomplishing many traditional and innovative tasks, allowing us to take different approaches to solving problems. It is a method of accomplishing geographic tasks.
- *Discipline:* Because of its wide range of applications and integrated methodologies, GIS is an academic and professional "interdisciplinary discipline," a special branch of integrated knowledge and field of study. Normally associated with the study of geography, GIS is found in numerous other disciplines, such as geology, anthropology, computer science, business, and many others.
- *Profession:* As a new field, there is a demand for GIS specialists, thereby making it a new profession and career. Because of the wide range of applications and breadth of technology involved in GIS, the professional must be much more than a mere technician, but must be competent in integrating the geography, the data and information, and systems. The first generation of trained GIS specialists is beginning to appear and the profession is growing very rapidly.
- *Business:* Many new businesses have taken advantage of the rapidly growing commercial aspects of GIS to offer hardware, software, data, consulting, services, and other marketable items. Today, GIS is a multi-billion-dollar (U.S.) business, growing at a very high rate.
- *Paradigm:* All of these roles, the many applications, and the global use of GIS have developed into an overall GIS paradigm. GIS has become greater than the sum of its parts (a synergistic evolution). The illustration shows two simple "include bracket" symbols demonstrating collection of a group of elements into a single entity. The top bracket "collects" the five roles above, and the bottom bracket reaches out beyond the page, into the constantly evolving world to gather many other components and aspects not covered here. Clearly, there is more to GIS than a few simple items. It has become a way of thinking about the world, and it is changing our ideas about data, technology, professions, and many other aspects of geography and the world in which we live.

CHAPTER 2

GEOGRAPHIC DATA AND THE DATABASE

Introduction

GEOGRAPHIC DATA AND INFORMATION ARE THE HEART OF GIS. To understand GIS, it is necessary to understand geographic data and the way it is structured, stored, used, and presented. Addressed here are the two fundamental components of geographic data: space (expressed as spatial data) and qualities (attributes). Both of these are stored in databases, the central working component of GIS. Discussed in this chapter are the nature of geographic data, types of geographic relationships, and the GIS database.

Data and Information Definitions

The terms *data* and *information* are often conveniently interchanged without real loss of meaning, but an important difference can exist. It is best to use the terms correctly, particularly in GIS work. The following, depicted in figure 2-1, point out important differences among several terms associated with data and information.

Data and Information

Data Definitions

Sample site	Measurement quantity	
A	3	Datum
B	6	Data
C	4	
D	4	
E	8	
F	5	
Total	30	Information
average	6	

Data Versus Information

Map A: Sample Points

Data

Sample Sites
Vegetation Types

1 - Red mangrove
2 - Black mangrove
3 - Coconut palm
4 - Ivory nut palm
5 - Lime
6 - Papaya

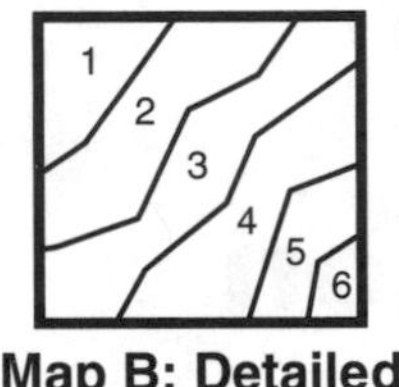

Map B: Detailed Vegetation

Information and Data

Collective vegetation types
Area Interpretation
Vegetation zones

1 - Red mangrove
2 - Black mangrove
3 - Coconut palm
4 - Ivory nut palm
5 - Lime
6 - Papaya

Map C: General Vegetation

Categories

M - Mangrove
P - Palm
F - Fruit

Fig. 2-1: The difference between data and information.

- *Datum:* A single number or fact; a single entry in the database. A datum can be a number, letter, or text. Although the term is correctly used as a singular of *data*, it is not commonly used this way. The more common use of the term is in reference to control points for establishing a geographic reference for the world sphere, an important aspect of accurate GIS coordinate systems.
- *Data:* A collection of facts in the database; multiple entries. In the illustration, the list of numbers constitutes the *data*. The word *datum* is singular and *data* is plural. Proper grammar of the terms is "datum is" and "data are." However, the word *data*

is typically used as both the singular and plural form, and "data is" seems to be easy on the ear because of common use. As noted, datum has a specific technical meaning. Here, data is used as both the singular and plural.

- *Data set:* A data set is a collection of related data, usually associated with a specific topic, such as population. Various measures and types of data may exist in the data set, but normally they should be related to the central theme.
- *Information:* The meaning or interpretation of data. Information is the knowledge obtained from data and implies explanation or significance of the collective facts or numbers. For example, a statistical measure is information concerning a string of numbers (data). The total and average measures of a group of data offer meaning, and are therefore the informational aspects of the small database in figure 2-1. These are not necessarily exclusive definitions. Sometimes, even a single datum can have meaning, so it too can serve as information.

The maps in figure 2-1 show how data and information are used and how geographic information can be developed, as follows.

- *Map A:* The top map shows a study area containing sample sites of vegetation identification (data points indicating individual observations). The numbers are codes for the type of vegetation found at each site. These are the data.
- *Map B:* When lines are drawn around identical data points, information emerges. The detailed vegetation landscape can now be interpreted and given meaning, namely areas of common tree cover. Collective data makes information. In turn, the map can be used as data input for further information, such as general tree types, as shown in map C. Thus, the map can serve as both data and information.
- *Map C:* Information derived from map B. The Red and Black mangroves have been combined into the category Mangroves (M). Coconut and Ivory Nut palms have been combined into Palms (P). Lime and Papaya have been merged into Fruit (F). Information can be data for other information.

To the GIS user, all three data sets and maps may be useful. Each map is a data set that has unique data and information; therefore, all are saved as part of a larger GIS data set. Note that GIS uses *information* as the core, not *data* (it is GIS, not GDS). Information is the primary purpose of GIS, not just data. Data is the input; information is the output.

Spatial Data

Spatial

TALIKA 12	TALIKA 14	DRANA 16
OXFORD COURT		
11	13	A ST. 15

MAP

Spatial Address	***Nonspatial*** Name	***Nonspatial*** Value	***Spatial*** Area
12 OXFORD CRT.	TALIKA	5000	600
14 OXFORD CRT.	TALIKA	7000	600
16 OXFORD CRT.	DRANA	6000	700

DATABASE

Spatial: Geographically referenced data, identified according to location

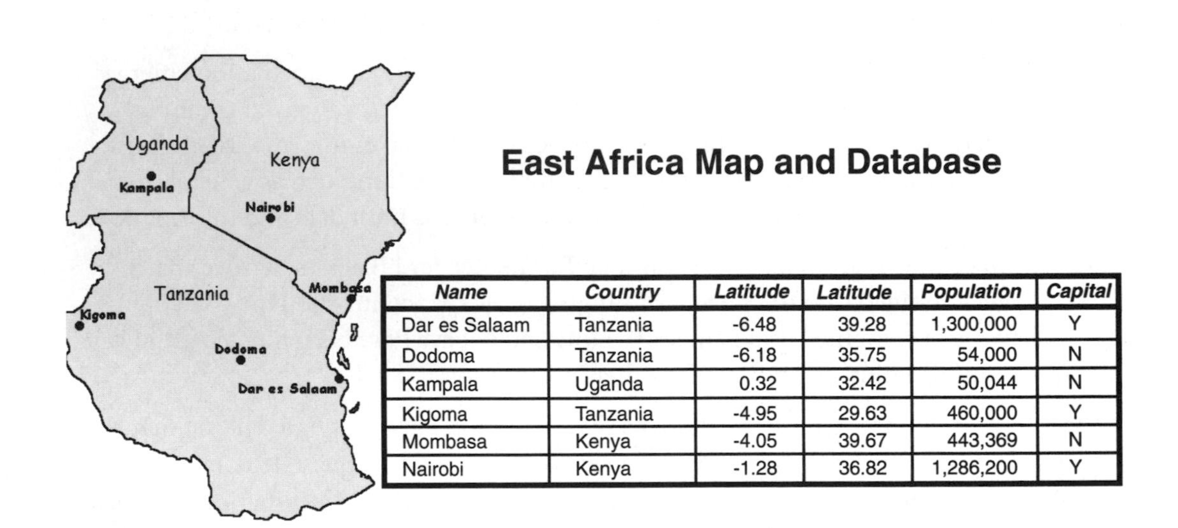

East Africa Map and Database

Name	*Country*	*Latitude*	*Latitude*	*Population*	*Capital*
Dar es Salaam	Tanzania	-6.48	39.28	1,300,000	Y
Dodoma	Tanzania	-6.18	35.75	54,000	N
Kampala	Uganda	0.32	32.42	50,044	N
Kigoma	Tanzania	-4.95	29.63	460,000	Y
Mombasa	Kenya	-4.05	39.67	443,369	N
Nairobi	Kenya	-1.28	36.82	1,286,200	Y

Which attributes are spatial?

Fig. 2-2: Geographic data in maps and databases.

Geographic Data

GIS uses spatial data as its primary component, but the term *geographic data* is more appropriate, as explained in the following. The sections that follow explore the nature of spatial and geographic data, such as the types and characteristics, spatial data relationships and their use in GIS, proximity relationships, and other aspects that establish the foundations for GIS.

What Is Spatial Data?

Spatial data occupies geographic (mappable) space. It usually has specific location according to some world geographic referencing system (such as Latitude-Longitude) or address system. Figure 2-2 shows a property map, with its database containing addresses. Both the map and database use spatial data.

The addresses in the database have specific locations and are therefore considered spatial data. The owner name and value of the property are nonspatial data. They are descriptive characteristics called *attributes.*

Many attributes are not location specific but can be found anywhere, even in more than one place (the owner can possess multiple properties). They are descriptions about the spatial data. For example, Talika is the owner of one address, but she could own, or move to, another address. The address does not change, but the owner and value might. GIS data sets usually contain spatial data and associated nonspatial data, both of which are termed *geographic data.*

Spatial data can be expressed in terms other than location. "Spatial" refers to "mappable" characteristics, such as size and shape. Size is calculated by the amount of area, and shape is defined by the position of the shape points (such as corners). The area of each property is given in the database, which is spatial data. The geographic data in a project, therefore, can consist of location, shape, and size as *spatial components*, and descriptions and associated data as *nonspatial attributes.* Various types of spatial data are discussed in the section that follows.

The map and database of East African cities is similar to the properties example, except with the use of latitude and longitude as the location references (expressed in decimal degrees, with negatives noting the Southern hemisphere, south of the Equator). Which attributes are spatial and which are nonspatial?

Spatial Data Features

Image 1: Spatial Data Examples

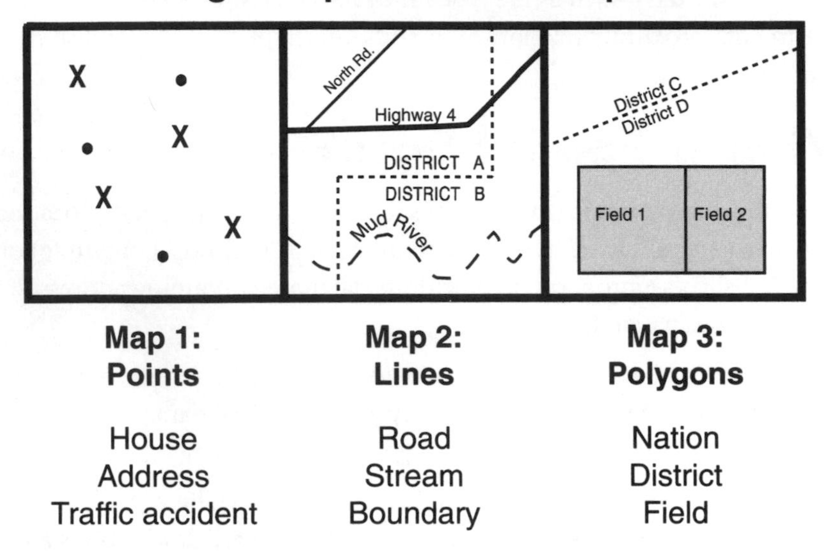

Image 2: Kenya Point, Line, Polygon Features

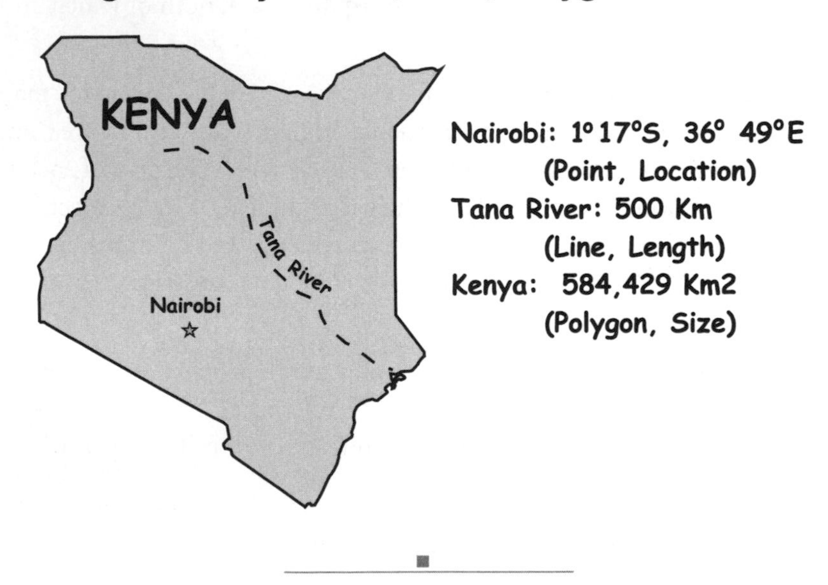

Fig. 2-3: Point, line, and polygon data types.

Types of GIS Spatial Data

In GIS, spatial data is classified as three main types: point, line, and polygon (area). Features, or landscape elements, are almost always depicted in one of these forms. Spatial data can take other, advanced forms, such as trend surfaces, but those are beyond the scope of this book. Image 1 in figure 2-3 shows examples of these three primary data types.

- *Point:* A *point* feature is a spot (or location) that has no physical or actual spatial dimensions, but does have specific location (map 1). That is, the point is shown as a convenient visual symbol (an X or dot, for example), to locate the feature it represents. However, the point does not indicate the actual length or width of the feature. Points generally designate features that are too small to show properly at the given scale (such as a house), or they depict locations of events (traffic accident) or nonphysical features (survey site).
- *Line:* A *line* is a one-dimensional feature having only length, no width. It has a beginning and an end (though not necessarily within the mapped area). Lines are linear features, either real (roads or streams) or artificial (administrative and property boundaries). Map 2 shows various line symbols that indicate different types of geographic features. The long-dash lines show a river, which is easily contrasted with the short-dash lines of the district boundary. Different line thicknesses ("line weights") can distinguish different classes of a feature type, such as road types. Here the heavier line is a highway and the thin line is a minor road. Line weight can also indicate a measure, such as amount of traffic.
- *Polygon:* A *polygon* is an enclosed area, a 2D feature with at least three sides and that has area and perimeter, such as parcels of land, agricultural fields, or political districts. Shown in map 3 are two polygons of agricultural fields that are inside a district polygon. The fields are shaded but the districts are not, a contrast that easily identifies the categories of polygons. Like roads, the two types of polygons have different line symbols (solid and dashed), which also help to establish basic classes.

GIS themes are usually stored as point, line, or polygon data files. In data entry, the type of feature usually must be declared so that the GIS can properly identify and manage the data. For example, a line feature should not be confused with one side of a polygon.

The map of Kenya (image 2) shows point, line, and polygon features and the type of measurement each has. A point has only a coordinate spatially, but it can be given an attribute symbol, such as a star denoting the national capital. A line has length spatially, and can be symbolized to note the type of feature it represents. Here, the river is shown as a dashed line. Polygons have area (size) and perimeter. They can be shaded or colored to show classification, if needed.

Discrete and Continuous Data

Map 1: Discrete Data

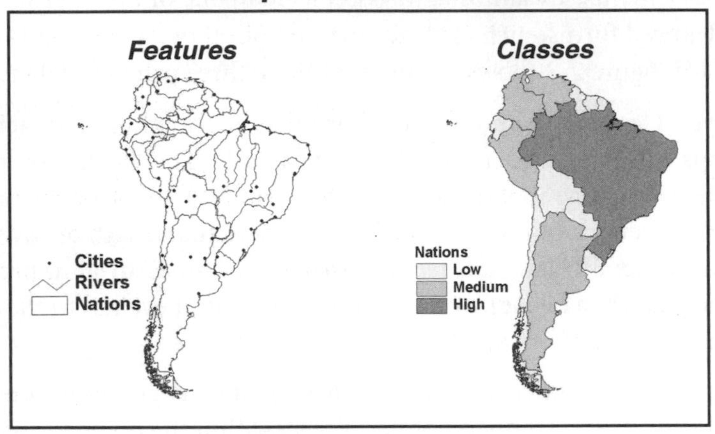

Continuous Data
Temperature Mapping

Map 2-A: Data Points

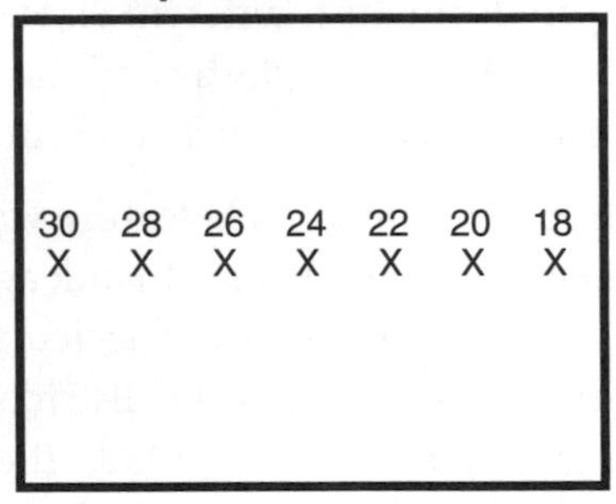

Map 2-B:Continuous Tone

Map 2-C: Discrete Zones

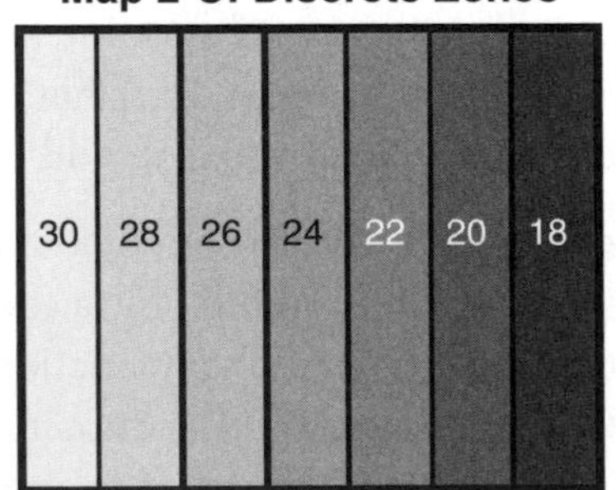

Fig. 2-4: Discrete and continuous data.

Discrete and Continuous Data

There are two basic types of measures of the geographic landscape, both of which characterize the environment, but in different ways. One is *discrete* data, the standard features and classifications recognized as normal map units. These are distinct features that have definite boundaries and identities. For example, a district has a definite location and shape, as do houses, towns, agricultural fields, rivers, and highways. They constitute separate entities; each is spatially well defined.

Map 1 of figure 2-4 shows examples of discrete features. The Features map presents individual cities, rivers, and nations, each of which is spatially and identifiably explicit and unambiguous. The classifications on the thematic Classes map are also discrete because they show the measure of a definite area.

Continuous data does not have definite borders or distinctive values. Instead, a transition from one measure to another is implied. Temperature from one point to another is not in discrete units, of course; there is a gradual gradation or transition between them. Map 2-A shows temperature data points for a region and map 2-B attempts to show the continuous nature of temperature change from one point to another.

The use of temperature zones as depicted in map 2-C is a convenient mapping device, but can be misleading because it implies discrete temperature regions that do not actually exist. Proper interpretation is easy for experienced readers, but others may not understand the true nature of continuous data. Other types of data can be treated as either discrete or continuous, depending on the project. Population distribution is often gradual from a dense core (such as a city) to more sparse rural areas, although not always in a smooth or regular progression. Slopes can be shown in uniform profiles or in discrete elevational units (stair steps). There are various methods of presenting geographic data and some will be explored in later chapters.

GIS Data Characteristics

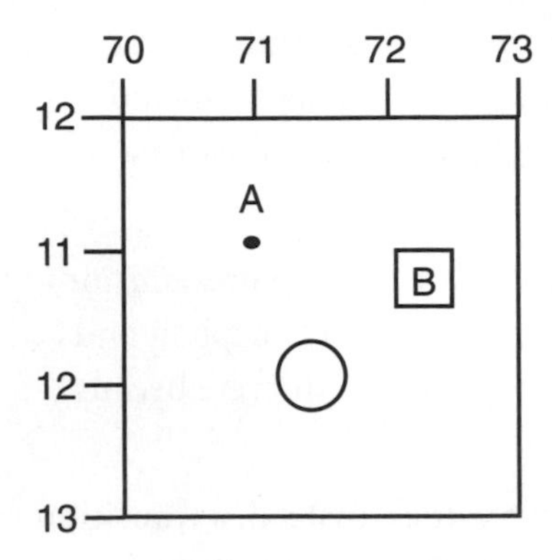

Map A: Location

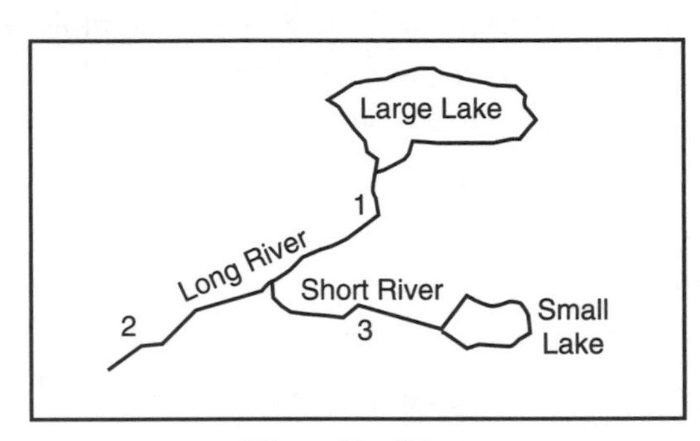

Map B: Size

River Database

Shape	ID	Name	Length
Line	1	Long	28
Line	2	Long	32
Line	3	Short	30

Lake Database

Shape	ID	Name	Area	Perim
Poly	1	Big Lake	506	77
Poly	2	Little Lake	160	35

C-1: Points

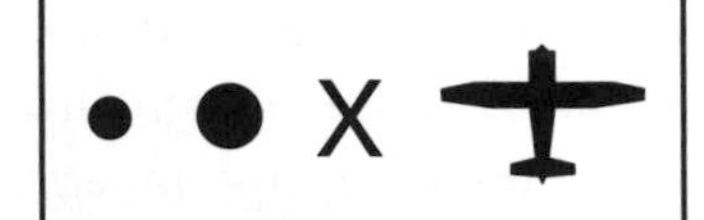

C-2: Lines

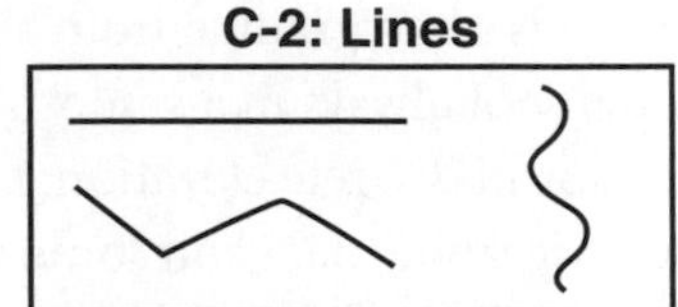

C-3: Polygons

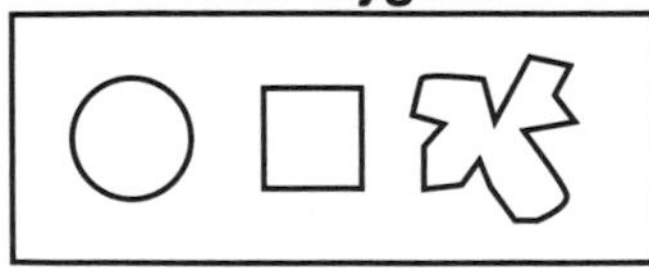

Map C: Shape

Fig. 2-5: Data characteristics in map and database form.

GIS Data Characteristics

At the heart of GIS applications is geographic data, which has numerous properties that make it much more functional and valuable than serving as mere features on a map. It is important to understand these major characteristics in order to take full advantage of GIS. The next several sections explore factors such as location, size, shape, spatial relationships, and time. Some of these primary characteristics can be easily seen on maps, whereas others require GIS.

Location, or *position*, is a major starting point of spatial measurement. Location can be descriptive, as in the old days of verbal descriptions ("10 meters north of the big rock, and then 150 meters east along the stream to the old house..."). Today, of course, better means of location are required. Standard geographic coordinate systems are normally used, such as Latitude-Longitude, which give X-Y (Cartesian) positions.

Map A in figure 2-5 uses a "Lat-Lon" system (sometimes "Lat-Long") that ranges from 10 to 13 degrees South latitude (the numbers on the Y scale) and 70 to 73 degrees East longitude (X-scale numbers). Point A feature on the map in the illustration is at approximately 11 degrees South, 71 degrees East (commonly noted as 11° S, 71° E). Where is polygon B feature? It is centered at 11.5° S, 72.5° E. What is at 12° S, 71.5° E? It is a circle feature. Sometimes a third spatial coordinate is used, referring to elevation or depth (typically called the Z coordinate). X-Y-Z are the major locational characteristics of GIS data.

Map B shows *size* characteristics. The lakes (polygons) have area and perimeter, whereas the rivers (lines) have length. Most GISs use databases or tables to store feature information, usually containing the feature type (shape), an ID number for each feature, and other descriptions, such as name and spatial characteristics. Normally, each type of spatial data (point, line, polygon) has an individual database, in this case the rivers and lakes.

Read the databases and note the spatial characteristics; namely, area and perimeter for the lakes and length for the rivers. For reasons that will be clear later, each segment of the river is represented in the standard database, identified by an ID number. The total length of a given river can be obtained simply by asking the database to add all Long River segment lengths.

Maps C-1 through C-3 show *shape*, an important descriptive element used in map and image interpretation. The shape of a feature often indicates its identity and role on the landscape. Point features have no real shape or spatial dimension, noting only the position of objects or occurrences (map C-1). They are represented by symbols, such as dots, geometric shapes, or icons.

Two settlement types (village and city) are shown as point features, using different size dots to designate population magnitudes. The dots are not the actual size or shape of those settlements. The X symbol specifies the location of a nonspatial feature, such as a riot location. Some maps use icons that have visual meaning in representing features; for example, airplanes indicating airports.

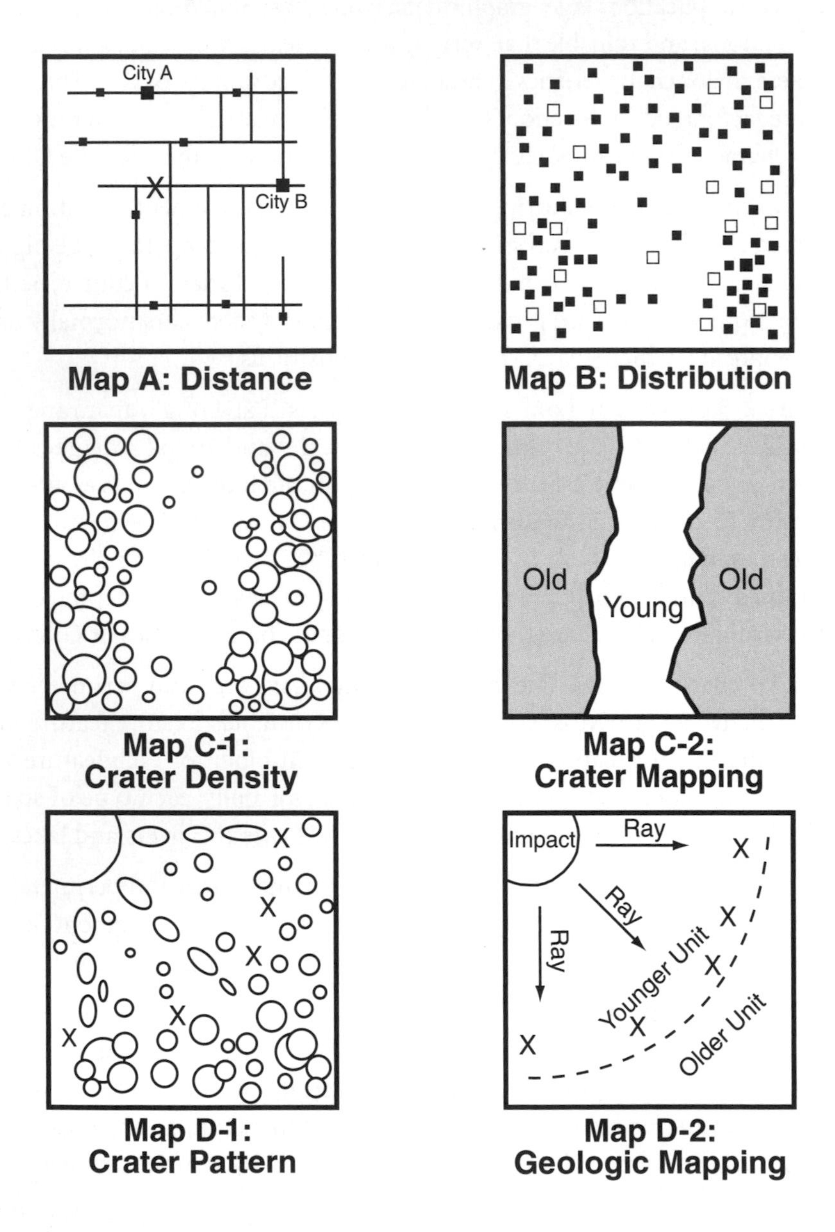

Fig. 2-6: Types of spatial relationships expressed in maps.

A line feature has length from beginning to end. The straight line feature in map C-2 (of figure 2-5)is a simple road, with only a beginning and end. The second line is complex, with various segments, each of which has individual length (as the rivers in map B). The curvy line feature is a river. It, too, has distinct segments or can use mathematically derived curves.

Polygon features (map C-3 of figure 2-5) have a wide variety of shapes, from easily interpreted circles and squares to complicated shapes that defy description. Spatial configuration can be important in understanding the landscape. Circles in agricultural areas, for example, suggest a certain type of irrigation technology (pivot sprinklers). Perfect squares or other geometric shapes typically signify either a human origin or an unusual natural process.

Shapes are quickly appreciated by the eye, but computers need special programs to make spatial form descriptions. Databases usually contain the feature type (point, line, or polygon), but they cannot adequately describe line or polygon shapes directly. Special mathematical techniques must be applied to describe and quantify odd polygon shapes.

Spatial Data Relationships

Features on maps have spatial relationships; that is, how they relate to each other in space. Spatial relationships can be very important in many applications, and GIS is an excellent tool for determining such characteristics. Of course, the human must decide the meaning of those relationships, but the GIS offers insight to the nature of spatial properties. Both the human and the computer are needed to interpret spatial relationships. Discussed in the following are distance, distribution, density, and pattern.

Distance

Distance from one feature to another is an elementary but important relationship. It is available through simple measurement. Map A in figure 2-6 shows point feature X and two cities. X is an accident site and the emergency response coordinator must know which city is closest. Distance can be measured as a straight line for helicopter flying or along roads for surface transportation. City A may be closer in air distance, but longer on the roads. GIS is very useful and very quick (almost instantaneous) in this type of measurement.

Determining the distance between the cities is easy, usually just by pointing to them with a measuring tool and reading the number. Many GISs offer other ways of measuring distances, including operations to determine distances from each small town (dots) to all other towns and cities on the map, either on a straight line or along the road network.

Distribution

Distribution is the collective location of features; the geographic dispersal or range. There are two basic ways of perceiving distribution: features among themselves and their spatial relationship with other features. Map B shows point features (small squares) distributed evenly in the upper half (though with no apparent pattern), and clustered to each side in the lower part. What do these distributions mean? Further investigation is needed, but some definite controls appear to be in operation on this landscape.

The point features are also considered in association with the square polygons. Is there a recognizable distributional relationship between the points and polygons? For example, there are more squares in the southern half of the map, where the point clustering to each side occurs. Is this a coincidence, or is there a functional relationship? More work is needed, but a basic look at spatial distributions has been useful.

Density

Density is the number of items per unit area; how close features are to each other. Maps C-1 and C-2 depict two basic densities of impact craters and what can be done with the information. Map C-1 shows lunar craters densely clustered on the east and west part of the map and very few in the center.

Impact crater density is an indicator of landscape age on some planets; that is, more craters indicate an older geologic unit because of the longer exposure to meteor impacts than the units with fewer craters. This area can be mapped into two ages, as shown on map C-2: the older units with many craters and relatively young lava plains with few craters. The significant difference in cratering densities indicates substantial time between their formations. The spatial characteristics of features are important clues to geologic processes on the moon.

Pattern

Pattern is the consistent arrangement of features, similar to (and can include) distribution and density. Where discernible patterns exist, specific processes may be in effect on the landscape. Map D-1 shows another part of the lunar surface, with a large crater in the upper left, and an interpretation map in D-2.

Note the three strings of craters at the upper half of the mapped area in D-1. Their pattern is linear and apparently radial from the large corner crater, suggesting an association and origin. These are probably secondary craters, made from material thrown out from the larger impact. Their pattern is the first clue, but the oval shape of the craters supports the thought of low velocity and low-angle trajectory (called "rays").

The lack of pattern in the lower part of the map area indicates a different origin for those craters; that is, they are not part of the larger impact. The higher density of craters suggests an older geologic unit (as in map C-1). Also, the X point features appear to be in a pattern, evenly spaced from lower left to upper right. They are gravity anomaly measurements (out of the ordinary), which may suggest a hidden geologic feature under the surface, or perhaps some relationship with the large impact crater, given their slightly semicircular pattern and even distance from the crater. Pattern and lack of pattern can be helpful in landscape interpretation.

Proximity Relationships

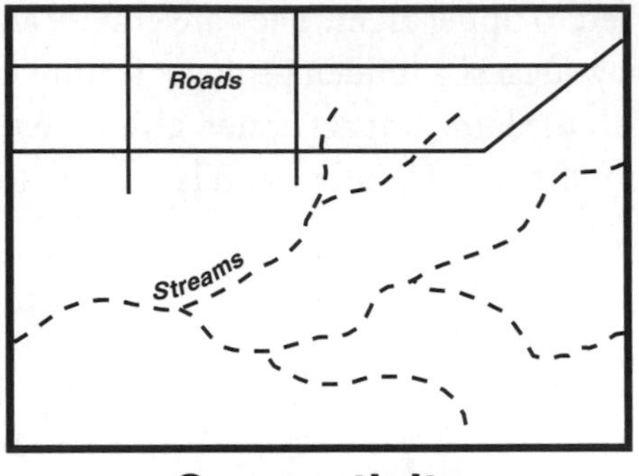

Connectivity

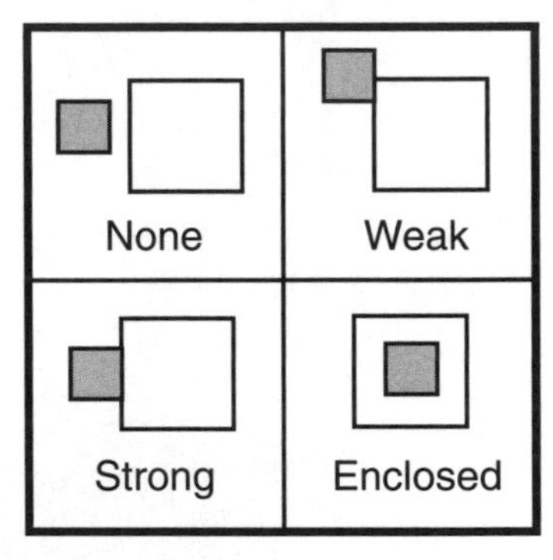

Contiguity

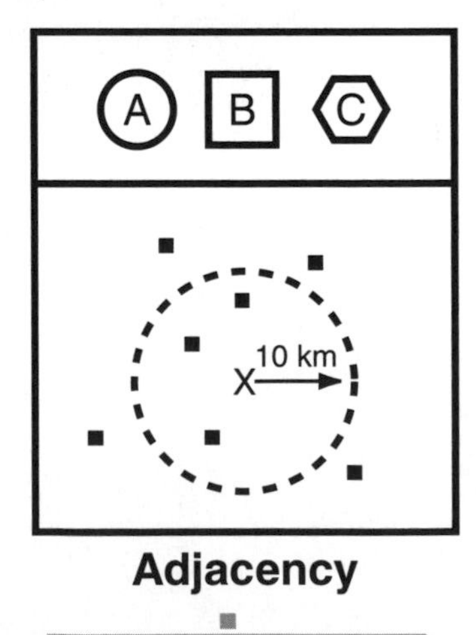

Adjacency

Fig. 2-7: Three types of proximity relationship.

Proximity Relationships

Proximity relationship refers to closeness; features that may have association because they are spatially near each other. These are also termed "neighborhood" characteristics; the elements of neighborhood or proximity analysis. The basic concept is that the closer features are to one another, the more potential relationship they have (and the further apart, the less likelihood of relationship). Depicted in figure 2-7 and described in the following are three common proximity characteristics that have potential meaning regarding functional relationships. Chapters 4 and 8 through 10 explore these characteristics further.

- *Connectivity:* Considers features that connect, or at least touch. It is a logical assumption that connected features may have some meaningful association. Roads that connect are probably part of a network system for transportation. Streams (dashed lines on the map) that connect are part of a single hydrologic system; part of the watershed that probably has common physical influences. A stream and road connection can indicate a bridge or drainage pipe.
- *Contiguity:* One aspect of connectivity is *contiguity*, the degree of connectivity. Polygons with shared borders probably have functional relationships proportional or corresponding to the amount of border that is shared. That may not be true, of course, but it is a logic that should be explored. Illustrated are four degrees of connectivity: none (no physical link), weak, strong, and enclosed. Can you think of examples each of these may represent?
- *Adjacency:* Considers nearness, or the features that are close to each other. In the illustration, feature B is adjacent to A, and C is adjacent to B, but C is not adjacent to A. Actually, a "near" distance is usually defined, and features within that distance are considered in proximity. The use of zones (buffers) around selected features is a common and very useful GIS operation. Determining the number of villages within 10 kilometers of point X is one example. The meaning of adjacency must be interpreted by the user for any given situation.

Time and GIS Data

Population Database 1

District	Pop.	Date
A	2000	1960
B	4000	1980
C	5500	1960
D	1500	1970

Unacceptable

Population Database 2

District	Pop. 1980	Pop. 1990
A	3500	4500
B	4000	5500
C	6000	8000
D	2000	3000

Acceptable

Geologic Database 3

District	Data Date	Age in Mil. Years
A	1955	11.2
B	1950	10.0
C	1970	12.3
D	1960	11.4

Acceptable

Map 1: River Dynamics

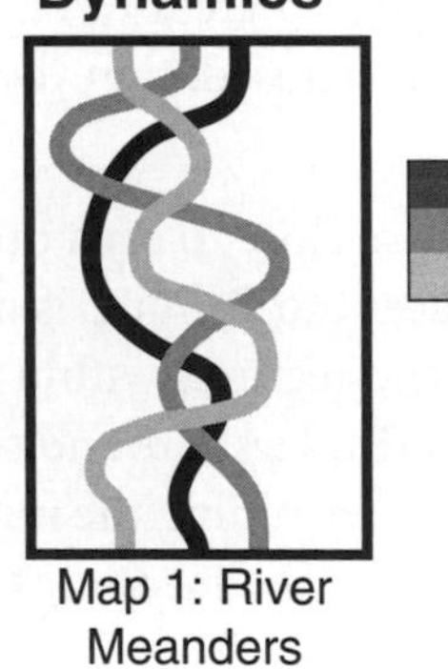

Map 1: River Meanders

Map 2: Spatial Evolution

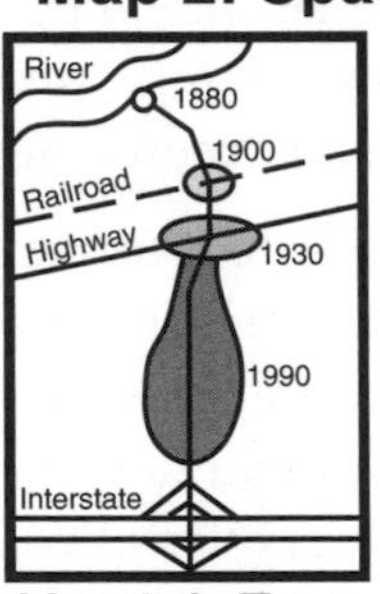

Map 2-A: Town Evolution

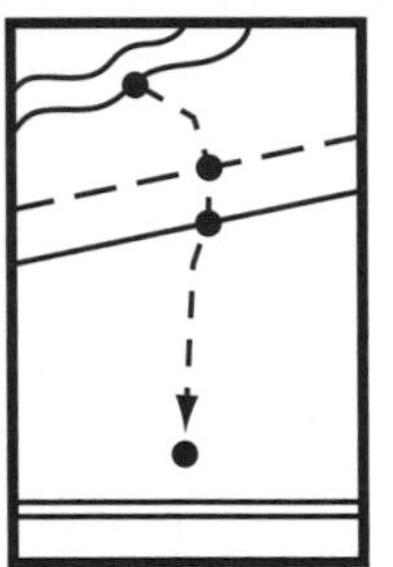

Map 2-B: Town Center Model

Map 3: Deforestation 1970-1990

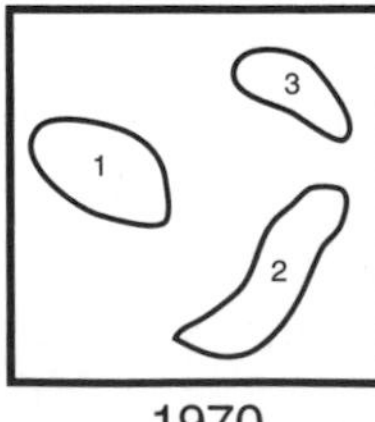

1970

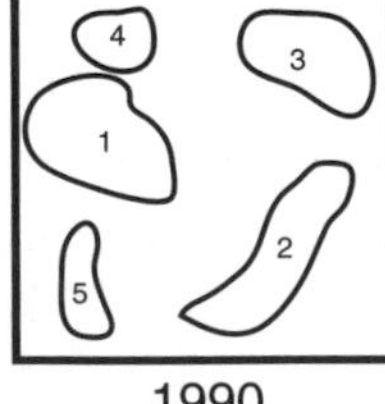

1990

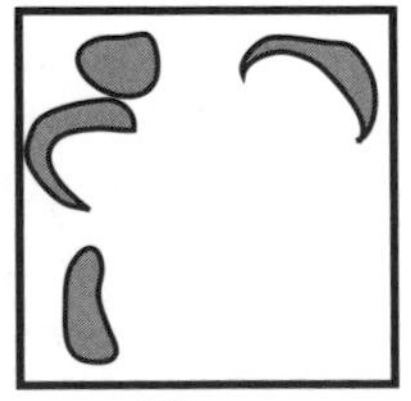

Change 1970-1990

Fig. 2-8: The effect of time on GIS data in database and map forms.

Time and GIS Data

Time is an important *data quality* element that can be a critical descriptive characteristic of spatial data. It can be used or expressed in terms of data quality, dynamic data, and trends. These are *temporal* components of geographic data that can be very important in understanding the processes affecting and shaping physical and cultural landscapes.

Although not always evident, the date is essential in some applications. A population project needs up-to-date and consistent data, and old data may cause confusion and error. The lifetime of data can be significant. Population numbers change constantly, and in many countries demographic data can be expected to have an official 10-year lifetime, the interval between censuses. Geologic data normally has long life, but urban land use changes rapidly.

Illustrated are two population databases at the top of figure 2-8. Database 1 shows district populations from various years (1960 to 1980). These dates are too far apart to be consistent for population analysis. The database is unacceptable for this project. Population database 2 presents two fields (columns) of dates, 1980 and 1990. The population of each district is recorded for those dates and there is consistency in the data. It is acceptable for demographic comparison and analysis.

The Geology database 3 has three geologic units with their age-dating measures. Even though the analyses were completed in widely varying dates (from 1950 to 1970), the time gap will have no effect on data credibility or comparison (unless there are questions about the differences in techniques used over that span of time).

Dynamic data is changing data. The universe is continually evolving, and static data does not actually exist. Geographic data is dynamic and often time is an important attribute to consider. Features can change shape, size, position, and attributes over time. In fact, many geographic phenomena are naturally dynamic and can be expected to show significant changes over a short span. These changes include such processes as natural seasonal changes, population growth, vegetation growth, and city evolution. This is particularly true of natural hazard events, such as floods and volcanic eruptions.

Rivers are very fluid and erosive features, changing shape constantly. Map 1 presents river meanders over time. Even though each channel can be displayed separately in individual frames, the single-frame presentation of the collective features allows comparison and a better visual appreciation of the process and its effects. Advanced technology will soon permit easy animation of change, making these standard static views obsolete.

Many settlements are located on transportation routes, from rivers in pre-mechanical days to super highways of today. Map set 2 shows two versions of an American town changing position and size throughout the years as transportation links evolved. Map 2-A shows the changing locations and growth of settlements as new transportation lines were developed,

Trends

Map A: Deforestation Trends (Spatial changes)

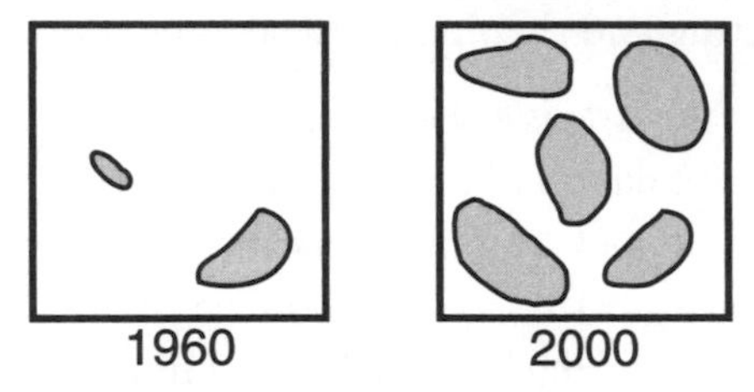

Map B: Population Trends (No spatial change)

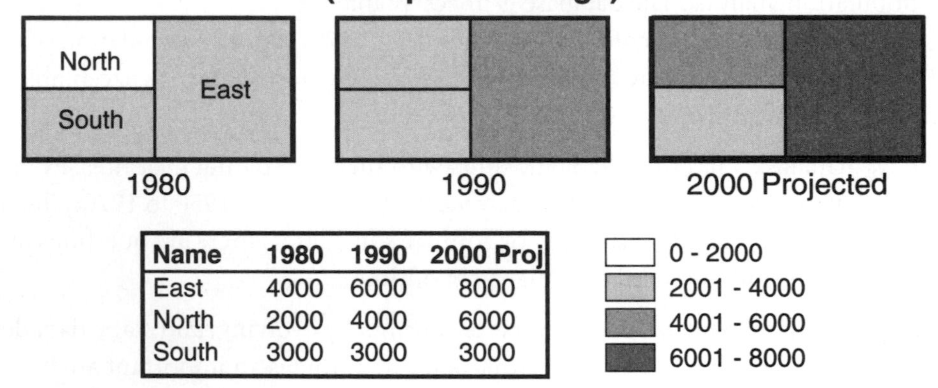

Name	1980	1990	2000 Proj
East	4000	6000	8000
North	2000	4000	6000
South	3000	3000	3000

Map C: GIS Generic Questions: Patterns and Relationships

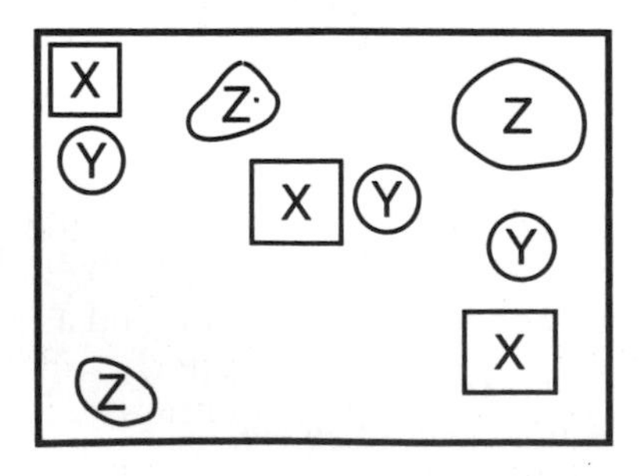

1. Does Feature X occur in a pattern?
Yes, in a line from northwest to southeast.

2. Is there a relationship between X and Y?
Yes. Y is always close to X and seems to change position around X in a counterclockwise motion from west to east.

3. What other spatial patterns exist?
Feature Z is always near a border and increases in size west to east.

Fig. 2-9: Comparison of data over time and space.

each period represented by a polygon. Map 2-B of figure 2-8 illustrates the points representing town centers (a model of the shifting locations). These presentations are a bit difficult to understand easily and the on-coming advances in animation will be most welcomed.

Deforestation is a major issue in many parts of the world, and a sequence of maps for periodic years illustrates forest changes (polygons). The figure 2-8 map 3 set shows forest coverage for 1970 and 1990, and then the actual changes. This type of GIS display is very useful for understanding landscape processes.

Trends

Trends are determined from comparisons of data over time (temporal trends) and over space (spatial trends). A few changes may show or indicate a trend. Geographic trends involve spatial directions or characteristics that are moving or changing in a particular manner, suggesting an active, orderly process affecting the landscape (otherwise, there is random, unpredictable change).

Trends are, in effect, patterns of geographic change that have some predictability. Geographic trends can be purely spatial (changing size, shape, or position of features over space and/or time), nonspatial (evolving attributes, such as growing population or changing value of some type), or both.

Map set A in figure 2-9 shows the predicted forest before (1960) and after (2000) the times shown in map 3 of figure 2-8. Of course, the actual results may be different, but in absence of observed data, the extrapolated trend is the best information available. Note that area 2 shows no change from 1970 to 1990, so no change is depicted for 1960 or 2000. If this is illogical to the foresters, the limitations of spatial statistics are demonstrated.

Areas 4 and 5 are new in 1990 and thus show very large growth for 2000. That is, the change was greater for those areas between 1970 and 1990 than for areas 1 and 3, so larger increase is projected for 2000, even though all of the areas may grow the same. Pure statistics may be used for a preliminary view, but the user must employ regional knowledge for the final analysis.

Map B illustrates a change of population attributes in a set of districts from 1980 to 1990 (there are no spatial changes in the districts). Population change (loss or gain) is used to predict for the year 2000. Study the database and map legend. Note that the classifications are consistent, and therefore comparisons and predictions can be made. In addition, a distribution of high and low population can be seen in each map—a purely spatial trend without the time factor. Map C demonstrates how changing shapes, locations, and associations can make patterns over space (across the mapped area). The questions address possible patterns and relationships. Perhaps several landscape processes are in effect here and further investigation is in order.

Project Data Files

South Asia Mapping Project

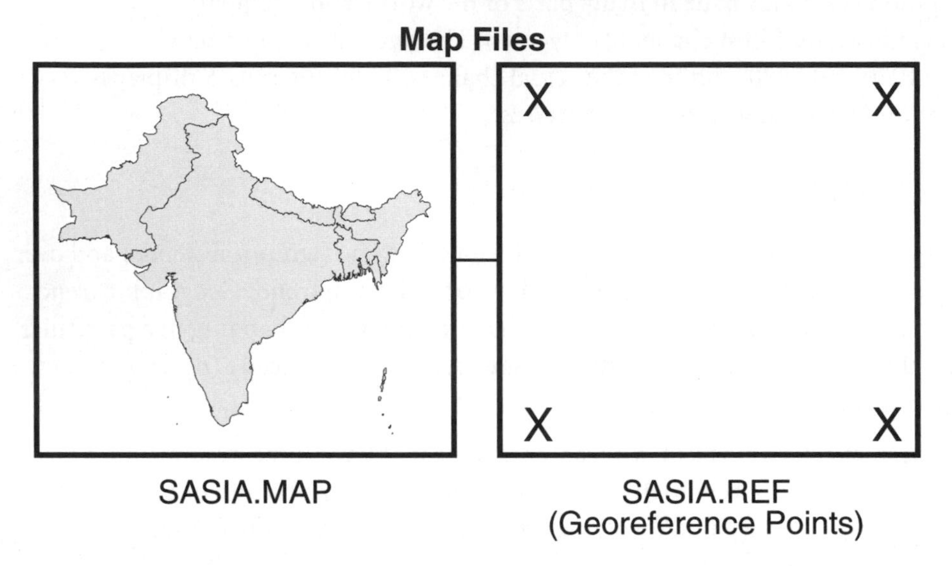

Database Files

Country	Population	Area (km^2)
Bangladesh	120732200	138507
Bhutan	1586631	39927
Sri Lanka	18321920	66580
India	894608700	3089282
Nepal	19927280	147293
Pakistan	126693000	877753

SOUTHASIA.DB

Metadata
Source: ESRI Corp. permission.
ArcWorld Supplement Map data
Date: 1998 compilation
Country name: unofficial
Population data: 1994 est.
Area: Square kilometers,
excluding maritime area
Original display scale: 1:26,000,000
Coordinate system: Latitude/Longitude
Datum: World Geodetic System, 1984
Coodinate precisions: Double
[Continued in Metadata.txt]

METADATA.TXT

Fig. 2-10: Data components of a GIS project.

GIS Projects, Data Files, and Themes

As illustrated in figure 1-8, a GIS project contains various types of data. It is a collection of data, analysis, presentations, support graphics, and associated files that address a particular application or problem. A GIS project usually contains at least several themes, all relating to an overall application or purpose.

A GIS theme includes one or more data sets dealing with a particular topic. Each theme in a project is defined as a point, line, or polygon, and each is stored in separate data files to avoid confusion of different data types. Discussed here are the nature of data files (what they contain), the basic types of data files, and several considerations GIS users have when dealing with data. These are the important project components; others will be seen throughout the following chapters.

As indicated in figure 2-10, most GIS projects use at least two types of working files (heavy border boxes in the illustration): map files that present the visual graphics, and the database with attribute data. The map file contains the visual data and presents the basic geography of the theme. A database has descriptive data in spreadsheet format (the following section discusses databases in detail).

Although some projects may function with the database alone, GIS links it with the graphics to provide data synergy for the project. (Synergy is the interaction of two or more agents or forces so that their combined effect is greater than the sum of the individual effects. In a GIS project, integration of visual data with the database makes a powerful GIS capability that is much greater than the contribution of the data files alone.)

Illustrated is a simple South Asia mapping project, containing the two primary files and two associated files. Depending on the software, various support files may be used. The map can have a separate file containing the primary location points defining its specific world area (usually entered at the data entry stage), called *georeference points* (discussed further in Chapter 5). The database often has a metadata file that explains the terms, measures, quality, sources, and other useful information about the data. Note that each type of file has a particular name and extension (*name.extension*) that helps to identify its function. The South Asia map file is "Sasia.map," whereas the georeference data is in the "Sasia.ref" file. The database has a ".db" extension, whereas the metadata resides in a common text file (".txt"). These are generic designations and specific GIS may use custom name extensions.

Data structure and data quality are two critical components for the GIS project. Data structure refers to the organization and format, or arrangement of data, used to define data so that the software can use it. Like books, data sets have many forms, or styles (formats), even though the text is English. For example, a satellite image is structured as a grid of cells, with each cell having a single data value. This is called a raster cell format. A GIS vector format, on the other hand, consists of lines, just like a hand-drawn map (discussed in the next chapter). Both may show the same area, but in different data formats for different software.

Data quality deals with the traits or conditions that determine how useful the data set is for a project. Acceptable data quality is accurate, credible, complete (no gaps or missing data), properly configured and defined (correct theme and correct labels), and timely (correct and appropriate date). A good data set has sufficient meta documentation for the user to judge the quality and how appropriate it is for the intended project. Having high-quality data is always a special concern for GIS projects.

The GIS Database

South Asia Attribute Mapping

South Asia Database

Country	Population	Area Km^2	Pop. Density	Currency	Landlocked	Capital
Bangladesh	120732200	138507	872	Taka	N	Dhaka
Bhutan	1586631	39927	40	Ngultrum	Y	Thimbu
India	894608700	3089282	290	Rupee	N	New Delhi
Nepal	19927280	147293	135	Rupee	Y	Kathmandu
Pakistan	126693000	877753	144	Rupee	N	Islamabad

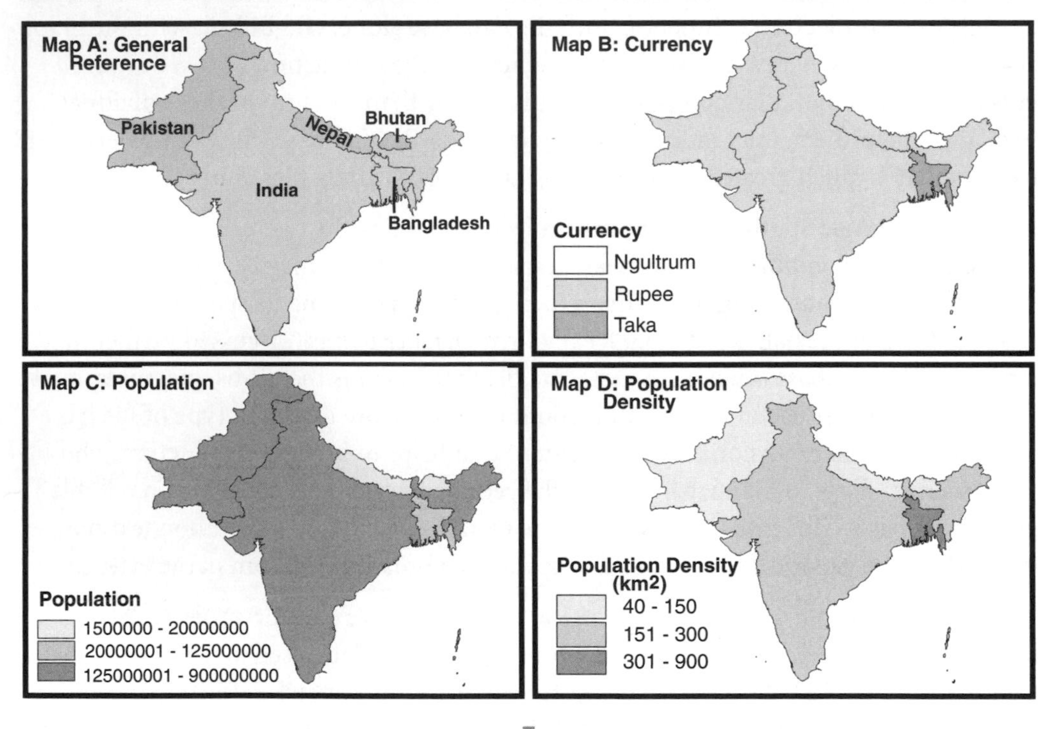

Fig. 2-11: Attribute components of a GIS database and their map equivalents.

The GIS Database

The database is essential to modern GIS; without it, GIS would be little more than computer cartography. This section explores the role and functions of the GIS database, beginning with a basic description, followed by the use of relational database, the link between maps and databases, and ending with the concept of the database approach in GIS.

The Database in GIS

As noted in Chapter 1, GIS data comes from many sources, such as maps, remote sensing imagery, CD-ROMs, and the Internet. These diverse sets of data are not always easily integrated. The central data integrator for GIS is the database. A major strength of GIS is that it accepts and merges diverse databases and different types of data, giving the user a flexible and powerful set of data for project work.

Database in GIS is a simple concept—a list or table of data arranged as rows and columns, as shown at the top of figure 2-11. Rows are the *records*, or each observation entered into the database. Columns are called *fields*, which present the *attributes* or descriptions of each record. Attributes are the data descriptors, such as color, ownership, magnitude, and classification. Databases can be simple, such as this one, or very large, with dozens of fields and hundred of records.

Field data, for instance, are records of descriptions or measurements of each surveyed site or feature. Attributes can be in various forms, including text description, numbered measurements, classifications, or any other way that is useful to the project. Achieving high accuracy and careful observations are necessary for quality attributes. The difference between qualitative and quantitative attributes is explained in the following.

- *Qualitative attributes:* Have no measurement or magnitude; they are non-numeric descriptions. Names, explanations, and labels serve as descriptions. Qualitative descriptors have no numeric meaning, even if they are code numbers (e.g., 1 = category 1), and therefore cannot be used in statistical analysis. (Students sometimes mistake code numbers for actual measurements, and careful reading is necessary.)
- *Quantitative attributes:* Are numeric and have mathematical meaning; the numbers serve as measurements or magnitudes of the feature to which they refer. The numbers might be a measurement for a single feature, such as a city's population. Mathematical meaning allows for statistical analysis.

The database for South Asia at the top of figure 2-11 (expanded from the database in figure 2-10), consists of an initial identity of each country and various attributes. Note that only the Area column is spatial and the others are nonspatial and descriptive. Population, area, and population density are quantitative, and the others are qualitative attributes. Each attribute can be made into a theme and a map.

Illustrated are four mapped themes derived from the database. The Country Names theme (map A) is shown as individual nations, but the Currency theme (map B) has only three classes because the rupee is used by four countries. Population (map C) and Population Density (map D) have been grouped into classes rather than mapping each number individually. Classification is usually easier to understand than unique numbers, and provides meaning (e.g., logical groups or associations).

Relational Database

Central America Database

Country	Population	Area (km^2)	Pop. Density	Currency
Belize	207586	22175	9	Dollar
Costa Rica	3319138	51608	64	Colon
El Salvador	5852470	20697	278	Colon
Guatemala	13021270	109502	94	Quetzal
Honduras	5367067	112852	48	Lempira
Mexico	92380850	1962939	47	Peso
Nicaragua	4275103	129047	33	Cordoba
Panama	2562045	74697	34	Balboa

South Pacific Database

Nation	Population	Growth (%)	Land (km^2)	Capital
Fiji	716000	2.0	18272	Suva
New Caledonia	146000	1.2	19103	Noumea
Solomon Is.	286000	3.3	29758	Honiara
Tonga	97000	1.1	697	Nuku'alofa
Vanuatu	140000	3.1	12189	Port Vila
W. Samoa	162000	0.7	2934	Apia

Query: Show all nations having:
Population > 100,000
Growth Rate > 2.0
Land Area > 15,000 Km ^{2}
Print Capital and Nation

Answer: Suva - Fiji
Honiara - Solomon Is.

Fig. 2-12: Relational databases have linked input categories.

The Relational Database: Using Related Data

There are several types of databases, each with special advantages and disadvantages. Some offer only basic options, such as a "flat-file" database system that is little more than a simple data table. That type of database is a rather elementary container of records, but is nevertheless one a GIS can use to perform meaningful work.

One of the best types for GIS is the relational database. It is an integrated table constructed so that each record and its attributes are linked and related (cross-referenced) to all records and their attributes. It is possible to ask questions that reference two or more attributes, looking for specific conditions in each. This is called a *conditional query.*

For example, in the Central America database in figure 2-12, a conditional query is entered to find the nations that have populations under 5,000,000 and population densities over 30/km^2. Read the data and do the work manually. Costa Rica, Nicaragua, and Panama meet these conditions. You can do this visually and slowly, possibly with error, but the computer requires almost no time to complete the task accurately. Consider a much more complex conditional query on a database with hundreds of records and dozens of attributes, and the power of database technology can be appreciated. As will be seen, once the nations have been selected in the database, there are additional options in GIS.

Try another conditional query that is slightly more complicated: What is the type of currency used by the nation(s) between 100,000 and 150,000 square kilometers of area and population densities over 50 people per square kilometer? Notice how the work is done manually, typically by selecting the appropriate area nations first, then the population density nations from the initial list. The answer is the Quetzal of Guatemala.

Computers are good at filtering attributes and entries in relational databases. Multiple conditional queries are very useful in GIS work, often forming the initial data reduction procedure (reducing the number of records to be analyzed). The second database illustrated in figure 2-12 is a set of South Pacific island nations and a few attributes. A search for nations having the specified conditions related to population, growth, and land area is shown below the database.

The program searches the first field for population requirements, selecting all records with populations over 100,000 (follow along: see the boxes in side each column of the database). From those selections, it moves to the second field and searches for growth rates over 2.5. Then, from the three records it looks for countries having land area over 15,000 square kilometers, finding two, Fiji and Solomons. The results are printed with the country name and its capital, as requested.

With a well-ordered and well-constructed relational database, the only limit to queries is the imagination. Most powerful GISs employ the highly useful relational database to perform this type of task. The next section discusses additional GIS capabilities using the relational database.

Database–Map Link

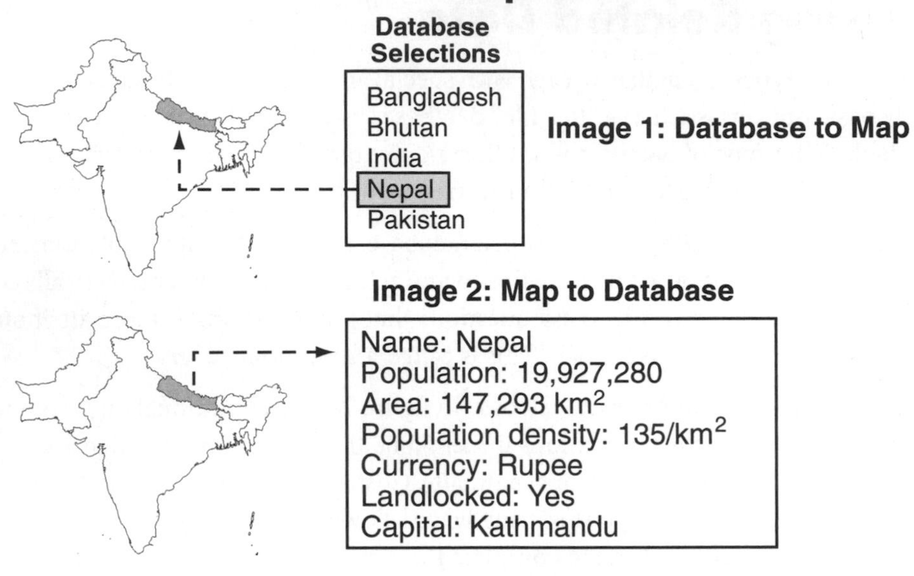

Image 3: Map to Database Multiple Selection

Country	*Population*	*Area*	*Currency*
Argentina	33796870	2781013	Peso
Chile	13772710	742298	Peso
Paraguay	4883464	400089	Guarani
Uruguay	384641	178141	Peso
Brazil	151525400	8507128	Cruzeiro Real
Colombia	34414590	1141962	Peso
Peru	24496400	1296912	Nuevo Sol
Venezuela	19857850	916561	Bolivar
Ecuador	10541820	256932	Sucre
Bolivia	7648315	1090353	Boliviano
Guyana	754931	211241	Dollar
Suriname	428026	145498	Guilder
French Guiana	130219	92911	Franc

Selection box

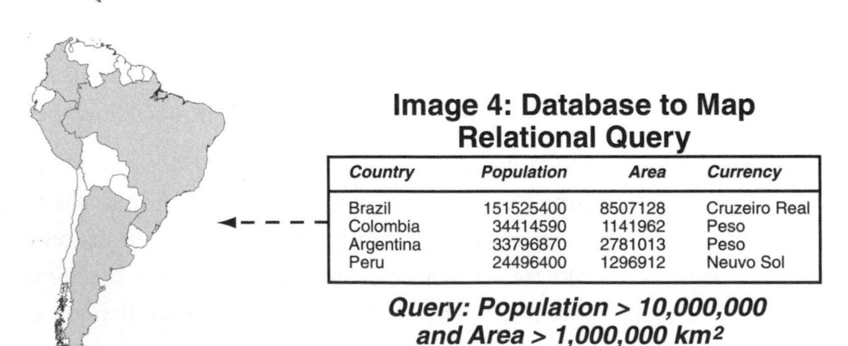
Image 4: Database to Map Relational Query

Country	*Population*	*Area*	*Currency*
Brazil	151525400	8507128	Cruzeiro Real
Colombia	34414590	1141962	Peso
Argentina	33796870	2781013	Peso
Peru	24496400	1296912	Neuvo Sol

Query: Population > 10,000,000 and Area > 1,000,000 km^2

Fig. 2-13: Maps and databases can be interactively linked.

Database–Map Link

A major strength of GIS is the interactive link between the database and the map. When selections are made in the database, the selected records are highlighted on the map. When features are selected on the map, they are highlighted in the database.

Image 1 in figure 2-13 shows a simple database-to-map selection. By selecting Nepal in the database or from a list, it is highlighted on the map. Image 2 demonstrates a map-to-database selection, usually by pointing and clicking on the map, and the information for Nepal is displayed. These are simple but very useful operations.

Image 3 is another map-to-database operation, this time using a click-and-drag box to select nations in southern South America (being careful to avoid Brazil with the box). The selection highlights the nations on the map, as well as in the database. All of the selections are displayed at the top of the database for easy reading, and are organized by population in descending order—two additional standard GIS database options accomplished by a few clicks of the mouse.

Image 4 is a relational query from the database, with the four nations meeting the specified conditions highlighted on the map and presented in a separate display box. This database-to-map function is very useful for many GIS applications and shows the strength and flexibility of GIS's map and database links.

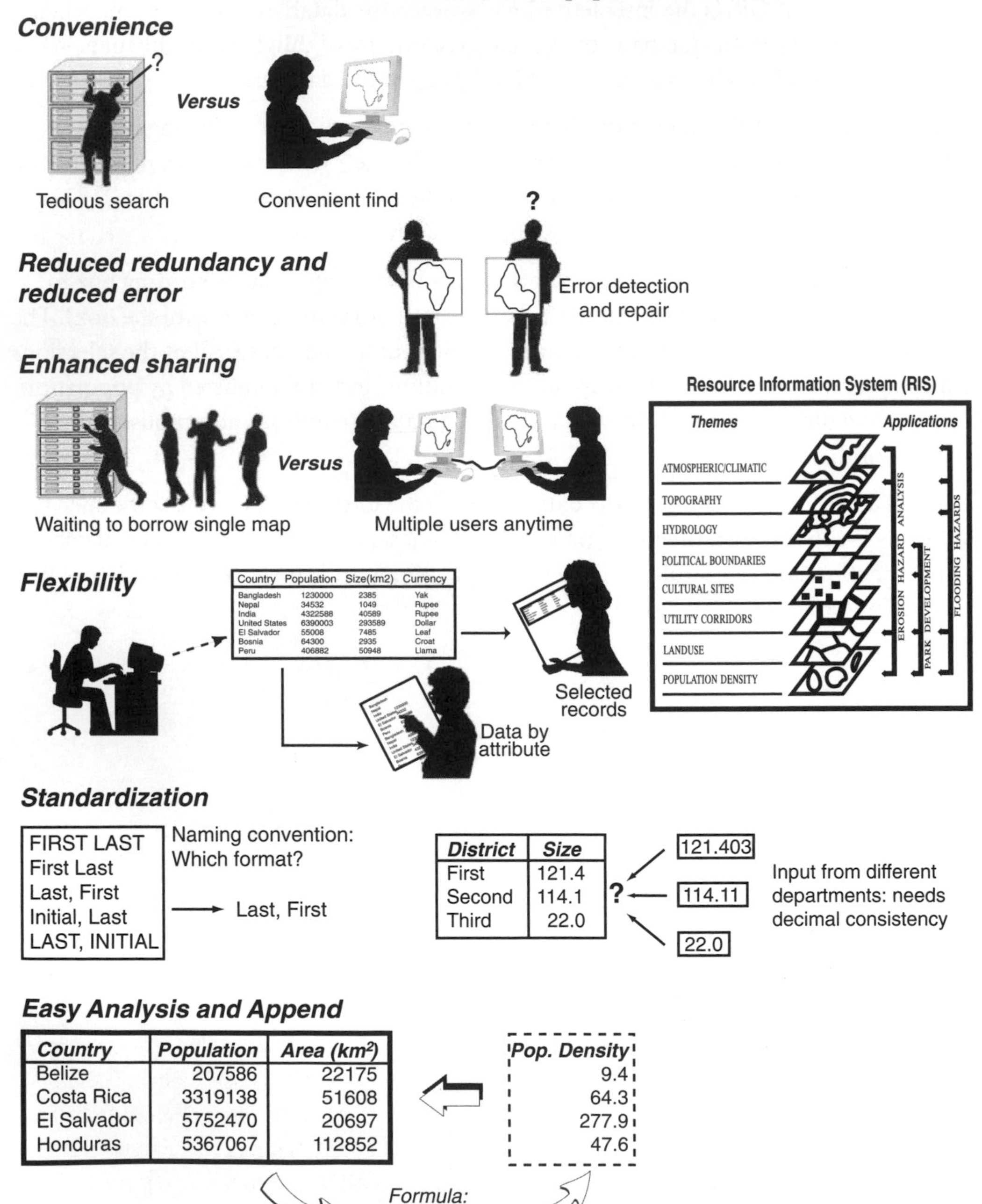

Fig. 2-14: Benefits of the database approach.

The Database Approach

It is easy to see how powerful and important the database is in GIS. Without it, mapping would be largely a drawing process, with very little analysis. Much of the analytical work in GIS is accomplished at the database, with the graphics performing valuable data visualization presentation. Visual attention may be on the sophisticated map and charts, but the core operations are often within the database.

The GIS *database approach* is a simple concept with two basic aspects. The first is recognition that the database is central to GIS projects and to organizations using GIS. Databases offer several benefits to projects and organizations, as presented in figure 2-14.

- *Convenience:* A central project database makes data accessible to everyone. (Access can be controlled by authorization programs.) Rather than searching map cabinets, reports, and statistics in various locations, a quick computer operation produces the desired data. Project and organizational efficiency is always welcomed.
- *Reduced redundancy and error:* Duplication of data is not necessary for individual parts of a project or organization when a centralized database is used. Data management costs are decreased and the potential for error is lower. There are fewer conflicts when everyone uses the same data. Inconsistencies can be discovered and repaired.
- *Enhanced sharing:* Several users in different offices or agencies can use the same database, thereby maximizing the efficiency of management and quality control. One database can serve multiple needs. No longer do workers have to wait for maps or data that have been borrowed by others (which is another factor in reducing redundancy). The illustrated Resource Information Systems (RIS) shows that various applications can use a central database.
- *Flexibility:* A good database offers a great deal of flexibility in the use of data, from a variety of retrieval options to diverse analysis and various reporting formats. It can "grow" to accommodate new users and new applications, while remaining stable for existing users. The two workers are viewing the same data with different designs. The RIS also demonstrates database flexibility.
- *Standardization:* Because a project or an organization can include many data contributors and users, control of the data format and quality is important. Usually there must be agreement on how the database is structured (such as naming conventions), what types of data go into it, how the data is structured (number of decimal places, for example), and how modifications are to be made. In short, GIS promotes control and standardization of data so that every user understands what is included. Often, the database approach introduces this type of agreement and standardization as a new idea to an organization. This also helps commit management to a specific direction, which in turn makes arbitrary changes more difficult.

A second important aspect to the database approach concept is that much of the data processing and analytical operations take place in the database when possible. Many of the early GIS systems used the map graphics for many tasks, such as overlay analysis and recoding, basically because of the poor links between the weak database (flat file systems) and graphics. Today's powerful systems permit the user to work with the data directly in the database and to link to the map graphics easily, making the project and organizational operations more efficient and effective.

Illustrated at the bottom of figure 2-14 is a simple operation that makes a new database column showing population density, which is calculated by applying the formula "Population/Area." The new number is entered automatically into each record in the new field. This type of operation is conveniently faster and easier than the old method of graphic overlay and tedious manual recoding.

The database approach says to use the database when possible, supported by the visual capabilities. This approach does not diminish the role of map graphics. Without accurate mapping, for example, the database is unreliable. Some operations are best performed with the graphics, particularly various raster modeling steps. And, of course, the data visualization is essential for many applications. With an integrated database and map graphics, GIS is a robust technology and methodology.

CHAPTER 3

RASTER AND VECTOR DATA

Introduction

SPATIAL DATA IN GIS HAS TWO PRIMARY DATA FORMATS (the arrangement of data for storage or display): raster and vector. Raster uses a grid cell structure, whereas vector is more like a drawn map. Only a few years ago, GISs were dedicated to one or the other, but today's systems integrate both. Each data format has its advantages and disadvantages, and the GIS professional recognizes that many projects have need for both.

Vector and Raster Formats

Image 1: Vector and Raster Data Structure

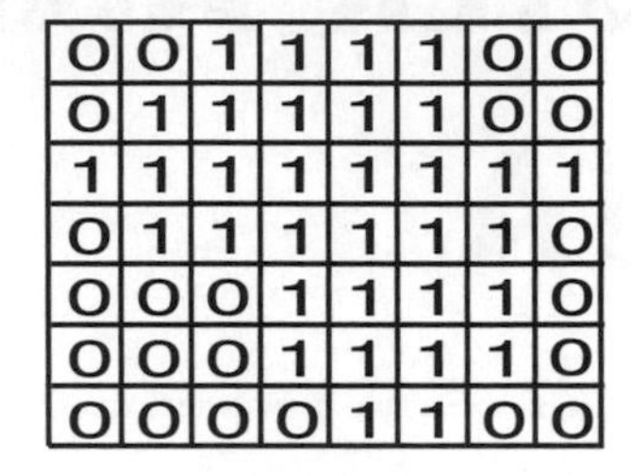

Vector
Points, lines, polygons
Map analog

Raster
Grid structure
Generalized reality

Image 2: Raster Cell Coding
Multiple land uses in one cell

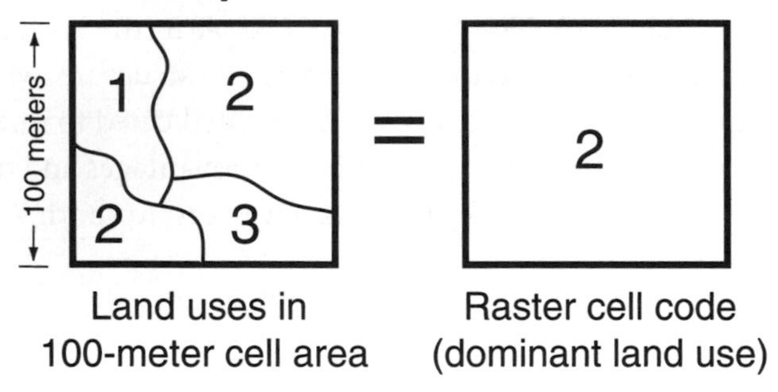

Fig. 3-1: Raster and vector formats.

Raster and Vector Data

The sections that follow discuss raster and vector data, the two important data formats in GIS. The descriptions of each type point out their respective data structures and uses, as well as their differences and similarities.

As a preview, image 1 of figure 3-1 demonstrates that vector format has polygons that appear normal, much like a map. Africa, coded 1, is easily recognized. Raster format generalizes the scene into a grid of cells, each with a code to indicate the feature being depicted. Africa, code 1 among the surrounding 0 cells, is difficult to recognize.

In the raster format, a landscape scene has a grid data structure. Each cell in the grid is given a single feature identity, usually a number (e.g., rainfall amount or a code number for a category such as land use) or a text label (a full name or a code letter). The cell is the *minimum mapping unit*, meaning that it is the smallest size at which any landscape feature can be represented and shown.

All of the features in the cell area are reduced to a single cell identity. This means that all of the geography in the area covered by a cell is accumulated and combined into one identification; it is a generalization of the landscape and its features. For example, if a cell covers an area 100 meters x 100 meters, all of the land information within is coded to a single value (although it is possible to use a code that notes two or more specific features). Image 2 shows a single cell that covers four areas of land use (coded 1 through 3), but because the cell can have only a single value, the dominant land use, code 2, is assigned (other coding techniques are discussed later).

Raster and Vector Data Models

Image 1: Raster and Vector Data Formats

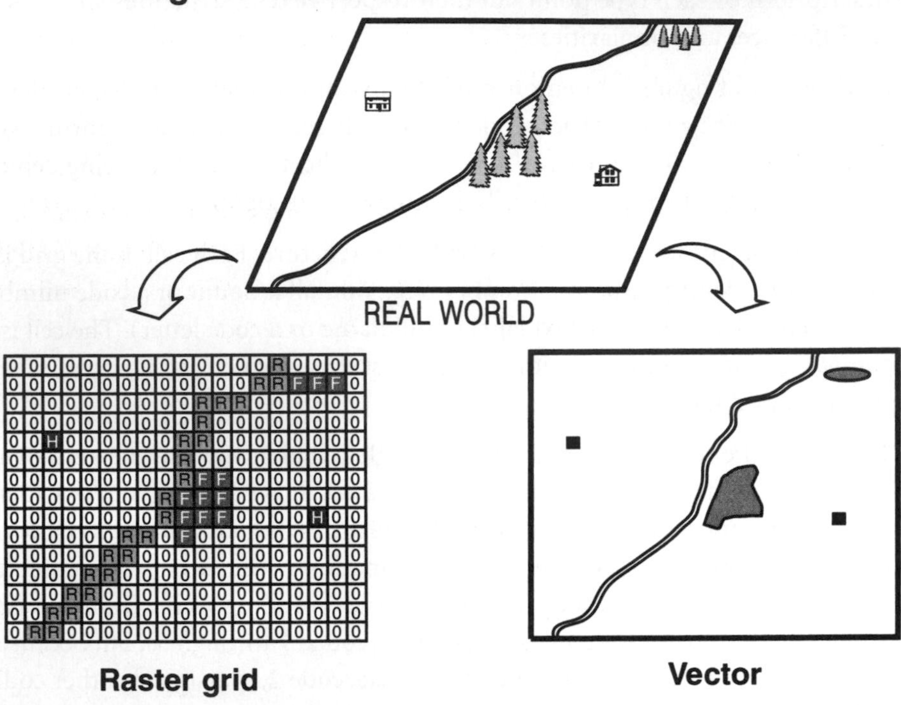

Image 2: Point Features

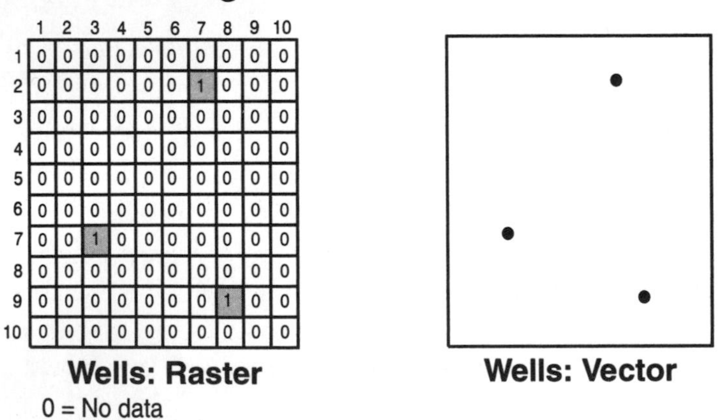

Fig. 3-2: Display effects of raster and vector.

Raster and Vector Data Models

Image 1 of figure 3-2 displays a real-world scene, with its raster version on the left and vector version to the right. The letters are codes for the features (R-river, F-forest, H-house, all shaded for visual convenience). The cells with no feature are coded 0. Like figure 3-1, the differences between raster and vector are evident.

Raster

Because the raster cell's value or code represents all of the features within the grid, it does not maintain true size, shape, or location for individual features. The river, for example, is actually more narrow than a cell, but only the entire cell can be coded as river, so the river appears wider than it really is. Also, note the river's change in shape—more geometric than sinuous (curvy) because of the square cells. Other features are generalized.

Even where "nothing" exists (no data), the cells must be coded. Most GIS themes depict only the necessary features in an area; showing everything on the landscape would be confusing. For example, there are only three wells in the image 2 area, and the raster format shows the three cells coded 1. However, the rest of the area must be coded 0, indicating No Data, or no wells.

Normally, a GIS map uses several thousand cells (too many to see individually without magnification), making the features more recognizable. For some projects, the spatial generalization is not important, but others need very accurate shapes and locations.

Vector

By definition, vectors are data elements describing position and direction. In GIS, vector is the map-like drawing of features, without the generalizing effect of a raster grid. The lines are analog, which means they are not broken into cells or fragments, but continue from start to finish in a continuous manner. Therefore, shape is better retained, much like an actual map. In fact, vector is much more spatially accurate than the raster format.

In image 1 of figure 3-2, the river maintains its curves and the forest areas have a realistic shape. The only limitation is the thickness of the line used to draw the features. Also, houses are not usually shown in actual size (except at very large scale), and they are made into point features, which must be represented by a symbol. A small square is used here.

Wells in image 2 are designated by dots in the vector version. Where no data occurs, nothing is entered and "No data" or no wells are presumed. Vector seems to be much simpler and easier than raster for data visualization.

Raster Coding

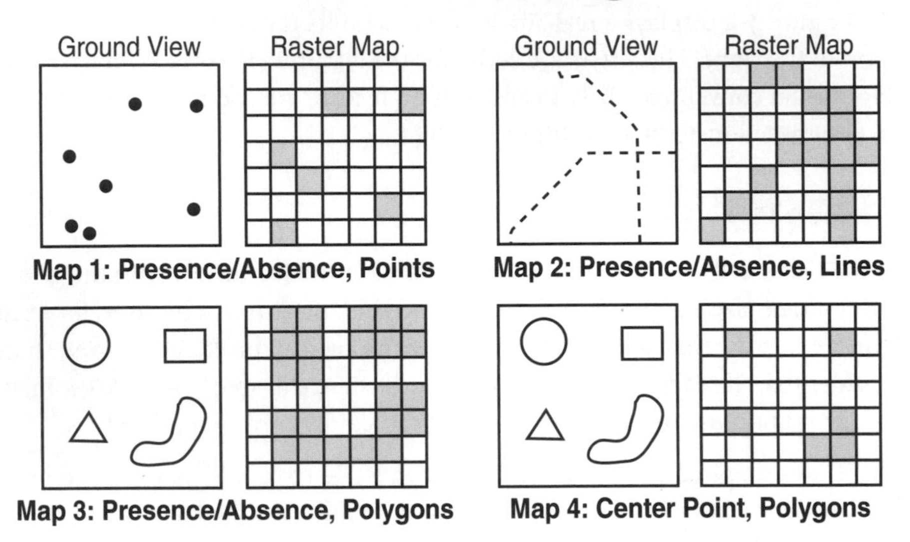

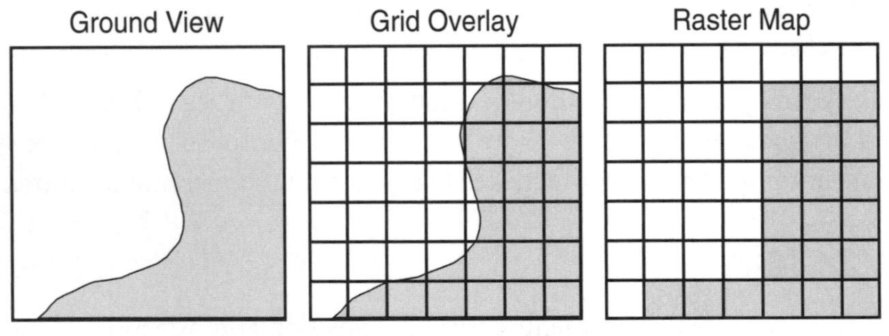

Map 5: Dominant Area

Forest Percentage

0	0	0	0	10	15	0
0	0	0	0	80	100	90
0	0	0	10	100	100	100
0	0	0	0	85	100	100
0	0	0	0	75	100	100
0	0	10	45	95	100	100
15	75	100	100	100	100	100

Water Percentage

100	100	100	100	90	85	100
100	100	100	100	80	0	80
100	100	100	90	0	0	0
100	100	100	100	80	0	0
100	100	100	100	75	0	0
100	100	90	50	0	0	0
85	75	0	0	0	0	0

Map 6: Percentage Cover

Fig. 3-3: Cell coding in the raster format.

Raster Data

The sections that follow discuss raster coding, raster coding problems, raster mapping, resolution, gridding and linear features, and raster accuracy and precision. These are important considerations and concepts needed by GIS users in order to understand the differences between raster and vector data formats.

Raster Coding

In the data entry process, maps can be digitized or scanned at a selected cell size and each cell assigned a code or value. The cell size can be adjusted according to the grid structure (100 x 200 cells) or by ground units (each cell = 30 meters), also termed *resolution.* The required resolution depends on the needs of the project, discussed in material to follow. There are three basic and one advanced scheme for assigning cell codes. These are discussed in the sections that follow.

Presence/Absence

The most basic method is to record a feature if some of it occurs in the cell space. This is the only practical way of coding point and line features, because they do not take up much area of a cell. Maps 1 through 3 in figure 3-3 show the ground view and then the raster map coding for point, line, and polygon features. Note the significant generalization and shape changes produced. For example, diagonal lines are coded into stair-step cells (map 2).

Coding the presence of a feature is complicated when two or more occur in the same cell space, as seen in the lower left of the point feature map, where two points are so close together that they share the same cell. The cell indicates the presence of a point, but not more than one. The ground has seven points, but the raster map shows only six. Even more complications arise when the two point features are different types. For example, if one were a water well and the other a gas well. Two different GIS themes or data layers may be necessary.

Cell Center

The second coding method involves reading only the center of the cell and assigning the code accordingly. When there are multiple features in a cell area, the one in the center "wins" the code. This is not a good scheme for points or lines, because they do not necessarily pass through the exact center. Map 4 presents polygon coding. Note the significant difference between the Presence/Absence and Cell Center methods. Obviously, there is considerable difference in total area calculations between the two results, which can be a possible major problem if accuracy is needed.

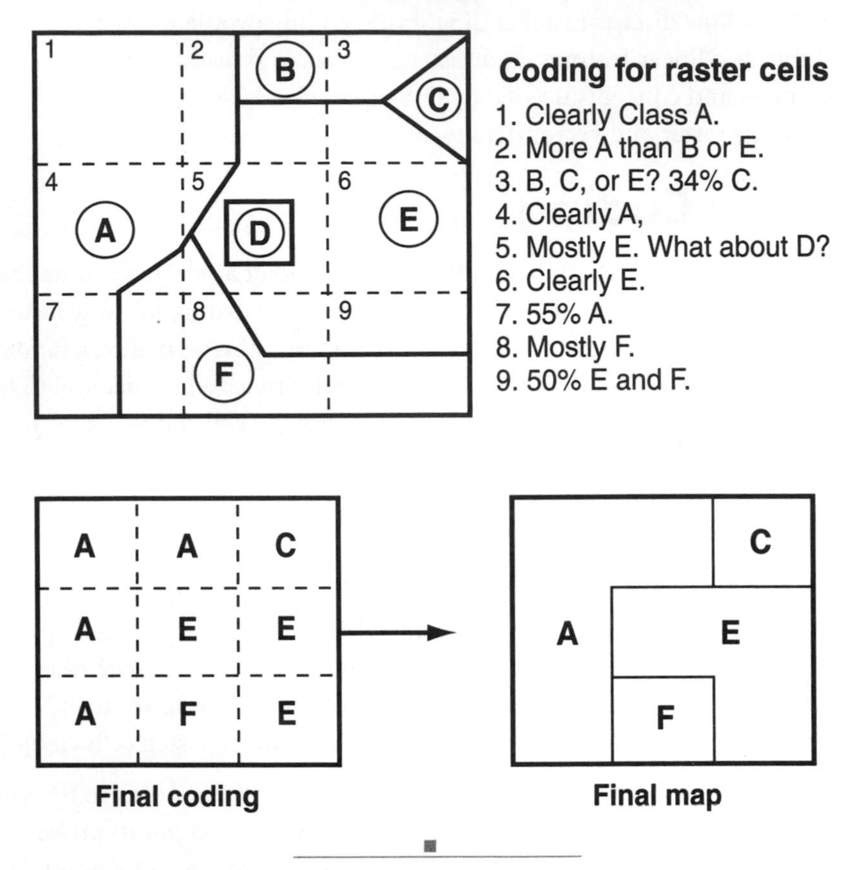

Fig. 3-4: Inaccuracies related to raster coding.

Dominant Area

A common method is to assign the cell code to the feature with the largest (dominant) share of the cell. This is suitable primarily for polygons, although line features could be assigned according to which one has the most linear distance in a cell. Map set 5 in figure 3-3 is a magnified portion of a landscape scene showing forest and water. The grid overlay demonstrates how features occupy the cells and the map view shows the final results. There is considerable difference between the ground view and the map.

Percent Coverage

Each of the three previously described methods results in spatial error. The dominant area coding, for example, can add more area for the dominant feature and delete the area for those features not coded. Although the error in each cell may be very small, the accumulated inaccuracy can be significant. A more advanced method is to separate each feature for coding into individual themes and then assign values that show its percent cover in each cell. Map 6 of figure 3-3 presents the forest and water area from map 5, with the percentages entered into each cell. A quick addition of all cells gives the area coverage of each feature.

The cells are given the percent coverage values. This is more difficult to work (requires more computation and time), but provides higher spatial accuracy. This method is especially useful for the transitional cells that contain several features. However, with a theme using 15 attributes (classes of land use, for example), 15 separate new themes must be produced.

None of these methods is perfect and some inaccuracy occurs in each. A few cells are typically incorrect or "unjustly" coded because of unfortunate positions. However, these are exaggerated illustrations of larger scenes that normally comprise several thousand cells; a few misleading cells normally do not present significant problems.

Raster Coding Problems

Raster coding produces spatial inaccuracies, though the loss may be acceptable, as discussed. Figure 3-4 illustrates some of the possible problems in converting a landscape into a raster data format. Land-use polygons are being coded in the nine cells (dashed-line boxes). Although this illustration is an exaggerated magnification of the spatial problems, these small errors can be compounded over many cells. The dominant cover method is used to assign land-use codes in this example.

The map shows land-use categories A through F (divided by thick lines and noted by circled code letters). Cell 1 is clearly part of class A, but cell 2 is divided into classes A, B, and E. Because a cell can have only one code, the decision to assign the dominant class gives code A to cell 2, despite the fact that A constitutes less than 50 percent of the cell. This decision results in a significant loss of categories B and E.

Cell 3 seems to have equal proportions of B, C, and E, which makes it difficult to classify. Exact thirds are probably very rare, but C is a fraction larger. Cell 4 is almost all A, so the decision is easy (but results in a loss of some F). Cell 5 is mostly E, but the small area of D will be completely lost. This is another special problem: D clearly exists in the area and may be important, but rasterization could eliminate it as a data category. On the other hand, if this is a large magnification, perhaps D is too small to be important.

Cell 6 is clearly E. Cell 7 is classed A because of the 55-percent coverage, but that means that the remaining 45 percent of land use F will be lost. Cell 8 is classed as F. Maybe the code E part that is converted to F in cell 8 will help to compensate for the lost F in cells 4 and 7. Cell 9 is 50 percent E and 50 percent F, and so is given the first letter (E), resulting in more loss of F.

Final coding maps are shown at the bottom. They appear quite different from the input land use. Note that land uses B and D have been eliminated, largely because of unfortunate minor positions of the cells; a slight shift in the grid could produce different results. Loss of some land uses can pose problems in the project, such as misleading spatial statistics. Using the cell center coding method, cell 5 would be assigned land use D, thereby eliminating the surrounding E and A.

Raster systems generalize a landscape and also yield spatial and classification inaccuracies. This might not be important for some purposes, but it could be essential for others. One possible solution is to increase the resolution by increasing the number of cells, making each one smaller and therefore more sensitive to accurate classification.

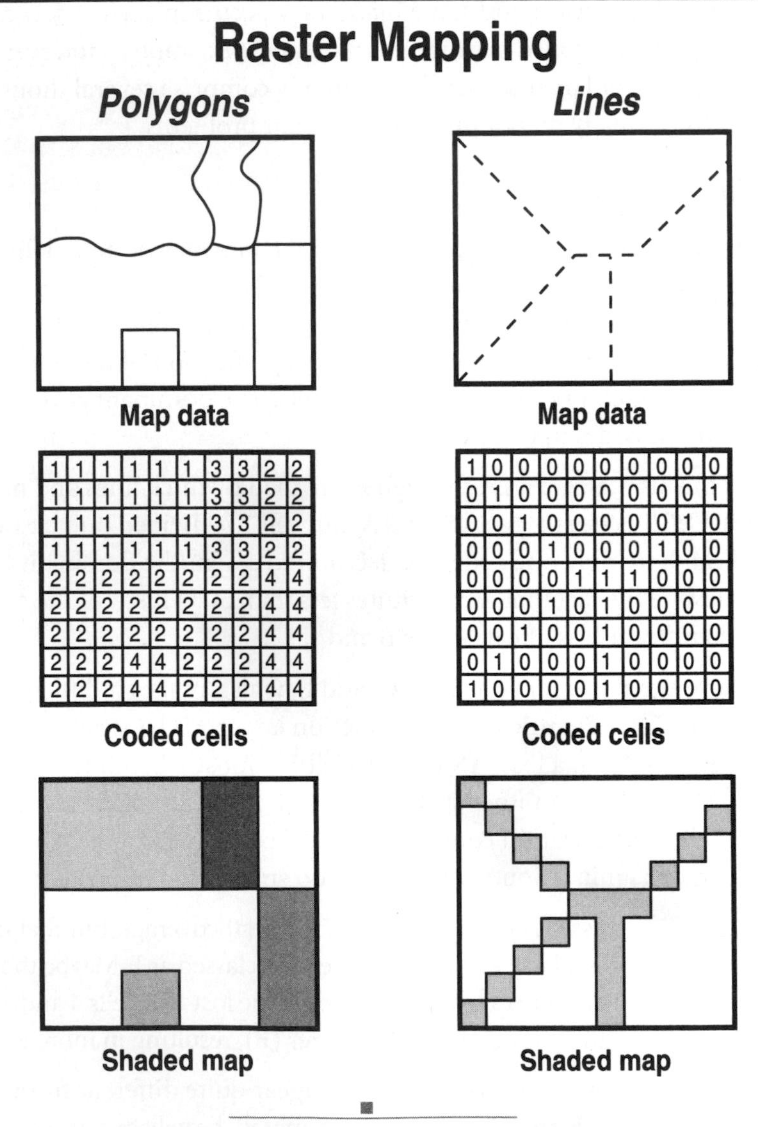

Fig. 3-5: Gridded nature of raster mapping.

Raster Mapping

A major problem with the raster structure is that the shape of features is forced into an artificial grid cell format. For right-angled features, such as square agricultural fields or rectangular political districts, this may not present a major problem. However, for many features, size and shape can become undesirably distorted.

Shown in figure 3-5 are two sets of map data: polygon and line features. The polygon set shows areas coded 1 through 4 (e.g., four different land uses). The polygon boundaries between 1, 2, and 3 are uneven and slightly odd-shaped (not geometric). The Coded Cells box (middle) represents raster cells, which force uneven lines into straight ones. Thus, land uses 1 through 3 have been generalized into right-angled, geometric features. This results in errors of spatial accuracy in terms of size and exact location.

Land use 4 features have square and rectangular shapes in the original Map Data display, so their raster versions are not significantly affected in shape. If lucky, the cell positions may match the real world and there will no loss of accuracy.

The shaded maps (lower boxes) show how codes are shaded or colored for better visualization (called a *choropleth map*). Obviously, a grid full of numbers is unreadable, so changing it into an eye-pleasing and readable map with colors, shades, and symbols is logical data visualization.

Linear features may be greatly affected by rasterization. Two important changes are made in the raster-cell shaded map: (1) shape change of the diagonal features and (2) width increase of the linear features because of minimum mapping cell sizes. Three of the roads in the line feature illustration are diagonals. The diagonals are turned into stair steps because of the effects of linking cells. The width of all linear features is generalized to the cell size because there is no smaller mapping size than the single cell (recall that it is the "minimum mapping unit").

As noted, most raster data sets contain thousands of cells. Each cell is usually too small to see without magnification, so features can appear realistic to the unaided eye, much like a newspaper photograph made of very small pixels (picture elements or dots). Often, the loss of spatial accuracy is insignificant at that scale. Each project must decide how much loss is acceptable.

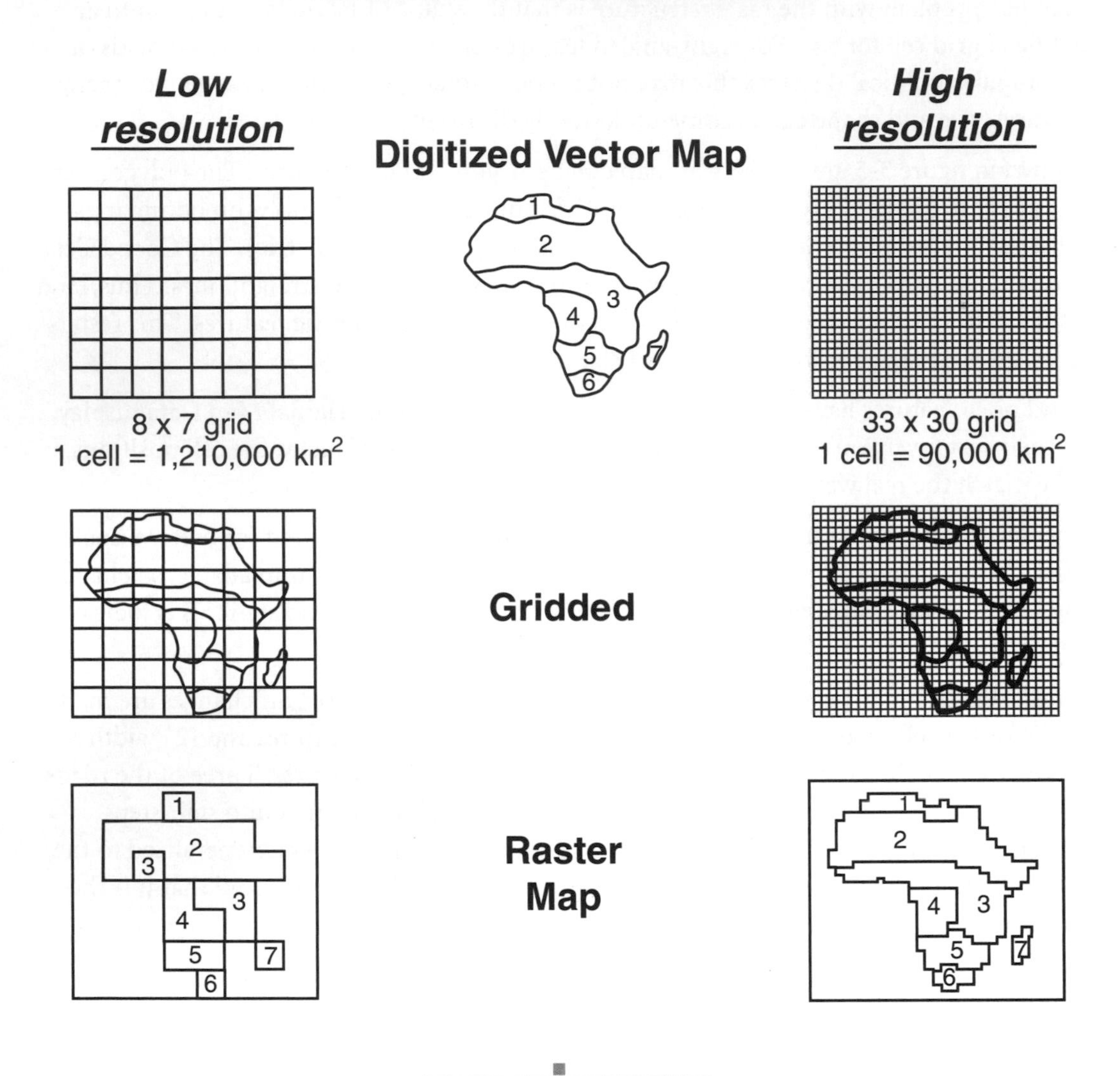

Fig. 3-6: Resolution and the raster format.

Resolution

Increasing the number of cells on a data set increases spatial resolution, which helps to increase spatial accuracy. This is usually done as part of the initial data entry stage for each map. The number of cells (100 x 200 grid) or the size of each cell (30 meters) may be set. Shown in figure 3-6 are a low-density, low-resolution grid on the left, and a high-density, higher-resolution grid on the right. The digitized ecosystem map in the top middle shows the data entry vector version.

The 8 x 7 grid on the left is rather coarse, with only 56 cells. When used to represent the African continent, the result is a shape nearly unrecognizable (bottom left). One advantage to using relatively few cells is the short processing time and ease of analysis. The true shape might be unimportant to a project in which the emphasis is fast analysis, and a general view is acceptable. The much more dense 33 x 30 grid on the right (990 cells) makes a recognizable shape, which might be preferred for mapping and analysis. This grid takes more storage space and has a slower computer processing time. Modern technology, however, is making storage concerns and processing time relatively unimportant.

Compare the cell resolution of each grid (1,210,000 versus 90,000 square kilometers per cell). Depending on the application, the 300-km cell may be too detailed, too general, or just right. Most GIS raster files contain many more cells and much better spatial definition than these examples.

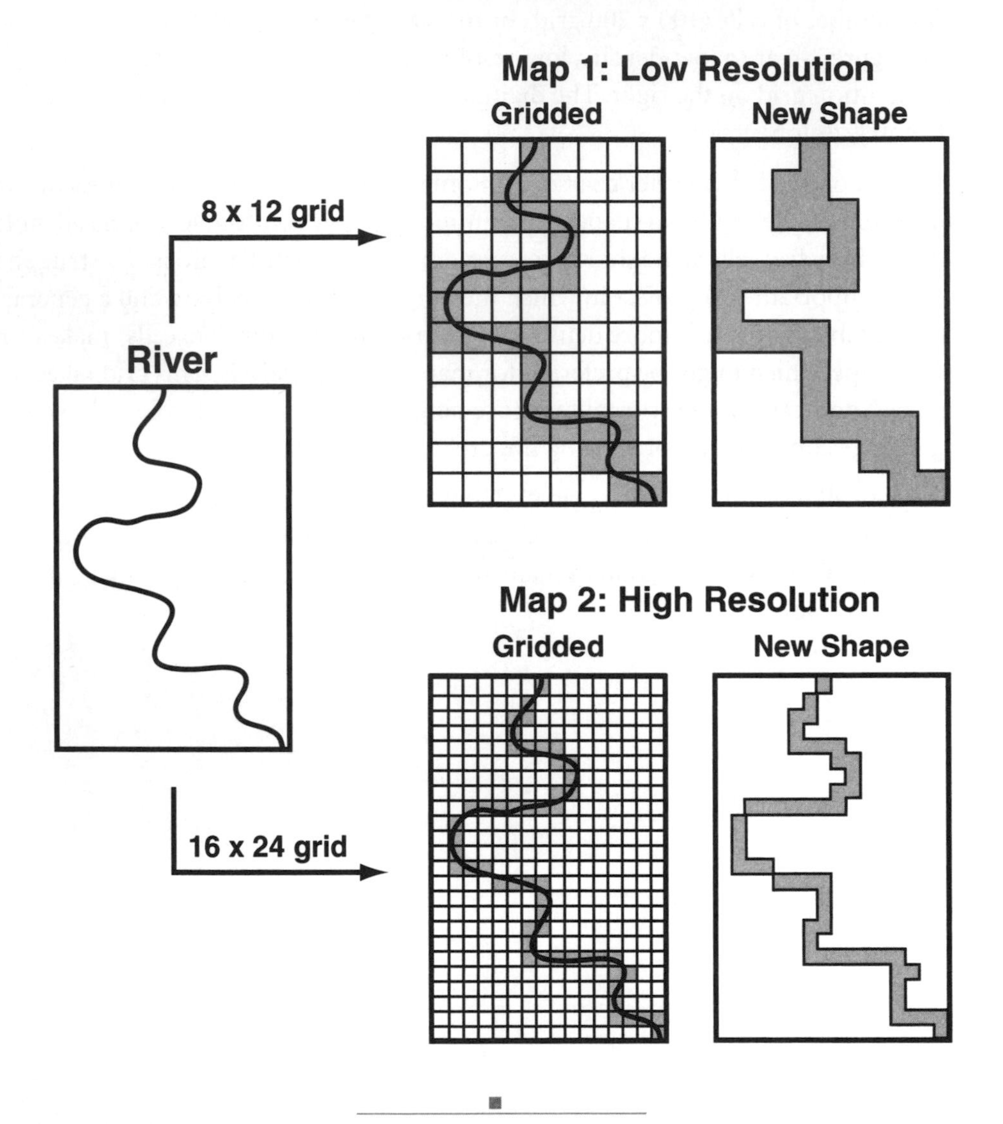

Fig. 3-7: Example of spatial inaccuracy due to the raster format and resolution.

Gridding and Linear Features

To see how the raster format can create spatial inaccuracies, note the actual shape of the river at left in figure 3-7 and then compare it to the two gridded versions (maps 1 and 2, which use the presence/absence coding method). Each cell touched by the river has been coded (and shaded) as "river." Map 1 uses a low-resolution 8 x 12 grid, resulting in a rather generalized and crude shape. Map 2 uses twice as many cells (16 x 24 grid), and the new shape appears more realistic, though still a long way from the vector shape and spatial accuracy.

For spatial measurement, suppose the side of each cell in the map 1 grid represents 1 kilometer. The actual river length is 24 kilometers, but there are 27 river cells in the new shape, thereby resulting in a 3-kilometer error. Contact with a cell is a matter of luck in this case; a very slight shift in initial placement of the grid might have resulted in more or fewer cells. However, shifting the grid is not necessarily an option, because it begins at the top left corner of the digitized area of the original map, which is difficult to adjust once it has been established.

Cells on the doubled grid in map C measure 1/2 kilometer, and the river count is 52 cells, yielding a length of 26 kilometers—a better measure, but not perfect. A higher density of cells in a raster system usually implies more accurate measurements (but not as good as vector). The size of the raster cells (resolution) is therefore important. However, the purpose of a project usually determines data accuracy needs and resolution.

Raster Accuracy and Precision

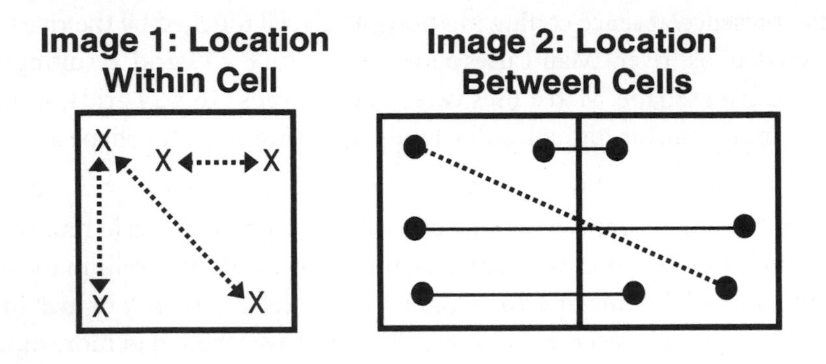

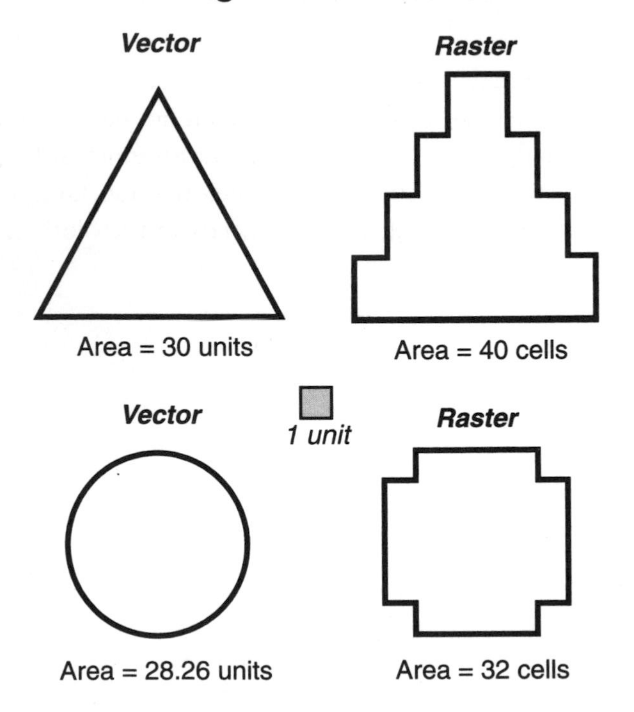

Fig. 3-8: Raster data precision and accuracy are not the same thing.

Raster Precision and Accuracy

Questions of raster data precision (the exact location) and accuracy (maximum spatial truth) are often a problem. Because the raster cell is the maximum resolution and the minimum mapping unit, there is no way to know exactly where any small feature occurs within the cell.

Image 1 of figure 3-8 shows that an X feature can be located anywhere within the cell area, but the location according to the raster format is simply the entire cell (or, basically, anywhere within the cell). This may not be very important for some applications, but locational uncertainty can be significant for others. Of course, smaller cells have less spatial error because the area of doubt is smaller.

Uncertainty becomes greater when measuring across cells (image 2). The actual real-world distance between two features in adjacent cells can vary considerably, either very close together (inside edges), to the outside edges of the cell, or even to opposite diagonal corners. Nonetheless, the raster distance is two cells because the cell is the minimum measuring unit.

Area measurements (image 3) are also generalized. A demonstration measure of one cell unit is shown in the center. The standard triangle measures 30 square units, but requires 40 cells in this low-resolution raster format. The circle is 28.36 cell units, but takes 32 raster cells to depict. The amount of error can be reduced by significantly higher resolution, of course, but clearly the raster format can create problems for some projects.

Vector Data

Image 1: GIS Features

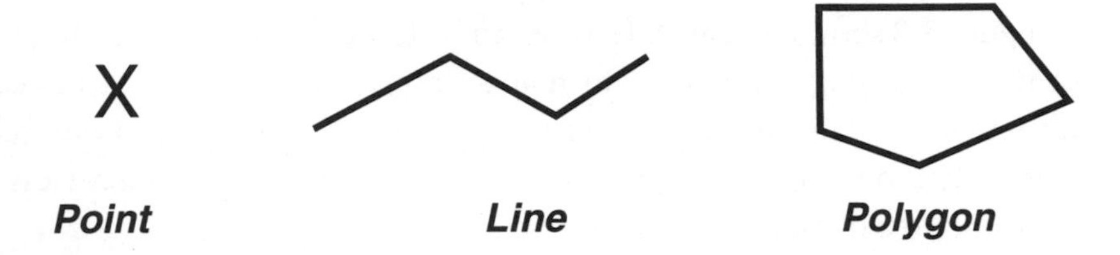

Image 2: Vector Data Structure

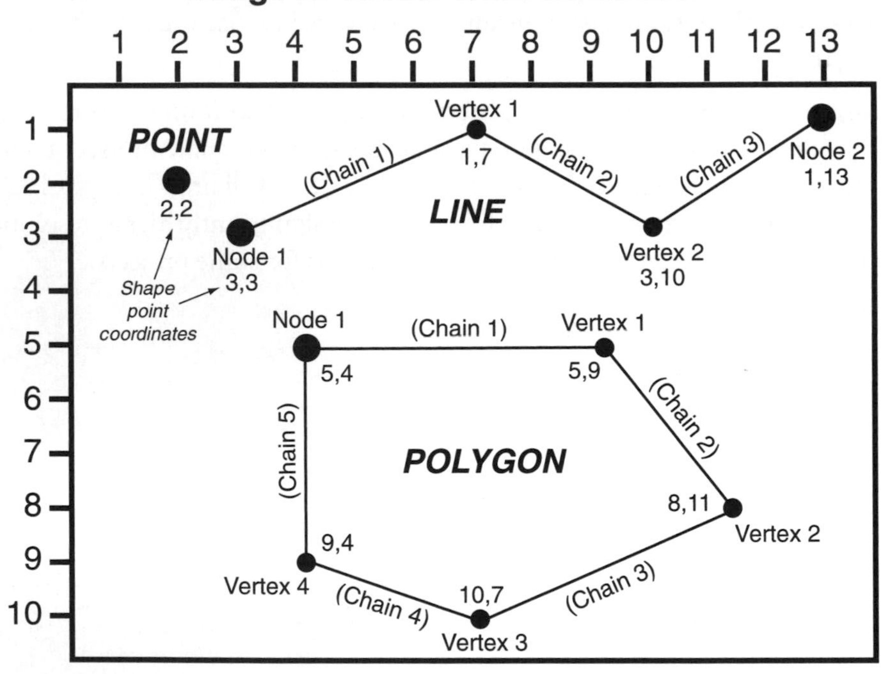

Fig. 3-9: The components and structure of vector data.

Vector Data

Vector features appear more realistic than raster features and have better spatial accuracy. Image 1 of figure 3-9 shows the three vector feature types: point, line, and polygon. As expected, the point is represented as a symbol (an X in this case), and the line and polygon appear as simple drawings, without the complications of a grid format.

Image 2 shows that vector data is structured with several elements, which are not seen in image 1. Vector features are defined primarily by their shapes, more specifically by the outline of their shapes (rather than a group of cells in the raster format). In GIS, the vector system is a *coordinate-based data structure*, meaning that each shape point of a feature is located by X-Y coordinates; for example, the point feature's location of row 2, column 2, in image 2.

Shape points are the ends and bends that define the feature's outline. At the beginning and end of every line or polygon feature is a *node*. At each bend (change of direction) is a *vertex* (plural: *vertices*). Nodes are end points and vertices are between, defining the shape. Point features are standalone nodes. A shape is recorded by using the coordinates of its shape points. Note the row-column numbers near each node and vertex in image 2 (simple coordinates are used here, but latitude-longitude or other systems can be used as well).

Chains connect the shape points to draw the feature's outline. Chains are vectors (hence the term *vector system*), or data structure paths that are not part of the actual stored data elements; they are not real lines, even though they appear on the monitor, but define and present the connection between shape points (they are also called arcs, edges, or links). Vector system data files store only the coordinates of each node and vertex; the hardware draws the connecting chain segments. On a monitor display, only the chains are seen, defining the feature.

Interestingly, nodes and vertices are the real, stored data structure elements, whereas the chain is a virtual component (existing in effect but not in actual form). The visual presentation is drawn for human convenience but is not needed by the computer. This is an efficient data storage format. Nodes and vertices can be inspected under special editing views. The vector data structure is also known as an *arc-node model* because it uses chains (arcs) and end points (nodes).

Because there are no chains attached to a point feature, the data structure notes it as a single node, such as row 2, column 2, in the illustration. A straight line feature is a simple chain (two nodes connected by a chain), whereas a complex line has several chains that define individual segments. The illustrated line feature starts at node 1, goes to vertices 1 and 2, and ends at node 2.

Polygons consist of three or more chains that enclose an area. Note that the illustrated polygon node is both the beginning and end node (common for single polygons). Connected polygons usually share several nodes. Study image 2 to understand the placement and function of the various vector system elements and how they define points, lines, and polygons.

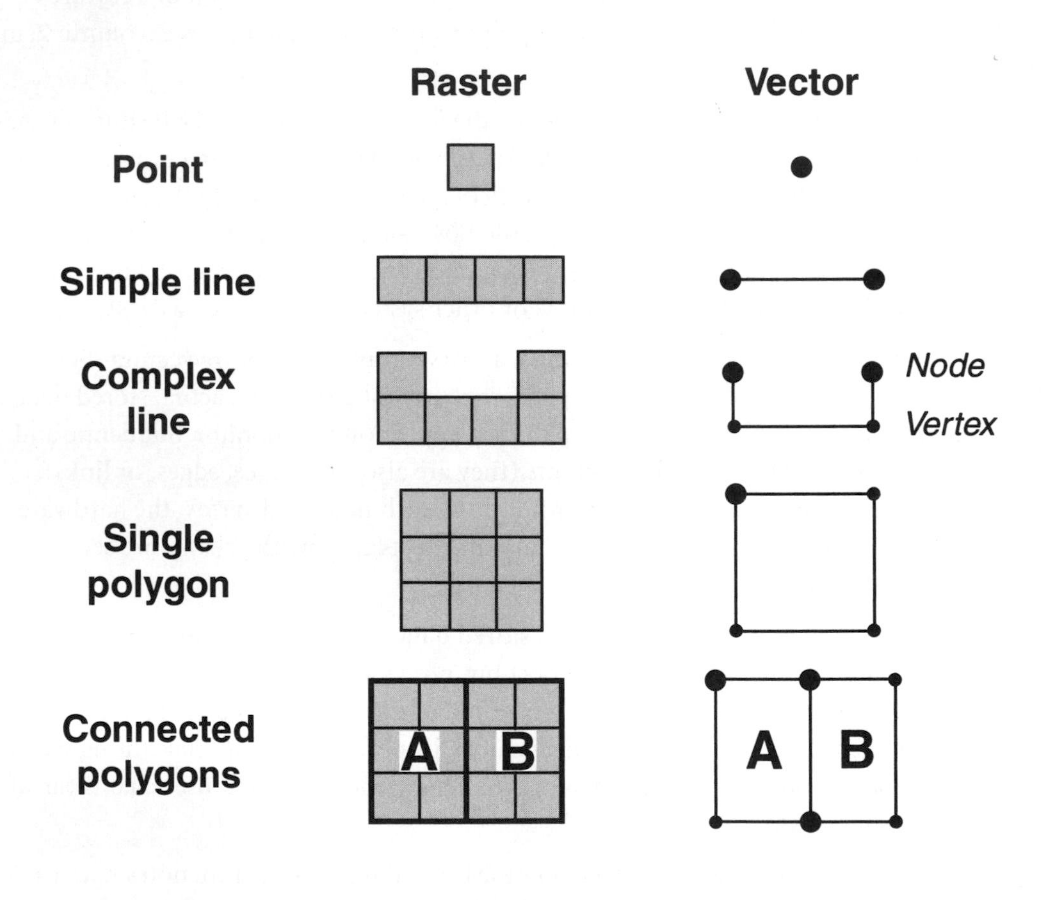

Fig. 3-10: The difference between raster and vector data structure.

Raster and Vector Structures

Raster and vector structures have different methods of storing and displaying spatial data. Figure 3-10 summarizes raster and vector features. Raster cells are stored and displayed as cells, but in the vector format only the nodes and vertices are stored (the chains are virtual elements, shown here for convenience). This results in considerable data storage differences.

A *point* in a raster system is a single cell, but in a vector system it is only a node represented by a symbol with its coordinate position noted. A simple *line* in a raster system consists of a sequence of cells, which mistakenly gives the impression of width. In a vector system, a simple line consists of two nodes (at each end) and a chain that connects them. A more complex raster line consists of connected cells, sometimes in stair-step fashion when they are diagonal. Complex lines in the vector format have vertices to mark changes in direction, with nodes at each end. (In the illustration, large dots are nodes and the smaller ones are vertices).

Raster *polygons* are filled with cells. For single polygons, the vector format usually has a single node and several vertices to mark the boundary direction changes. Connected polygons are simply two blocks of cells, side by side in the raster format, but in vector they share a common border and some common nodes. Illustrated are adjacent polygons A and B, two connecting series of cells in the raster format, but a chain separating them in the vector system (with nodes at each end).

For programming reasons, normally only one data type is in a GIS theme (a point, line, or polygon file). One reason is to avoid confusion between features that may appear alike but are different, such as four connecting streets (line features) that could be interpreted as a square polygon. For presentation and mapping, the final display can have a mix of feature types (i.e., point, line, and polygon), but they still come from different data files.

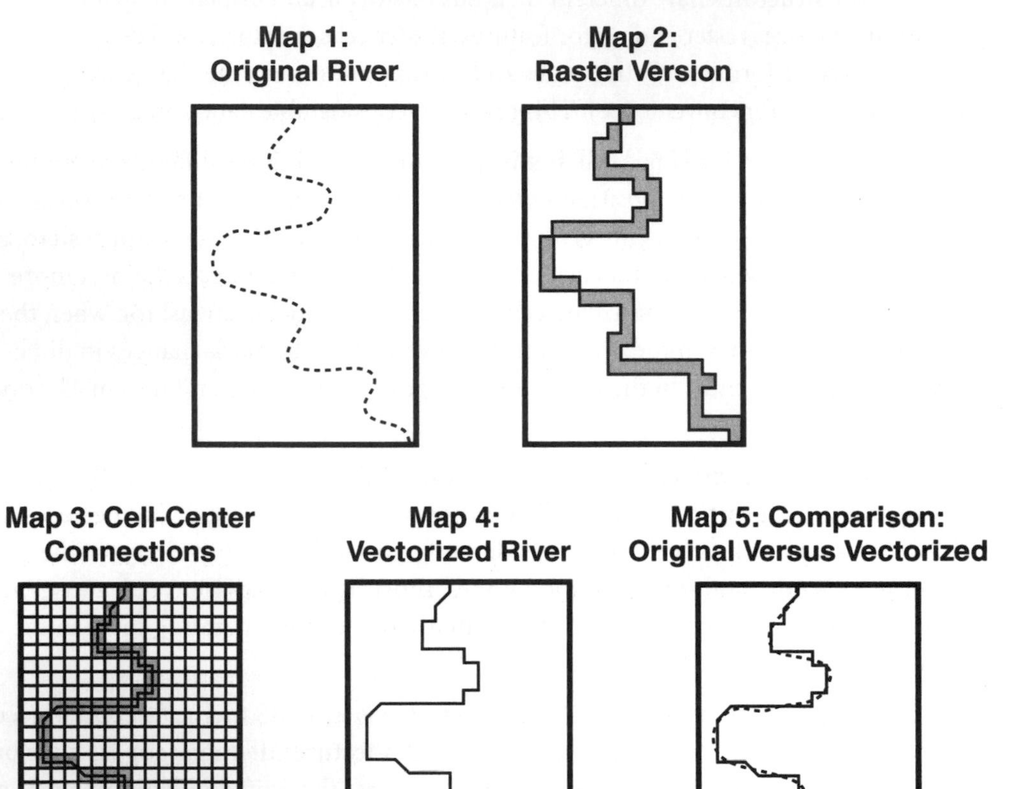

Fig. 3-11: Conversion of data from raster to vector format.

Although modern GIS systems can manage both raster and vector data, there are reasons to convert from one format to the other. There are special problems to consider, and this section discusses the process, with emphasis on raster-to-vector change.

Figure 3-11 presents the spatial evolution of a river from its original vector shape to a raster version, and then converted into a new vector version. The process of changing raster features to vector features normally involves connecting cell centers. Because there is uncertainty of the river's location within the cell area, the center is the logical location for a vector vertex.

Map 1 shows the original river, taken from figure 3-7. When imaged by remote sensing in a raster data structure, it is generalized and made into a blocky, unrealistic form (map 2). (This view is an exaggerated magnification and the actual raster data file may show the river in reasonable visual shape.) When a vector version is needed, the cell centers are linked to construct the best-guess path (map 3). The result is a rather jagged river, without the expected curves (map 4), creating doubt about the actual river shape.

This uncertainty causes problems in accuracy. The final product can *appear* to be very accurate because it is vector (though not in this magnified view), but the new river is nothing more than chains connected with spatially estimated vertices. Map 5 compares the original river shape with the new vectorized version. The difference is not great in this case, but it can be significant when high accuracy is required.

Some GISs can make the new vector version look smoother by "softening" the sharp vertices or sometimes creating artificial curves between vertices (called splines), but again, the user can be fooled into believing the nice appearance is more spatially accurate than the raster version, when in reality it is not.

There are at least four basic reasons to convert from raster to vector: (1) despite the accuracy problems, the better visual appearance of vector features may be preferred, (2) some plotters work only on vector data (easier to draw lines than to plot squares), (3) comparison with vector data is best when both data files have identical formats, and (4) although most modern GIS systems accept both vector and raster data files, some (especially older ones) have either raster or vector as the central operating data structure, and data formats should be identical to facilitate data management and analytical operations. This requirement is becoming outdated and may not be a significant reason any longer.

Rasterization of vector data is often called *gridding* because of the conversion from smooth vectors to a grid structure. The size of the raster cells can be established by the user, such as a certain ground size (100-meter grids), the number of rows and columns, or to match another raster theme. The process is demonstrated by maps 1 and 2, which are from figure 3-7, where raster gridding was discussed.

Raster Advantages

Image 1: Simple Data Structure

3	1	1	1	0	0
0	3	2	3	3	1

Image 2: Easy Analysis

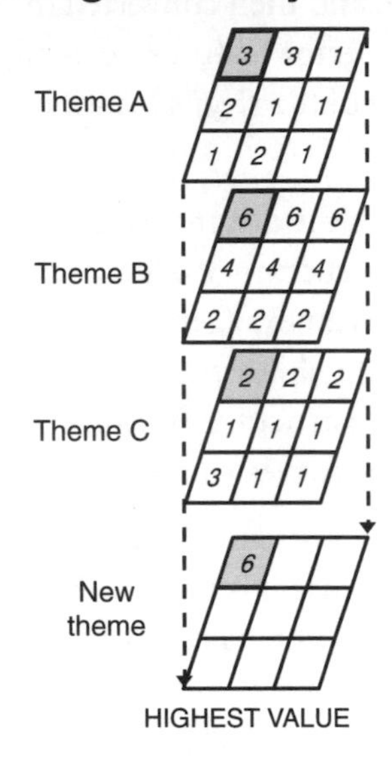

Image 3: Low-Tech Platforms

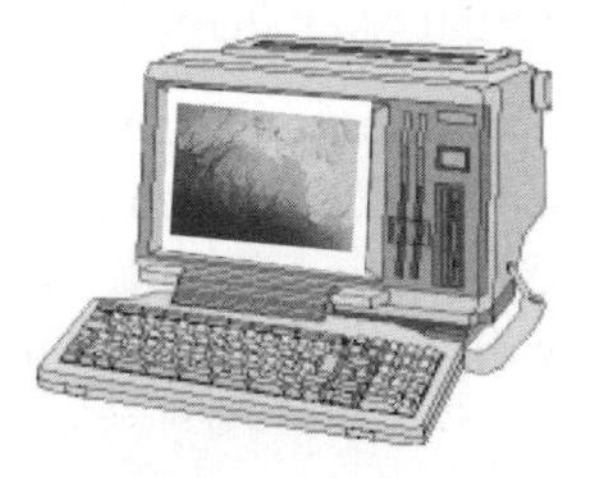

Image 4: Remote Sensing Data

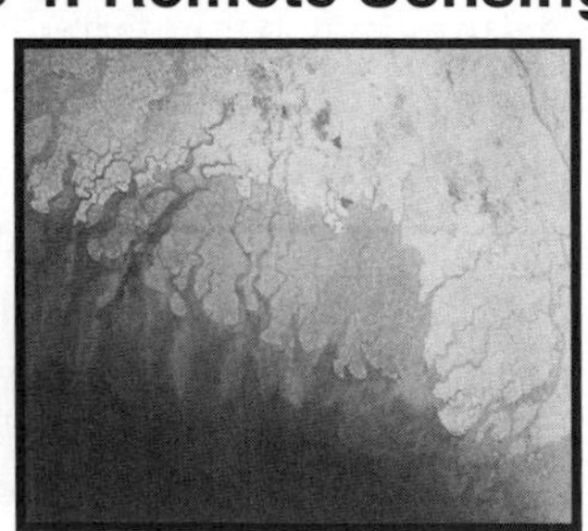

Image 5: Modeling

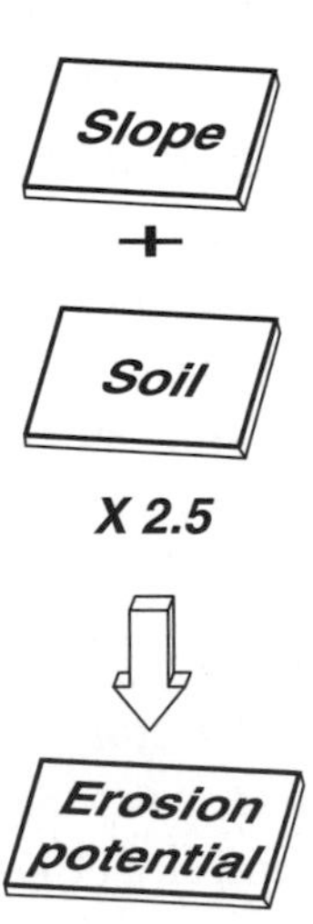

Fig. 3-12: The advantages of the raster format.

Raster and Vector Advantages and Disadvantages

Although data usually comes in vector or raster format, it can be difficult for the new user to know which one to request or when to convert. The sections that follow discuss the advantages and disadvantages associated with both the raster and vector formats.

Raster Advantages

Despite the accuracy problems, raster data has some advantages (depicted in figure 3-12 and discussed in the following).

- It is a relatively simple data structure: a grid with a single code in each cell (image 1).

It is easy to understand and use, even by beginners.

- Thc simple grid structure makes analysis easier (image 2).

Computers are very fast at comparing numbers, and if there is a stack of data files to be manipulated for analysis, the computer reads each grid cell position one by one and does the analysis on that cell for each data file, such as finding the highest value. For example, the illustration shows three themes (A through C), with the first grid cell in each one shaded (cell position 1). The program compares raster values for cell number 1 and enters the highest of the three in New Theme. This type of operation can be accomplished very rapidly over a large raster grid file.

- Because of the relative simplicity of raster formats, the computer platform can be "low tech" and inexpensive (image 3).

Older, slower, and limited machines can manage raster data easily, and the level of technology and expense can be kept fairly low. However, even today's moderate machines are capable of vector GIS work, and this advantage is becoming less relevant.

- Remote sensing imagery (from aircraft or satellite) is typically obtained in raster format (image 4).

This allows integration and comparison between imagery and GIS data. The image is easily integrated into a raster format GIS because of the identical data formats (discussed further in Chapter 5).

- Modeling is the creation of a generalized data file or a set of universal procedures to accomplish a certain GIS task (image 5).

If, for example, predicted soil erosion in an area is expressed as a formula, using slope and soil values, raster numbers in a grid system are easier to analyze than vector formats. The illustration shows a simple example of adding the slope value to the soil value and then

Raster Disadvantages

Image 1: Spatial Inaccuracy

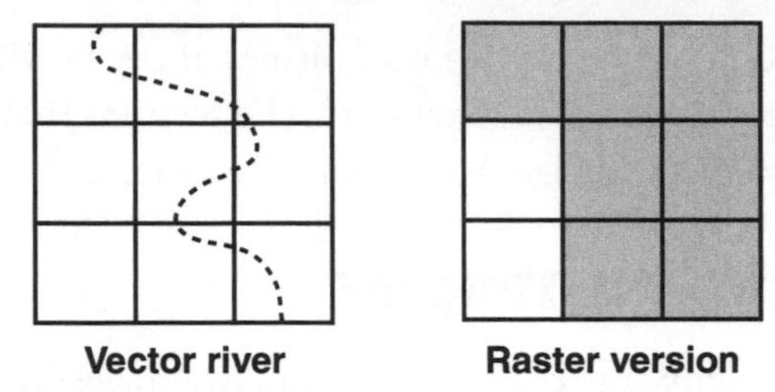

Image 2: Generalization

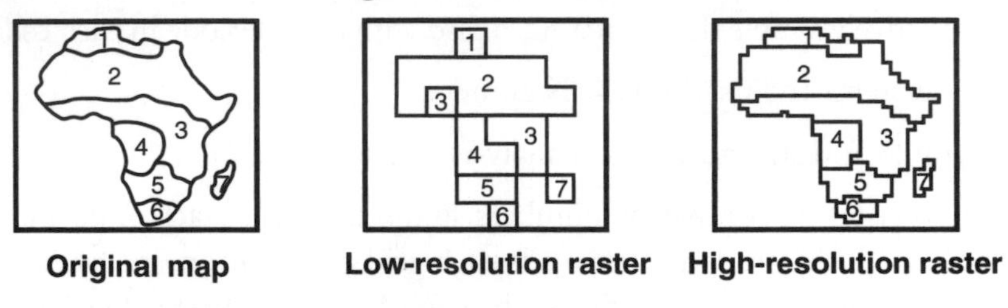

Image 3: Implicit Data

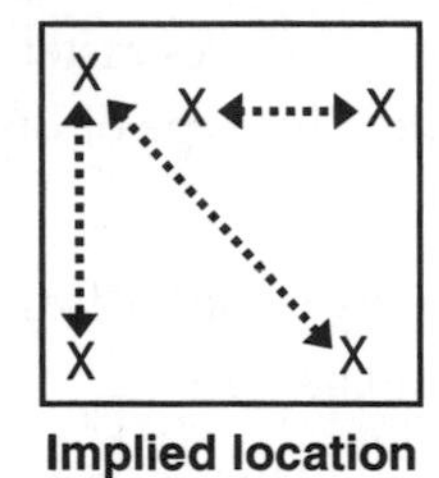

Image 4: Large Data Sets

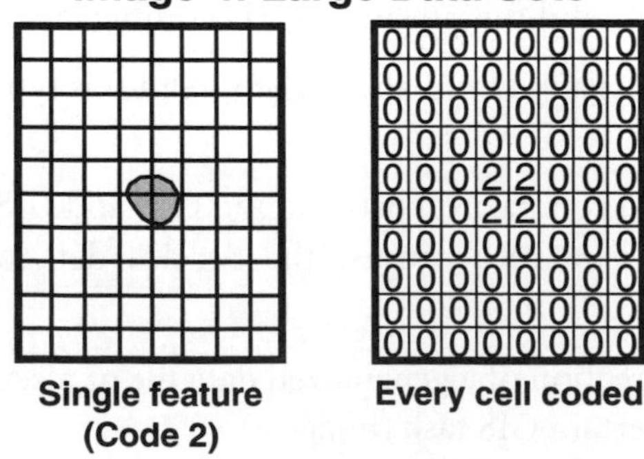

Fig. 3-13: Disadvantages of the raster format.

multiplying the sum by 2.5 to arrive at a number indicating the erosion potential. The calculation is on a cell-by-cell progression, as shown in image 2 of figure 3-12. Performed for each cell in the data set, the potential for erosion over the entire area can be mapped and analyzed easily and rapidly. Modeling is discussed further in Chapter 10.

Raster Disadvantages

Disadvantages of the raster format, depicted in figure 3-13, include the following.

- Spatial inaccuracies are common with raster systems (image 1).

It is usually hoped that losses are compensated by gains and that, overall, inaccuracies are canceled out. This may be wishful thinking, however, and projects that need high accuracy either have to use more cells (greater resolution) or use a vector format when possible.

- Because each cell tends to generalize a landscape, the result is relatively low resolution compared to the vector format (image 2).

Image 2 shows the original map and a comparison of low and higher resolution for Africa. Even the use of a very high number of cells can only guarantee better resolution, not necessarily satisfactory accuracy. More cells means larger files and slower computation and display.

- Because of spatial inaccuracies caused by data generalization, a raster format cannot tell precisely what exists at a given location (image 3).

Consequently, raster cells imply (suggest) truth—an implicit structure (understood though not directly expressed). This means that what is shown is not necessarily what actually exists on the landscape. The user must judge if the level of generalization is acceptable or not.

- Each cell must have a code, even where nothing exists. That is, even No Data must be coded, usually 0 value (image 4).

Every cell is coded, making computer storage needs high, especially for high-resolution grid cell formats (many cells). Thus, raster systems can have very large data sets, even when the useful data is low.

Vector Advantages

Image 1: High-resolution Map Analog
Scale-independent Resolution

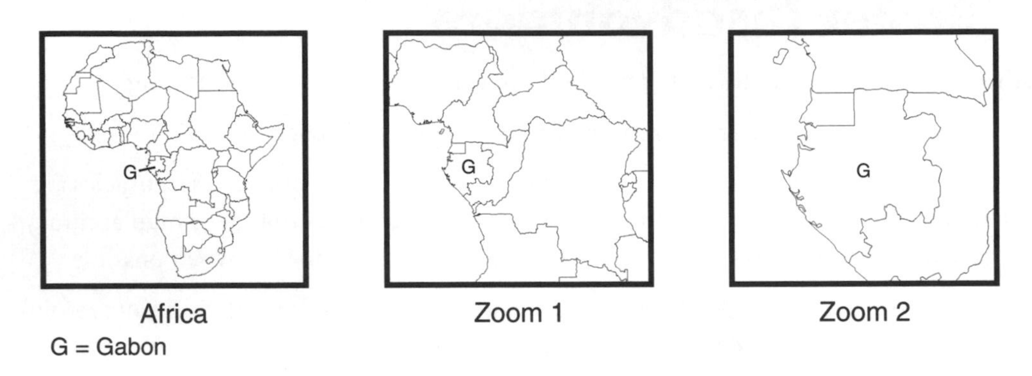

Image 2: Node-Vertex Storage

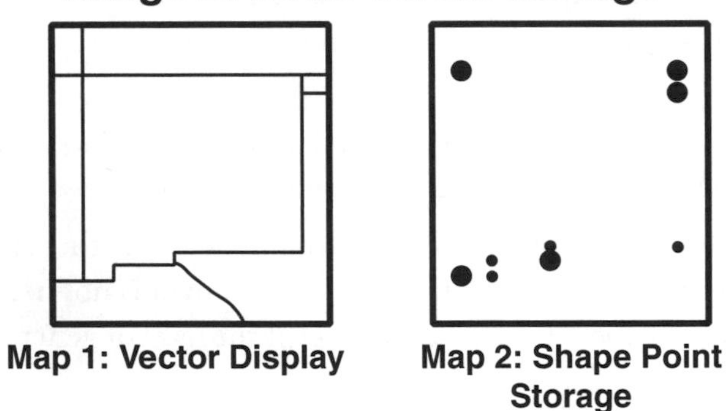

Image 3: Understandable

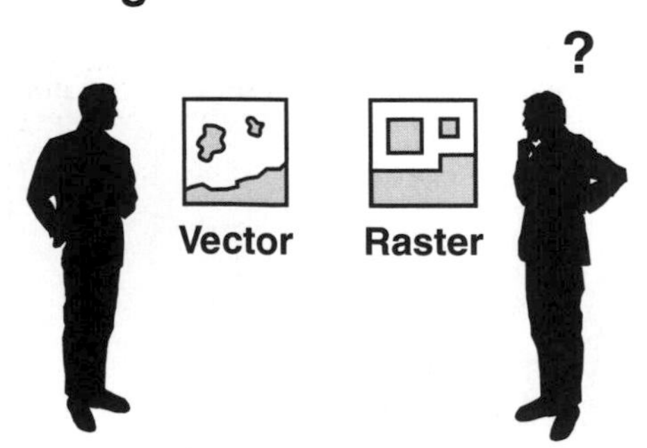

Image 4: Topology

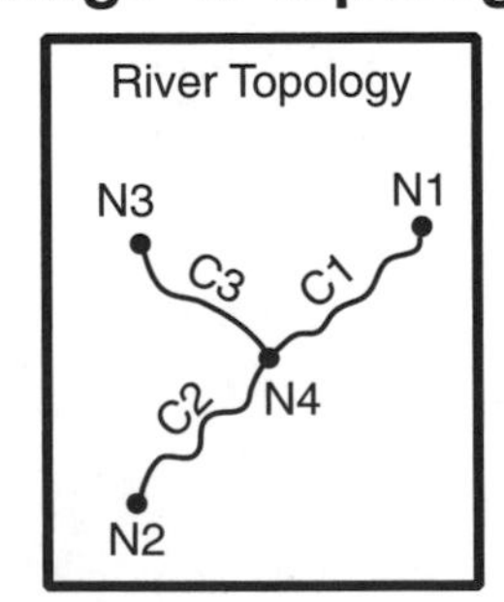

Fig. 3-14: Advantages of the vector format.

Vector Advantages

The advantages of using the vector format, depicted in figure 3-14, include the following:

- In general, vector data is more map-like than raster data (image 1). The geographic data is more accurate and credible (believable) than the raster format. Vector displays are also more pleasing to the eye.
- Vector data is very high resolution.

Magnification does not damage the display quality. In fact, the detail and accuracy may be higher than the display resolution of the computer monitor or mapping device; limitations can be due to hardware restrictions rather than the data. Image 1 shows a zoomed-in view of Gabon (G) without generalization of spatial data.

- The high resolution supports high spatial accuracy.

Because vector systems use coordinate structure data and vector chains, the display can be magnified to the limits of the original data collection standards. If ground teams gathered data at plus or minus 10 meters positional accuracy, the monitor display can zoom in until that limit is reached. Outlines of the features will not grow thicker or more inaccurate.

- Vector formats have storage advantages (image 2).

Systems and data managers are always concerned with data storage. Vector formats take less storage space and usually offer better storage capabilities than raster formats. This is because vector features are defined and stored only as nodes and vertices, whereas raster data sets have every cell coded. This means that vector data files can be smaller and faster than raster files. In vector, only the essential data elements are stored.

- The general public usually understands what is shown on vector maps (image 3).

Vector displays may be more acceptable than generalized raster presentation to the nonspecialist. Vector data structure seems to be the system of choice for many GIS users, particularly among projects requiring control of data accuracy and presentation. The general public, in particular, easily recognizes vector displays. However, presentation quality is less of an issue today, given the power of modern systems to present raster and vector data in pleasing and understandable designs.

- Vector data can be topological (image 4).

As explained in the next chapter, topology is a special data structure that has special advantages in GIS. Raster data does not use topology, and many users consider vector to be the system of choice.

Vector Disadvantages

Image 1: Complex Data Structure

Chain Topology

Chain	Start Node	End Node
1	1	2
2	2	3
3	3	5
4	4	5
5	5	6

Node Topology

Node	Chains
1	1
2	1,2,3
3	2
4	4
5	3,4,5
6	5

Chain Coordinates

Chain	Coordinates
1	70,22 71,23
2	71,23 73,23
3	71,23 71,51 71,21
4	70,21 71,21
5	70,21 72,21 72,22 73,22

Node Coordinates

Node	X	Y
1	70	22
2	71	23
3	73	23
4	70	21
5	71	21
6	73	22

Image 2: Demanding Teaching

Technical skill required

Image 3: Introductory Training

Decisions of cost, needed technical skill, applications, relevancy, and other considerations

Fig. 3-15: Disadvantages of the vector format.

Vector Disadvantages

Although vector may be the best all-around format for many users, there are a few disadvantages that must be considered.

- Vector data formats may be more difficult to manage than raster formats.

They are usually stored as long lists of coordinates for nodes and vertices—easy for the computer to understand but difficult for editing by the user. Knowing how to read and work a data file can be demanding. However, most users may not be concerned with this issue. Image 1 of figure 3-15 previews the topology data of the next chapter by presenting four data tables that usually accompany vector topological systems. These offer a degree of complexity to data management.

- Whereas very simple, "low-end" computers can operate many raster-based GISs, vector formats require more powerful, high-tech machines.

Management of computer equipment becomes more of a problem. Again, the power of modern computers is resolving this problem and it is less of an issue today.

- The use of better computers, increased management needs, and other considerations often make the vector format more expensive.

However, most users today have reasonably inexpensive, powerful technology available to handle the higher needs of vector-based GIS.

- Learning the technical aspects of vector systems is more difficult than understanding the simplicity of the raster format, particularly when topology is introduced.

This makes teaching more difficult, especially for secondary school or beginning college students. Image 2 symbolizes the demands on the teacher. Some of the knowledge is not intuitive and must be taught with care and skill.

Because of these considerations, some users prefer the more simple and less expensive raster format. For example, raster systems might be easier to use in introductory training for applications with minimal requirements or that have limited financial and human resources. The issue of raster versus vector must be decided for beginning students (image 3).

Computer technology has advanced so much in the past several years that very powerful systems are available at affordable prices for most users. There is more integration between raster and vector formats today, and distinctions between raster and vector are less problematic. The disadvantages of both formats are becoming less important. Perhaps in the near future raster and vector will be truly "seamless."

CHAPTER 4

TOPOLOGY

Introduction

TOPOLOGY IS ONE OF THE MOST USEFUL DATA STRUCTURE CONCEPTS in GIS. It is special programming on the data that creates powerful connections among features. This chapter discusses what topology is, how it works, and some of the common uses it provides for GIS. In effect, with topology, each feature has the following characteristics.

- "Knows" where it is: a feature's position is part of the data knowledge.
- "Knows" what is around it: the connected and surrounding features are recognized.
- Has recognized spatial relationships with other features.
- Has length, distance, perimeter, and area information.
- "Knows" how to get around: gets from one location to another using connections and paths.
- "Understands" its environment: by virtue of the connections and spatial relationships (its surroundings), topology identifies features and uses their attributes to accomplish various spatial analysis tasks.

Topology

Image 1: Topology Concepts

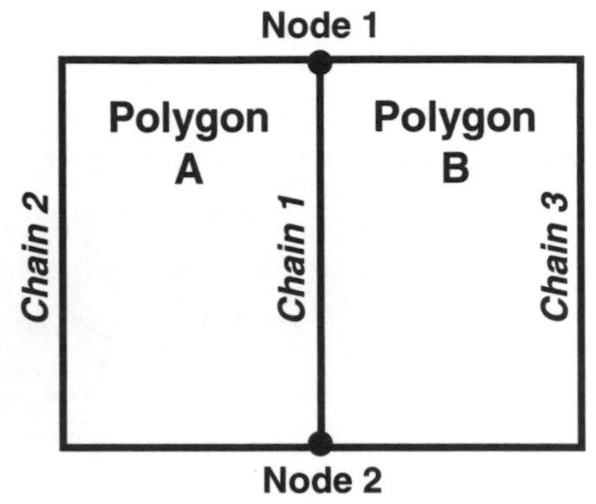

Polygon A is related to polygon B using chain 1.

Chain Topology

Chain	Left Polygon	Right Polygon
1	A	B
2	O	A
3	B	O

O = Outside

Image 2: Afghanistan-Pakistan Topology

Chain Topology Afghanistan-Pakistan

Chain	Left Polygon	Right Polygon
1	Afghanistan	Pakistan
2	O	Afghanistan
3	Pakistan	O

O = Outside

Fig. 4-1: Illustration of spatial relationships and topology.

Topology

For the beginner, topology can be a bit difficult to understand, so careful reading of the text and illustrations may be helpful. Basically, topology is the mathematical procedure for defining spatial relationships. What does this actually mean? Topology is special programming applied to GIS data so that the features have "intelligent" connections with other features.

There are several definitions of topology, but in simple terms it can be considered in two associated perspectives. First, it is defined as the mathematics of connectivity and adjacency for spatial features (how features join). Second, topology can be viewed as the programming that provides spatial relationships between nodes, chains, and geographic features. For GIS, this means that topology is a special data structure that establishes connections and links for the nodes and chains in order to recognize spatial relationships among the geographic features.

It is not difficult for people to find and trace a route on a map, from one point to another, and at the same time identify the features on each side of the road along the way. It is also possible to measure the distance of that route. That is easy for us, but computers do not have the natural ability, so special programming is needed. Topology gives GIS the power to perform some spatial operations that humans do intuitively, and sometimes to do it faster and better.

When topology is applied to GIS, data structure tables are built for features, nodes, and chains. The tables are used to determine various relationships, such as what is connected, what is adjacent (left and right), and the direction of chains. By reading these tables, GIS can track connections, make paths, measure distance and area, and determine a variety of spatial relationships. Features are not stored simply as standalone spatial elements but are now linked and integrated with other features. Essentially, topology consists of "intelligent vectors" that integrate features and the environment.

Image 1 of figure 4-1 serves as a preview of topology. Shown are polygons A and B, with chain 1 separating them. Chain 1 actually belongs to both polygons; it defines the right side of polygon A and the left side of B, just like the borders of two adjacent nations. Topology recognizes this defining property by reading the data structure tables (chain topology, illustrated), which show the left and right polygons of the chain.

Topology establishes chain 1 as part of both polygons and thus relates one to the other in terms of their connection. Polygon A is directly related to polygon B because chain 1 is their common topological feature. Each feature "knows" the other is connected and adjacent, because the spatial relationship is related in the table. A simplified data table is shown, demonstrating how topology builds the chain and node information used to recognize the spatial relationships. Fortunately, the user does not have to deal with these tables to make topology work; the tables are largely invisible unless there is a special need to display them.

Applying Topology

Image 1: Establishing Topology

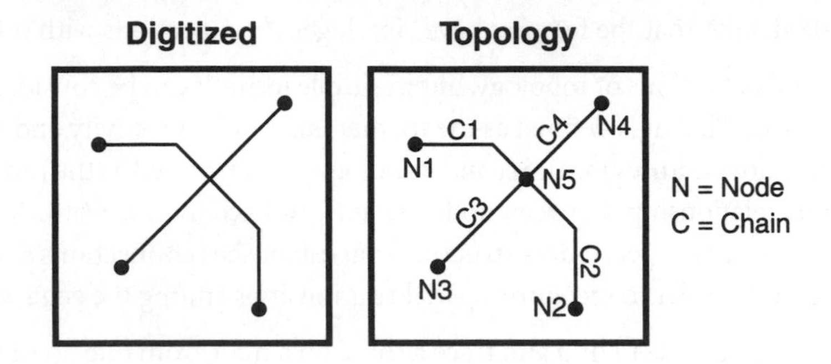

Image 2: River Topology

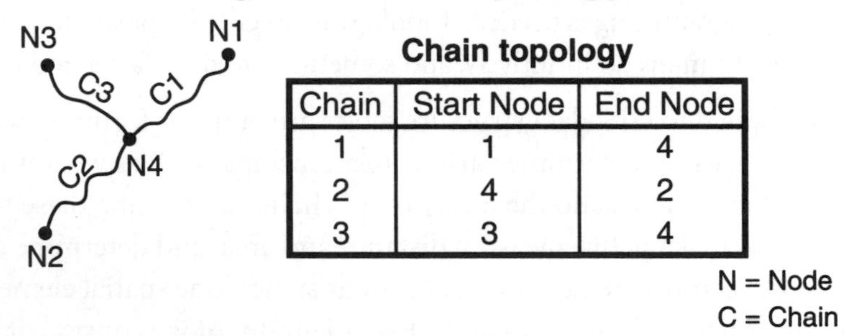

Chain topology

Chain	Start Node	End Node
1	1	4
2	4	2
3	3	4

Image 3: Complex Line Topology

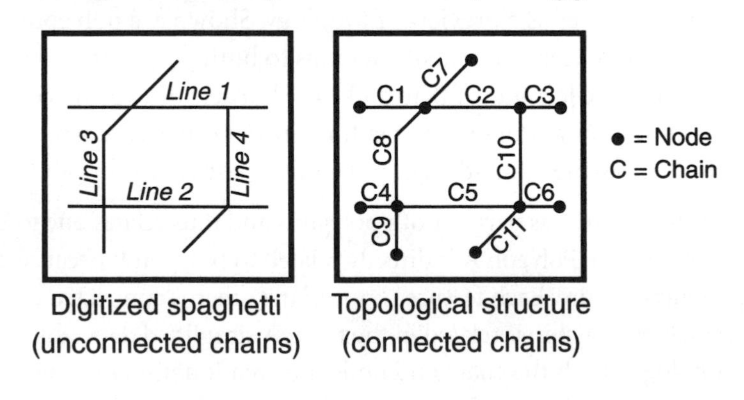

Digitized spaghetti (unconnected chains)

Topological structure (connected chains)

Fig. 4-2: Examples of applied topology.

Image 2 of figure 4-1 applies topology to Afghanistan and Pakistan borders. The table is almost identical to image 1, and there is very little difference between the conceptual model and the real-world application.

Applying Topology

Topology is actually applied (sometimes referred to as "built") by a simple operation (typically a single command), usually after digitizing. (Digitizing is the digital tracing of maps, a primary means of entering map data, discussed in Chapter 5).

Features are first digitized into the GIS as they are outlined from the map, and the resulting lines are without connection or "intelligence." This is the *spaghetti model* of data: lines that cross and form shapes without connection or relationship to other lines and shapes. There is no informational content to the lines except location. Like spaghetti, following the path of one line can be difficult; identifying and keeping track of the lines that cross is even more complicated. Topology establishes the needed information.

Topology builds the spatial information by recognizing nodes at the ends of each digitized line and creating new nodes at intersections where lines cross, thereby making true topological chains. Image 1 in figure 4-2 shows two digitized lines (with their nodes that define the start and end points). They are not functionally connected at this time but are merely two lines standing alone, with no recognized spatial relationship.

When topology is applied, each node is numbered (N in the illustration) and a new node is established at the intersection. Note that each digitized line feature now consists of two topological chains (C), basically because the intersecting new node 5 defines the ends of the chains. That does not present a problem to the GIS software because the line features are now defined as two chains each: one line consists of chains 1 and 2 (using nodes 1, 5, 2), and the other line consists of chains 3 and 4 (nodes 3, 5, 4).

Image 2 shows a river and its smaller tributary. They were digitized from the headwaters (upper right) to the mouth to maintain water flow direction. Topology builds the intersecting node and names the chains. The river is now two chains, with the new node 4 separating the two features. The table describes the chains and their nodes, identifying the first and last node (also known as Start and End nodes, or sometimes To and From nodes). Because direction is important, the starting and ending node identifications are very useful in defining water flow paths.

Image 3 shows a more elaborate version of digitized spaghetti and resulting topological structure for a road system. For example, after topology is applied, line 1 consists of chains 1, 2, and 3 (nodes are not numbered here). Study the illustration to understand how topology builds nodes and chains to construct recognizable spatial relationships. Several tables make up a complete topological data structure, as discussed in the next section.

Topology Tables 1

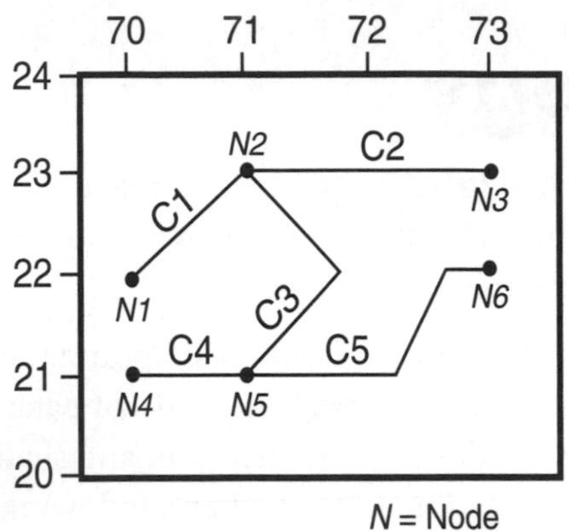

N = Node
C = Chain

Chain Topology

Chain	Start Node	End Node
1	1	2
2	2	3
3	3	5
4	4	5
5	5	6

Node Topology

Node	Chains
1	1
2	1,2,3
3	2
4	4
5	3,4,5
6	5

Chain Coordinates

Chain	Coordinates
1	70,22 71,23
2	71,23 73,23
3	71,23 71.5,22 71,21
4	70,21 71,21
5	70,21 72,21 72.5,22 73,22

Fig. 4-3: Various tables derived from topological database information.

Topology Tables

The power behind topology is the set of integrated "topology tables" containing the data structure information used in feature definition and in determining spatial relationships. The tables are created automatically when topology is applied to the data. As noted, nodes and chains define line and polygon features, and are the important data structure components for topology.

Figures 4-3 and 4-4 show various topological tables. Each commercial GIS has special names and table formats (separate tables or one large integrated table), but the basic idea is to have a set of tables that contains complete spatial information.

Topology traces features by finding nodes, chains, and polygons as needed. In figure 4-3, line features (property survey lines) have each node and chain identified on the map (they are normally not seen—an invisible part of the data except under special editing views). The Chain Topology table presents each chain and its start and end nodes. Then the Node Topology table lists the chains that connect to each node. Study and verify the tables. Find each connection.

These tables work together to define spatial relationships. For example, if the task is to follow a path from node 1 to node 6 sites, the program first finds the chains connected to node 1, which is only chain 1. Then it finds all nodes connected to chain 1, which is only node 2. The chains connected to node 2 are chains 2 and 3. Both are followed, with chain 2 ending in a dead end at node 3, but chain 3 ends at node 5, which has several possible paths. Continuing the path exploration, chain 4 also ends in a dead end at node 4, but chain 5 goes to the target node 6. The only path from node 1 to node 6 has been identified: N1-C1-N2-C3-N5-C5-N6. Fortunately, this is not necessarily the end of topology's capabilities.

Coordinate tables for the chains and nodes present the real-world locations for each node and vertex (positive numbers for northern latitude and eastern longitude in this illustration). Note that the coordinates for nodes and the vertices of each chain are given in order of occurrence.

If the distance between the start and end sites is needed, topology calculates chain lengths and adds the ones used in the path. The real-world coordinates can be translated into actual distances and then simply added to give total path distance. These types of operations are easily activated by the user, who does not have to know exactly how topology works. Typically, the process is very rapid and mostly transparent.

Image 1 of figure 4-4 closes the lines from the survey map of figure 4-3 to make property polygons A and B (circled). Polygon topology is only a bit more complicated. It uses node and chain tables similar to those shown previously, but with the new chains 6 and 7 as the enclosing borders. A Polygon Topology table is made, identifying the chains that make each polygon. The Chain Topology table identifies the polygon to the left and to the right, with O indicating the outside (sometimes called the "universe"). Follow some of the chain entries and you will see that it is a fairly simple but effective system.

Are there any polygons adjacent to A? The program searches the polygon topology table to find all chains that make polygon A. Then it goes to the chain topology table to see if other polygons are associated with those chains.

Note that only chain 3 lists a polygon that is not O. In this case, only B (obviously) is adjacent to A, but the system works identically even for highly complex multiple-polygon landscapes, such as continents or soil maps. The procedure would go from polygon to chain to

Topology Tables 2

Image 1: Polygon Features

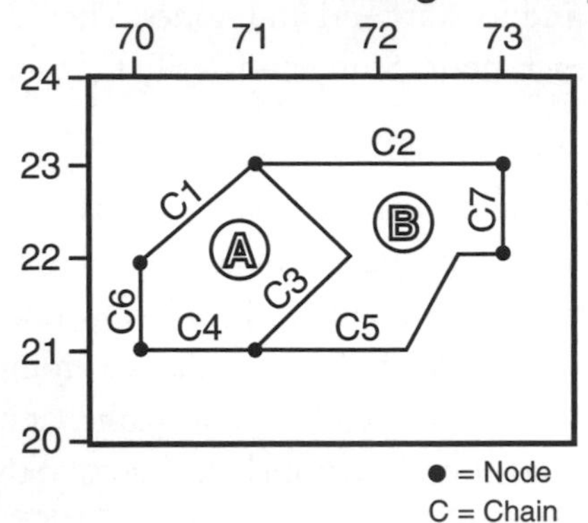

Chain topology

Chain	Left Polygon	Right Polygon
1	O	A
2	O	B
3	A	B
4	A	O
5	B	O
6	O	A
7	B	O

Polygon Topology

Polygon	Chains
A	1,3,4,6
B	2,3,5,7

Image 2: East Africa

Chain Topology

Chain	Left Polygon	Right Polygon
1	Ethiopia	Kenya
2	Kenya	Somalia
3	Kenya	Outside
4	Tanzania	Kenya
5	Uganda	Kenya
6	Sudan	Kenya
7	Sudan	Ethiopia
8	Ethiopia	Somalia
9	Somalia	Outside
10	Tanzania	Outside
11	Outside	Tanzania
12	Uganda	Tanzania
13	Outside	Uganda
14	Sudan	Uganda
15	Outside	Sudan

Polygon Topology

Polygon	Chains
Kenya	1,2,3,4,5,6
Ethiopia	1,7,8
Somalia	2,8,9
Tanzania	4,10,11,12
Uganda	5,12,13,14
Sudan	6,7,15

Polygon Attributes

Nation	Perimeter	Area (km^2)
Kenya	3446	584429
Ethiopia	5311	1132328
Somalia	2366	639065
Tanzania	3402	944977
Uganda	2698	246050
Sudan	7687	2490409

Fig. 4-4: Various tables derived from topological database information.

polygon to chain as many times as necessary to show linkages between polygons that are spatially separated but have functional, topological connections.

Image 2 shows East Africa, with the chains (numbered) and nodes (dots). The chain topology and polygon topology tables are presented to demonstrate the connections. These appear to be complicated, but upon second look, the system is fairly simple. The linkage from Ethiopia to Tanzania is Ethiopia to C1 (chain 1) to Kenya to C4 to Tanzania.

As with the line features, chain length is easy to calculate, thereby making polygon perimeter an easy attribute to provide (sum of the chains making a particular polygon). With relatively simple geometry (for the computer), area is also easily determined. Some GISs offer line length and polygon perimeter and area as automatic figures in the standard attribute database available to the user. Image 2 shows a polygon attribute table presenting perimeter and area for each nation.

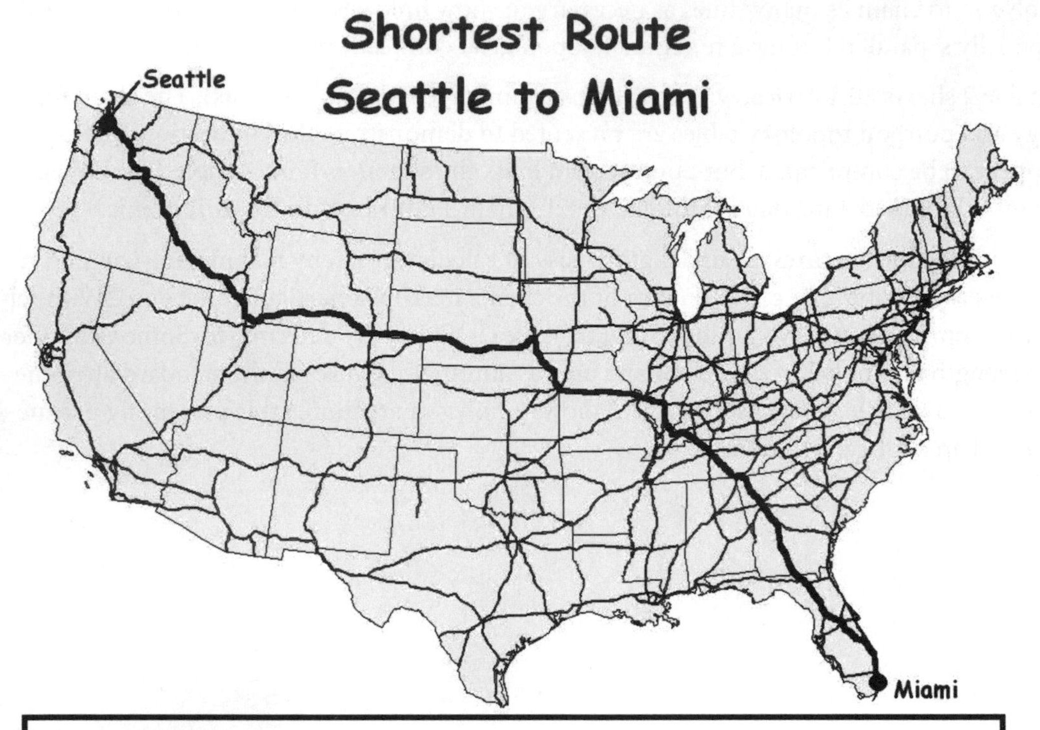

Directions
Starting from Seattle
Turn right onto Interstate 5
Travel on Interstate 5 for 1.88 mi
Turn left onto Interstate 90
Travel on Interstate 90 for 100.83 mi
Turn right onto Interstate 82
Travel on Interstate 82 for 140.94 mi
Continue straight onto Interstate (OR/ID/UT) 84
Travel on Interstate (OR/ID/UT) 84 for 569.78 mi
Turn left onto Interstate 80
Travel on Interstate 80 for 878.12 mi
Continue straight onto Interstate 29
Travel on Interstate 29 for 178.04 mi
Continue straight onto Interstate 70
Travel on Interstate 70 for 243.95 mi
Turn left onto Interstate 55
Travel on Interstate 55 for 3.18 mi
Continue straight onto Interstate 64
Travel on Interstate 64 for 69.06 mi
Turn right onto Interstate 57
Travel on Interstate 57 for 51.05 mi
Turn left onto Interstate 24
Travel on Interstate 24 for 311.85 mi
Turn right onto Interstate 75
Travel on Interstate 75 for 488.42 mi
Continue straight onto Florida State Hwy 528
Travel on Florida State Hwy 528 for 0.14 mi
Turn right onto Interstate 95
Travel on Interstate 95 for 10.93 mi
Turn right into Miami

Fig. 4-4: Topology used for routing analysis.

Multiple Connectivity

Topological connections can work on an integrated set of line features, such as a highway system, called a "network." Each highway often consists of a set of chains because there are numerous intersections with other highways. Topology provides the connections and routing analysis, as shown in figure 4-5. When lengths and total distance are included, the shortest paths can be found among the complicated network.

The illustrated task is to determine the shortest path along the U.S. Interstate Highway system from Seattle, Washington (in the northwest corner), to Miami, Florida (in the southeast corner). The search starts at the selected node (the Seattle city location) and searches all possible connections until it reaches the target city: Miami, Florida.

Using topological links, the search follows the first chain until a node is reached. It then reads the node and chain tables to follow other connections. Eventually, there may be numerous possible paths between the two cities, but because distances have been calculated along the way for each path, it is a simple matter of determining the shortest distance and its associated path. Then the theme attribute table is referenced to identify the highway names.

A final report can include visual highlighting on the graphics for easy identification, and a list of highways and individual distances. Some GISs even provide directions, such as "Travel on Interstate 90 for 100.5 miles. Turn right onto Interstate 82 and travel for 149.5 miles…"

Other attributes could be used in the search, such as one-way streets, barriers that prevent continued movement, speed limits, restricted streets, or the need to pass through selected cities on the route. For example, streams are basically one-way paths, and tracking the flow path and distance from a small headwater tributary down through the system to the mouth is just another example of network topology at work. These types of factors are discussed further in Chapter 8. Network topology has opened many new transportation applications in the business and pleasure fields. When combined with GPS, there are even more uses, such as in-car routing and directions, enhanced emergency response, transportation planning, and more efficient delivery systems.

Zambia-Mali Adjacency

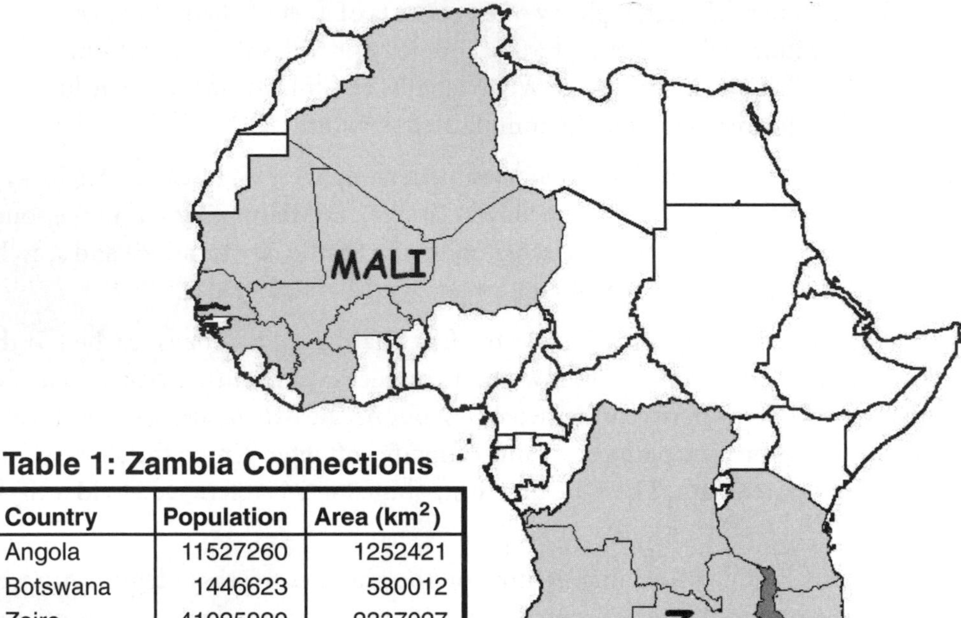

Table 1: Zambia Connections

Country	Population	Area (km^2)
Angola	11527260	1252421
Botswana	1446623	580012
Zaire	41025920	2337027
Malawi	10660480	119028
Mozambique	16604660	788329
Tanzania	28386270	944977
Namibia	1550917	825632
Zimbabwe	11106690	390804

Table 2: Mali Connections

Country	Population	Area (km^2)
Algeria	27459230	2320972
Guinea	62420070	246077
Ivory Coast	13498860	322216
Mauritania	2204077	1041570
Niger	8797739	1186021
Senegal	8116554	196911
Burkina Faso	10164690	273719

Fig. 4-6: Spatial analysis made possible using topology and relational query.

Topology and Relational Queries

Topology provides the connections and relationships for a variety of spatial analysis operations. Previously discussed were how length, distance, perimeter, and area are easy to obtain. Such spatial attributes are very useful for GIS analysis, particularly when combined with nonspatial information.

The map in figure 4-6 shows the countries that border (are adjacent to) Zambia (Z on the map). Although this could be accomplished by manually selecting the nations, a simple GIS command is faster, easier, and more accurate; let topology do the work. For example, overlooking Botswana would be understandable; only very large scale would reveal the 200-meter border with Zambia, but any connection is easily identified by the topology tables, regardless of size.

When topology is connected to a relational database, the power of GIS is greatly enhanced. After the topological work is done (finding the adjacent nations), the selected nations can be queried for specific information. Here the assignment is to find which nations bordering Zambia have populations between 10,000,000 and 20,000,000 and areas less than 10,000,000 square kilometers. This is a simple chore for the database operations. Final query results are shown in table 1. Topology and relational databases are powerful synergistic partners in GIS.

Compare Zambia with Mali, another landlocked African country. Which of the two nations has the most neighboring countries and which has the most surrounding population? Doing the adjacency first, followed by the database selection, the figures in table 2 show that Zambia has eight surrounding nations with a total population of 119,980,811, but Mali, with seven nations, has a higher surrounding population (132,661,220).

GIS can be very useful in helping to understand many different types of landscapes and the problems associated with spatial factors. This type of GIS operation demonstrates the potential complicated geopolitics of landlocked nations, in this case both of which have a large number of neighbors relative to most non-landlocked nations, making significant international political and economic concerns for maintaining good relations.

Topology Contributions

Image 1: East Africa Topology

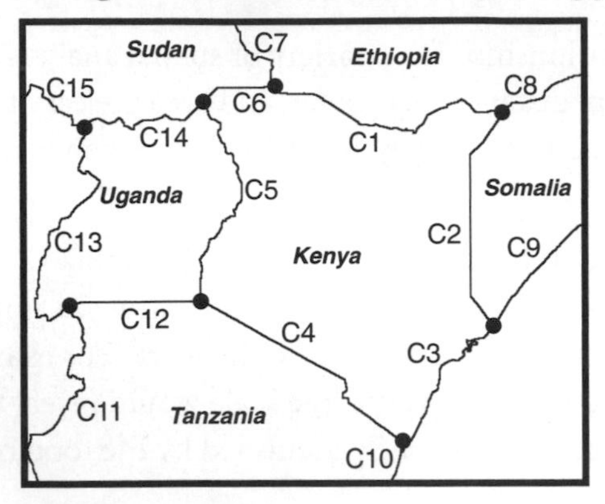

Chain Topology

Chain	Left Polygon	Right Polygon
1	Ethiopia	Kenya
2	Kenya	Somalia
3	Kenya	Outside
4	Tanzania	Kenya
5	Uganda	Kenya
6	Sudan	Kenya
7	Sudan	Ethiopia
8	Ethiopia	Somalia
9	Somalia	Outside
10	Tanzania	Outside
11	Outside	Tanzania
12	Uganda	Tanzania
13	Outside	Uganda
14	Sudan	Uganda
15	Outside	Sudan

Polygon Topology

Nation	Perimeter	Area (km^2)
Kenya	3446	584429
Ethiopia	5311	1132328
Somalia	2366	639065
Tanzania	3402	944977
Uganda	2698	246050
Sudan	7687	2490409

Spatial Relationships

Tanzania — C4 — Kenya — C2 — Somalia

Image 2: Network Analysis

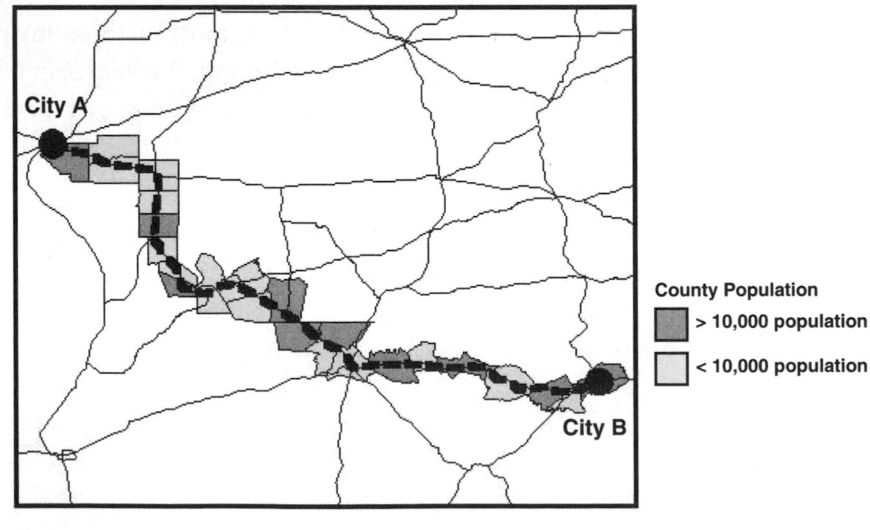

Fig. 4-7: Contributions of topology.

Topology's Contributions

Topology offers special information to the data structure and provides powerful functions for spatial analysis. It is a valuable component of GIS and is useful in a very wide range of applications. Topology offers a number of advantages and contributions to geographic data and GIS, which are depicted in figure 4-7 and described in the following.

- *Spatial information:* Topology provides length, distance, perimeter, and area. Image 1 shows East Africa and the database resulting from topological calculations.
- *Spatial relationships*: Topology creates connections, which functionally link features that are adjacent, such as the nations in East Africa. A query asking which nations border Kenya will be replied with Tanzania, Uganda, Sudan, Ethiopia, and Somalia, as read from the topology tables.
- *Multiple linkages:* Each feature is linked to other features, providing multiple connections (linkages) that join (unite) them. Image 1 shows the polygon adjacency connections from Tanzania through Kenya to Somalia.
- *Network analysis:* The functional connections, distance, and other spatial relationships, combined with the relational database, are ideal in interpreting network features (such as highway systems), and for performing specialized analysis (e.g., shortest route between features and use of associated nonspatial attributes). Image 2 shows the shortest route between two cities (large dots) and the intersected counties along the way, which are classed according to population (light and dark tones). All of these attributes are easily determined because of topological connections and database queries.

Chapter 5

Data Entry

Introduction

The data entry process is the most important and most time-consuming part of GIS operations. Careful attention is needed to locate and acquire data, enter it into the system, and prepare it for use. Without reliable and appropriate data, the rest of the GIS work is meaningless. This chapter examines the principal elements of the data entry process, such as the types of data GIS uses, digitizing (a common data entry process), georeferencing and projections, the database, and GPS.

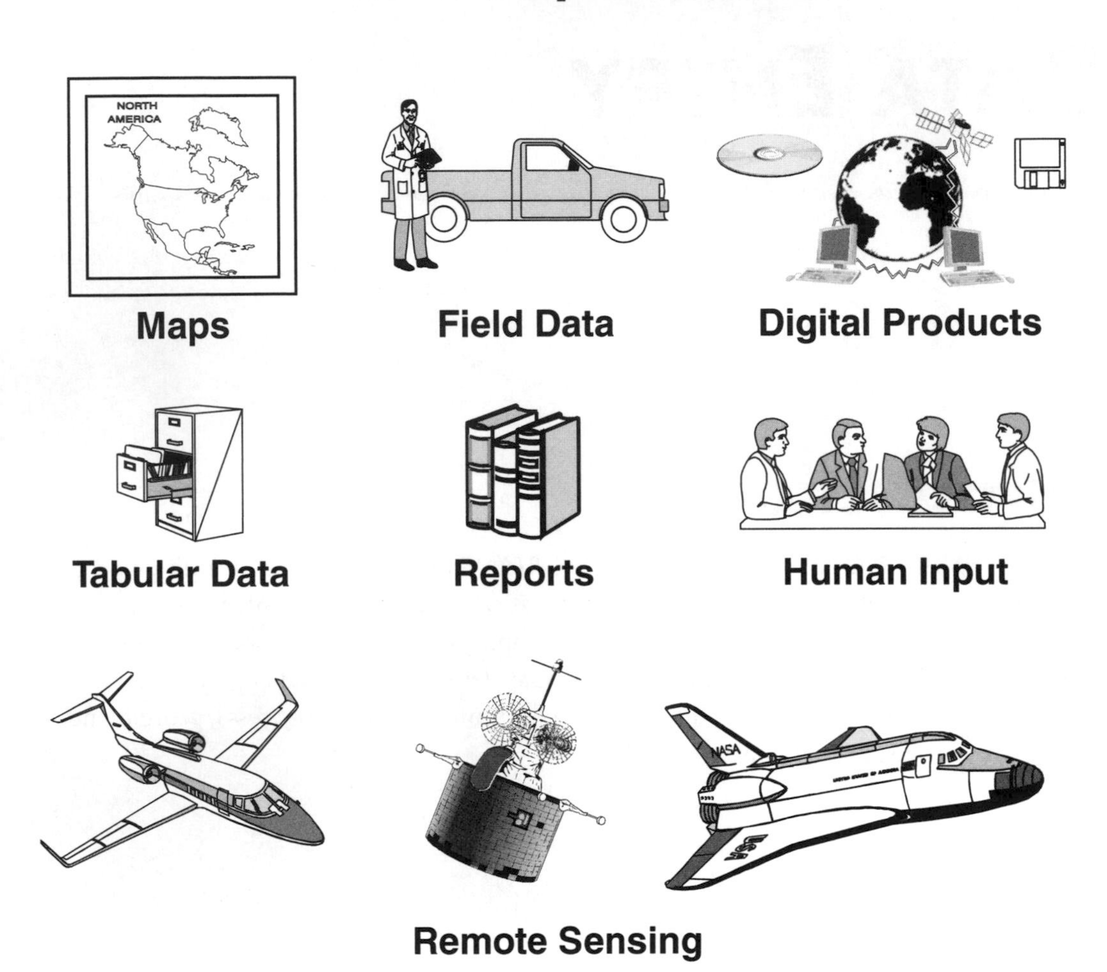

Fig. 5-1: Methods of GIS data acquisition.

GIS Data Acquisition

As indicated in figure 5-1, there are many sources of GIS data. Maps, the most common presentation of spatial data, are a primary source for most GIS projects. Just about every type of map can be *digitized* (electronic copying) into GIS. However, the process is not always easy or brief.

Data collected in the field (study area) is considered "primary" data, which refers to the first observations put into the database, typically unprocessed regarding classification or meaning. This usually involves researchers or survey teams taking measurements, notes, and photographs of the places under investigation and then entering the data into the GIS.

Digital products, which are sets of processed data ready to use, are available from various organizations and commercial sources. They may be databases, images, maps, or any type of data or information that can be integrated into GIS. Data may be from several media, such as CD-ROMs and the Internet. Digital products are becoming a major source of GIS data because of their efficient storage, ease of transfer into computers, and convenience. The Internet has developed into a major means of presenting and exchanging data.

Tabular data may be standard lists, such as census reports or business marketing information, formerly in printed form but rapidly transitioning into standard digital products. Typing or scanning is still a common method of entering nondigital data into GIS. Reports contain a variety of data that can be copied into the GIS, such as text, maps, and tabular data.

Text information can be important, though not easily reduced to GIS format. It can be translated into useful data by manual entry or scanning, to be stored as an associated part of the database. Today, many reports and their data are stored on CD-ROMs or even on the Internet, making them more accessible and useful. Hardcopy data will not disappear in the near future, but digital media are becoming more common.

Human input, such as decisions concerning classifications, can also be included in GIS. This is an important flexibility of GIS because judgments, interpretations, and even new data based on personal knowledge can be as valuable as field data. Commentary, such as explanations or elaborations of the data, may be included in order to add depth to the information.

Geographic remote sensing is the collection of landscape data from above, such as by aircraft carrying cameras and electronic sensors, and space vehicles (the Space Shuttle and satellites) equipped with special imaging devices. Remote sensing is a major source of GIS data, and with a little work imagery can be placed directly into the GIS database. Remote sensing is discussed in this chapter.

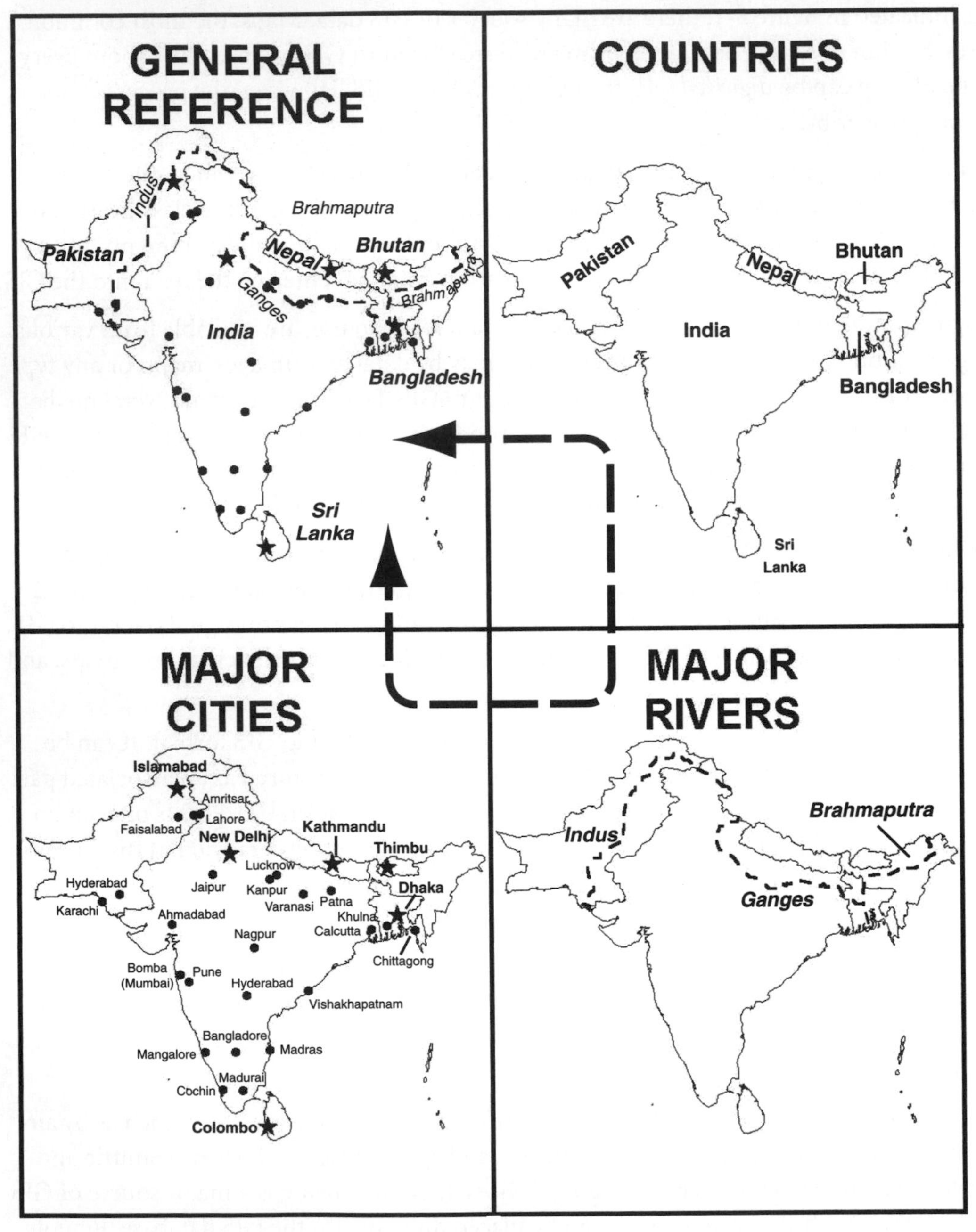

Fig. 5-2: Mapped thematic data.

General Reference to Thematic Data

A common source of data is existing maps, many of which are *general reference* types containing various themes, such as U.S. Geological Survey topographic maps, which show land cover, topography, roads, towns, and so on. Most countries have similar versions of these maps. GIS projects normally consist of separate themes instead of one multi-theme general reference map. Therefore, themes can be digitized separately from the general reference map and then stored as individual files.

Figure 5-2 is a simple example of a standard project, showing a general reference scene of South Asia, consisting of countries, cities, and rivers. Individual themes can be extracted and presented as thematic maps. In reverse, a set of thematic data sets can be combined into a general reference map, often as a good project summary map. Projects often work in both directions. GIS can be a convenient mapping tool.

Remote Sensing

Image 1:
Alps, Central Europe

- *Synoptic view*
- *Consistent data*

Image 2:
Hurricane Elena

- *Real time*
- *Disaster management*

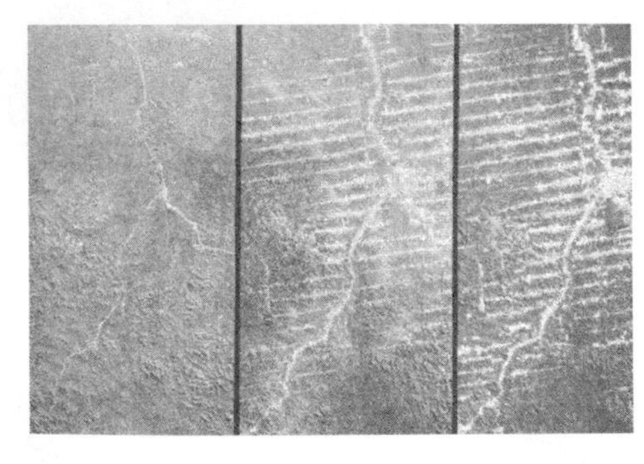

Image 3:
Rondonia, Amazon Basin, Brazil

- *Permanent record*
- *Inaccessible areas*
- *Development tool*

Fig. 5-3: Remote sensing is a major GIS data source.

Remote Sensing

As noted, geographic remote sensing is one of the major data sources for GIS projects. Image 1 of figure 5-3 shows a satellite view of the Alps and Central Europe. A large amount of geographic information is available. For example, the major physiographic regions of the area can be mapped. Much better data can be extracted by experienced image experts and sophisticated computer analysis. Remote sensing, which captures images such as this, has several advantages that are important in GIS, including the following.

- Although potentially expensive, images usually cover very large areas (the *synoptic view*), especially from satellites. This makes the cost per ground unit lower than time-consuming, labor-intensive ground crew collection. The cost of the satellite imagery for the Alps region coverage may be several thousand dollars. However, any other means, whether ground-based or aircraft photography, would be many times that amount.
- A satellite image is very consistent regarding data quality and conditions of collection (the entire image typically is a single view that is uniform all over), whereas field crews tend to produce uneven observations when a long time or several crews are required. Consistency of data collection is very important to change analysis over time because data needs to be as comparable as possible. Considering how long it would take to cover the large area of the Alps image by ground crews and how problems of data uniformity can result, satellite imagery collection would be a better method in this case.
- Remote sensing imagery can be delivered to users very quickly, sometimes in real time. Field teams take time to investigate a study region, which is a disadvantage for some applications, such as disaster evaluation. For example, real-time thermal imagery for forest fire mapping can be crucial in preventing tragedy. Hurricanes, such as Elena in image 2, are tracked numerous times a day to provide warnings to save lives and property. Without remote sensing, these storms can be tragic.
- Images provide a permanent record that can be checked and used for a long time. In image 3, the three images of Rondonia in the Amazon Basin (1975, 1985, 1995) show the changes due to human settlement programs and demonstrate the utility of consistent data over time. (The images were originally in color, which provides even better information.) Development programs, for example, can monitor and analyze landscape and environmental changes in order to make adjustments.
- Remote sensing can gather data from difficult or inaccessible areas, such as deserts, rain forests, or the polar regions. A satellite, of course, only requires a convenient orbit to cover any desirable part of the world (or another planet). The Amazon images cover an area not easily accessed by ground or aircraft.

Electromagnetic Energy Spectrum

Wavelength µm (micrometer)

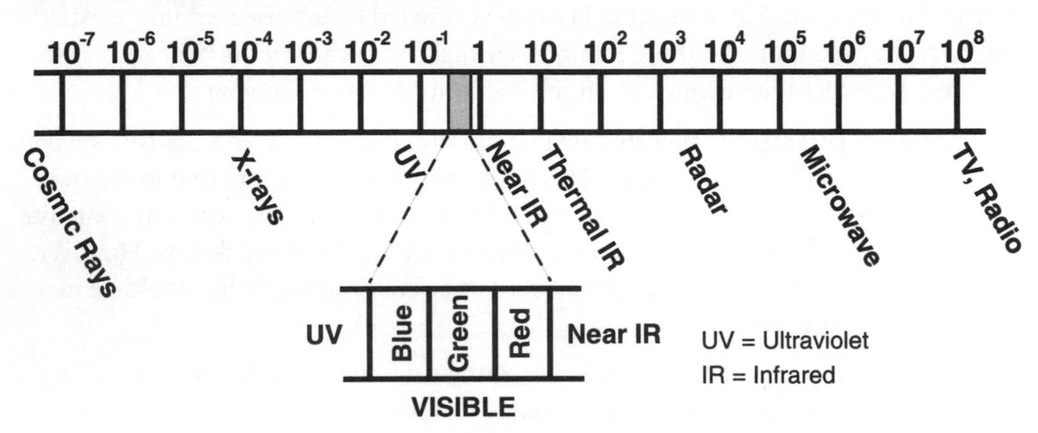

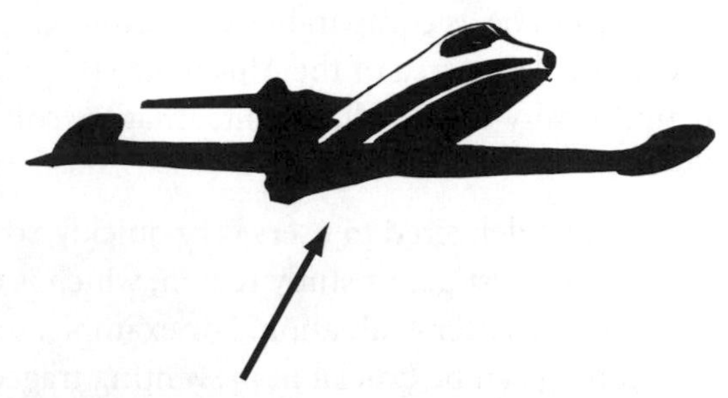

Remote sensing instruments:

- Visual photography camera
- Color infrared camera
- Thermal infrared scanner
- Radar scanner
- Electronic scanner for 10^{5} – 10^{6} wavelengths

Fig. 5-4: The electromagnetic energy spectrum and its association with remote sensing.

- For poor nations, remote sensing can provide valuable landscape data resources that may not be obtainable by other means. Satellites can cover an entire nation in one or a few images. Remote sensing has high value in development programs.

Other advantages will be seen. Perhaps you can think of a few.

Electromagnetic Energy Spectrum

Aerial photography was the first remote sensing data, providing visual imagery of landscape on film (a non-electronic medium). In recent decades, sophisticated electronics and high-tech sensors have been developed that gather data from parts of the electromagnetic (EM) energy spectrum that are invisible to the human eye.

Figure 5-4 shows that portions of the EM spectrum are expressed as wavelengths of energy, ranging from very short (cosmic rays on the left) to very long (television and radio waves on the right). Note that visible light is a very narrow part of the EM spectrum, meaning that humans see only a small part of the spectrum, whereas remote sensing equipment can obtain information from a much greater portion of the spectrum.

Each part of the EM spectrum defined in the diagram has particular advantages in remote sensing. Landscape features, for example, typically absorb some of the incoming solar radiation and reflect some, the portion depending on the nature of the feature. Beaches are very bright reflectors in the visible, and water absorbs near infrared. Some features emit their own radiation, such as heat from hot springs, which shows up bright on thermal infrared imagery. Knowing the responses of various features under various conditions is important for interpreting remote sensing imagery. Examples of remote sensing imagery are provided in the following section.

Some portions of the EM spectrum are not useful as GIS data sources, such as X rays (used in medicine and engineering). The short wavelengths of ultraviolet (UV) are affected by atmospheric scattering and are not effective sources of landscape data on Earth. UV remote sensing has been used successfully from the moon, however, and possibly could be practical for extraterrestrial studies (although largely Earth-based, GIS can be applied to the study of other planets).

The illustrated airplane carries a "multispectral" (multiple wavelengths) remote sensing package, including sensors that are sensitive to the visual (photography), near infrared, thermal infrared, radar, and one part of the microwave portions of the EM spectrum. Film and digital data will be produced for use in various projects.

Remote Sensing Imagery
Major Wavelengths

Image 1: Visual
Aerial Photograph

Image 2: Visible
Manaus, Amazon River

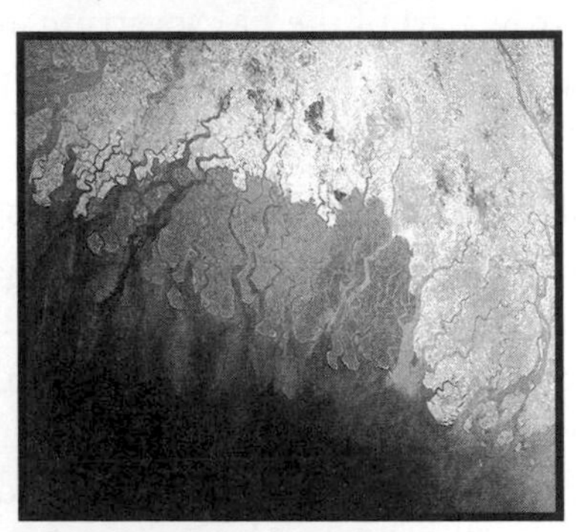

Image 3: Infrared
Sundarbans, Bangladesh

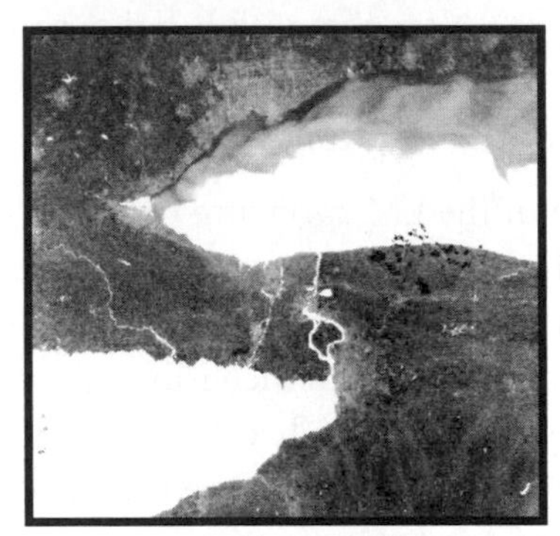

Image 4: Thermal
Lake Ontario, Lake Erie

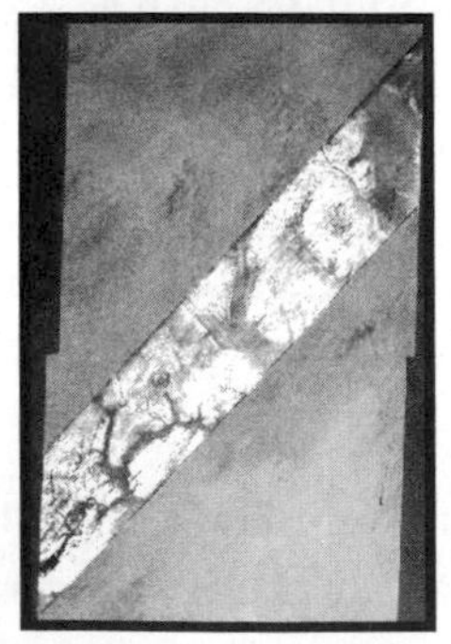

Image 5: Radar
Ancient Rivers, Sahara Desert

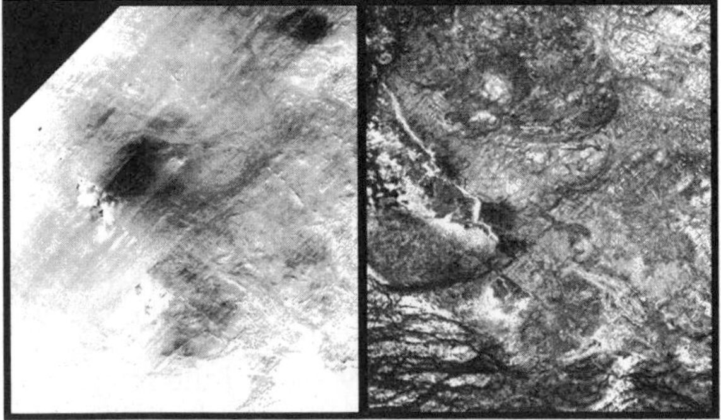

Image 6: Radar
North Africa

Fig. 5-5: Types of remote sensing imagery.

The EM Spectrum and Types of Remote Sensing Imagery

The middle range of the EM spectrum contains wavelengths that are visible to the human eye (the *visual*), ranging from blue to red. Ordinary camera film is sensitive to this spectrum, producing standard aerial photography. Image 1 of figure 5-5 is a standard visual film photograph of an urban area. Image 2 is a photo from the Space Shuttle, showing Manaus, Brazil. The original color photo is only a bit more informative than this monochrome (black-and-white) version, primarily because of the high moisture content of the atmosphere, which diffuses light.

The heavily silted Amazon appears light (reflection from opaque water), whereas the cleaner Rio Negro is darker. Manaus is the bright area near the junction of the two rivers. Very little vegetation information can be detected through Earth's atmosphere. This is an unusually clear day, without much cloud cover.

Electronics and some films capture the next longer wavelength, just outside the visual, called the near infrared. The near infrared reveals information not available in the visible spectrum, such as plant health and stress. Visual infrared can show damaged vegetation before it is apparent to the observer on the ground. This is highly valuable for many environmental GIS projects, and infrared—either color (CIR) or black and white (B/W IR)—is a common data format.

The image 3 monochrome version of a Sundarbans, Bangladesh, CIR photograph contrasts the darker mangroves with the surrounding lighter farmland and brush. Heavy siltation is evident into the bay (lighter tones, similar to the Amazon's siltation). Clean or deep water appears dark in infrared.

Thermal infrared sensors detect very small temperature differences and display them on film or electronic display, such as in image 4 of lakes Ontario and Erie, USA. Warmer water is brighter than cooler land. This is useful in thermal pollution monitoring, where, for example, industrial effluence can be analyzed in terms of heat characteristics. Urban heat sources can be detected easily, and landscape thermal mapping can reveal geologic and vegetation differences. Thermal data is also important in fire fighting.

Radar and microwave data are long-wave, producing land and water information much different from that of the visual. Radar is useful, for instance, in penetrating cloud cover, and can help to map topography in humid, cloud-covered tropical regions where aerial photography has been unsatisfactory. It can also penetrate some types of vegetation and can detect subtle geologic features, such as fault lines.

Radar imagery from the Space Shuttle has detected ancient stream systems under several meters of sand in the Sahara Desert (image 5). The two views of image 6 show a region in North Africa, with a visual version on the left showing mostly sand, and the remarkable difference from radar on the right, revealing a wealth of geologic information under the sand.

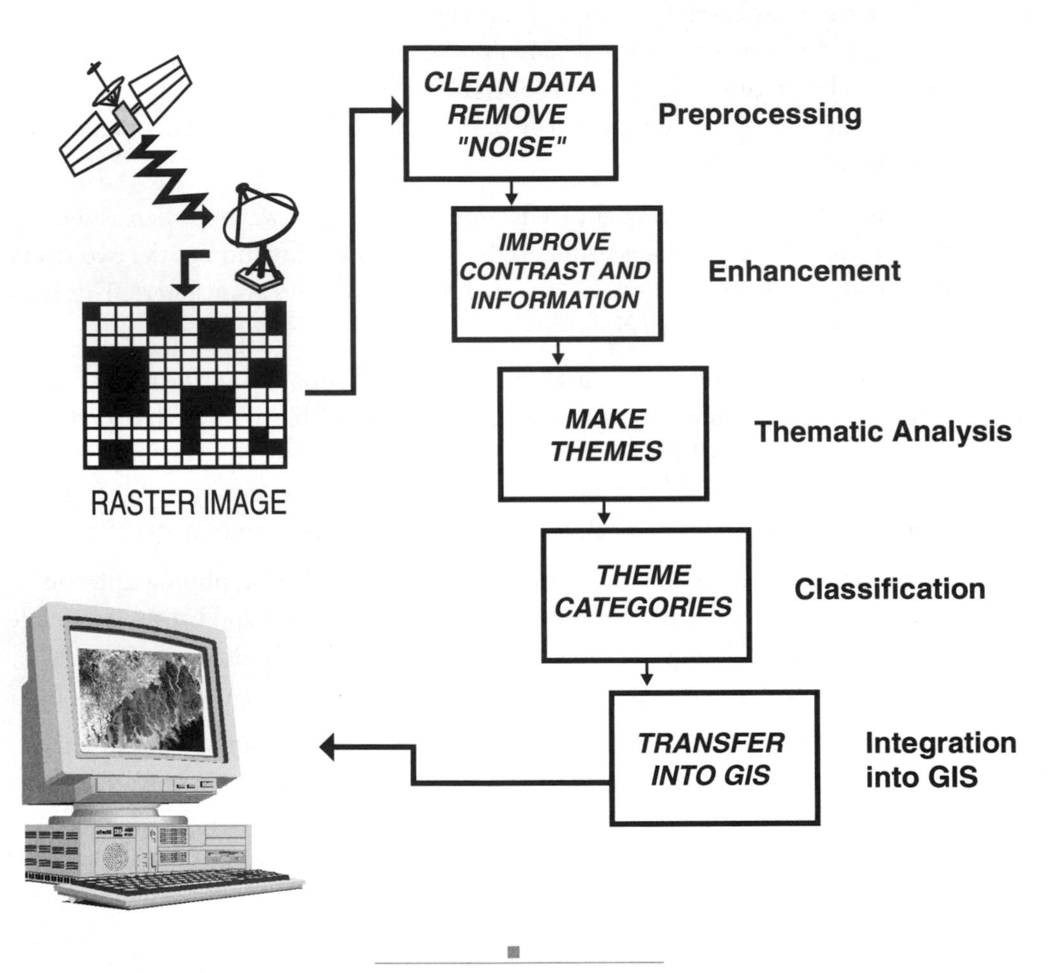

Fig. 5-6: The process for using remote sensing imagery.

The microwave region yields special geologic information (e.g., soil moisture), and can be very useful in some oceanographic applications, such as oil spill mapping and monitoring sea ice distribution. Microwave scanners can make rapid changes in the small increments of wavelengths, giving data from multiple parts of the spectrum. This offers many images of the same area at once.

Today's advanced remote sensing scanners can make imagery from many parts of the EM spectrum simultaneously, termed "multispectral imaging," which offers a synergy of data for the landscape. Because of the wealth of data, computers are needed to merge and analyze the data. As discussed in the following material, additional digital techniques are used to improve the imagery and to incorporate it into GIS.

Imagery and GIS

Remote sensing data, usually in the form of digital raster format images, can be incorporated into GIS, but preparation is required. Camera film produces photographs, which are analog, not digital, and thus must be scanned into a digital format in order to be placed into GIS. Most other remote sensing data is in digital form.

The sensor, on an aircraft or satellite, scans the landscape, recording electronic data from selected parts of the EM spectrum in a raster format. Each cell has a measurement value that is converted to a digital number and then relayed to the ground station. There, the raster cells are constructed into an image of the scanned area. Now it is an image data file, ready for processing into a GIS. The basic process, depicted in figure 5-6, works as follows.

- When first received, the imagery is not ready to be used. The first step is *pre-processing*, which cleans the data by removing electronic "noise" and correcting mistakes, such as missing scan lines. Data quality is improved and the image is then distributed to users.
- The next step is for the user to enhance the data so that better information can be obtained. This often includes improving visual contrast, such as changing subtle differences in gray tones into shades that are more distinctive, as an aid to interpretation. Special digital techniques are used for better information detection. Perhaps better vegetation data could be extracted from the visual Manaus photo in figure 5-5.
- Thematic analysis is the next stage in the process. This involves turning enhanced data into selected themes. For example, landscape images contain a variety of land covers (such as various vegetation types) that can be separated to provide a useful thematic map. Programs to find statistical groups of numbers (from the raster cells) are applied to help find common themes.
- Classification of themes into distinct categories is a logical next step, though it is not always necessary. A vegetation theme, for example, might include a sequence of classifications, starting at overall land cover (e.g., forest, grassland, agricultural land, and barren areas), to the community level (e.g., deciduous forest, coniferous forest, and mixed forest), and down to areas of particular plant species. Classifying data according to a project's specific purpose is an important "data reduction" step in keeping the amount of data manageable and focused.

Global Positioning System

Garmin Handheld GPS

Trimble Field GPS

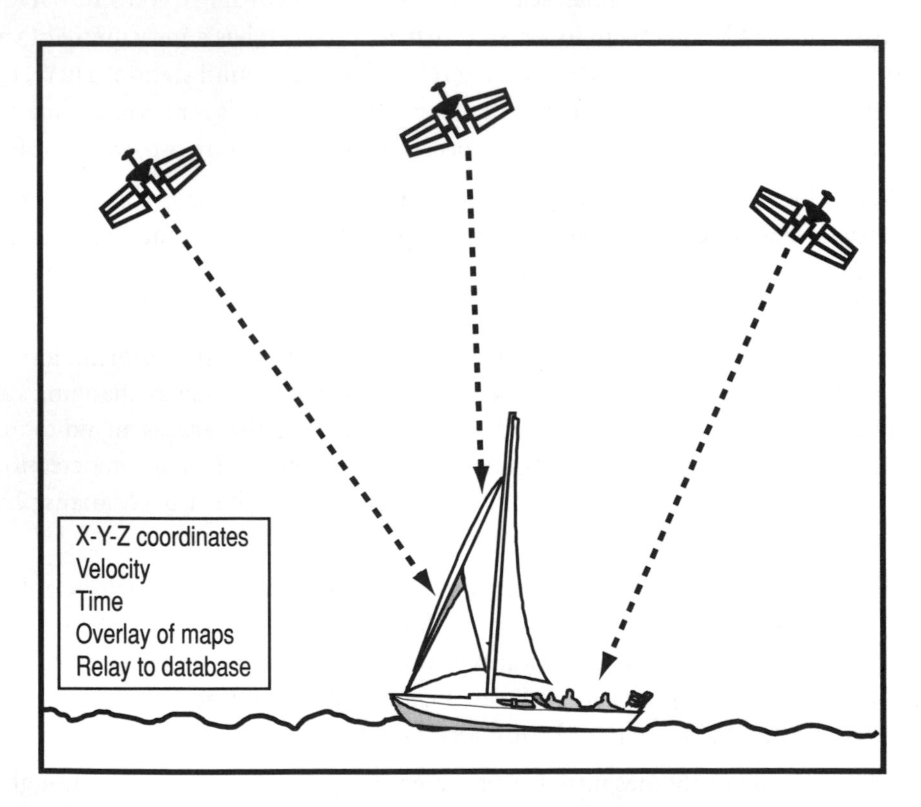

Fig. 5-7: GPS in GIS.

- Transferring imagery into GIS can be simple copying for use as a background or complete integration as a full data set. That is, images used as background for on-screen digitizing are merely digital pictures, whereas a full data set requires proper georeferencing, projection, and database development, perhaps even classification and theme creation steps.

Getting remote sensing data into a GIS requires several technical steps, with the basic aim of extracting new and effective data for a project. Special software may be necessary in achieving this goal, though many GISs now include some capabilities for integrating remote sensing data. Many vector GISs, for example, include programs that read and incorporate raster imagery.

GIS and GPS

The sections that follow discuss the Global Positioning System (GPS) as it relates to GIS work, as well as GPS attribute input and interactive GPS relay. GPS has become a major component of GIS and will continue to be an essential input for fieldwork.

What is GPS? A data source similar to remote sensing technologies is the Global Positioning System, a U.S. Department of Defense system of over two dozen satellites that transmit signals to special receivers on the ground for precise determination of X-Y coordinate position. The receivers triangulate incoming satellite data (measure the differences between three or more signals) and calculate position.

The receivers may be small, handheld units or larger instruments that can have sub-meter accuracy. Two commercial systems are shown at the top of figure 5-7. GPS data can also give elevation (Z coordinate), velocity (while moving), time of measurement, and even custom input of selected variables (see figure 5-8).

Figure 5-7 shows a sailboat receiving GPS data. In the middle of an ocean (or swamp, forest, desert, and other environments), there is an absence of easily recognizable landmarks for proper location reference. GPS provides highly accurate position almost anywhere on Earth. There are numerous GIS applications that require precision, such as mapping of urban streets and their components (e.g., water drains, manhole covers, and utility boxes). GPS continues to provide increasingly accurate data input for a wider range of GIS applications.

Because GPS data can be incorporated directly into GIS, update and modification of geographic data can be made in the field. For example, by overlaying the GPS signal on an existing road display, corrections to maps and entry of new data can be accomplished interactively, even while traveling. It is no wonder, given the importance of accurate input data, that GPS is becoming a major technology for many fields, including GIS.

GPS Attribute Input

Image 1: Highway Inventory

F1	F2	F3	F4	F5
DATA POINT (Location and Time)	2-LANE	4-LANE	REPAIR	CONTROL
	U - Unimproved	D - Divided	S - Surface	2 - 2-Way Stop
	I - Improved	U - Undivided	P - Pothole	4 - 4-Way Stop
			R - Repaint	C - Caution
				T - Traffic Light

Image 2: Interactive GPS Relay

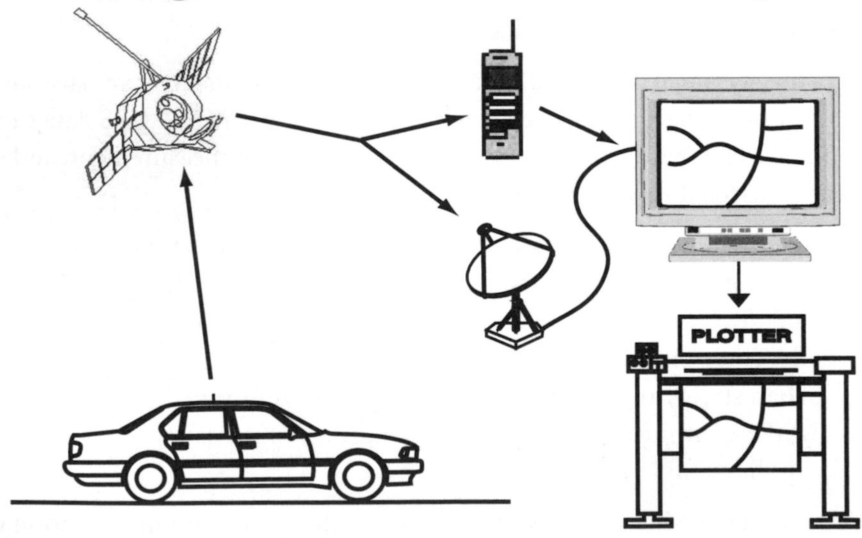

Fig. 5-8: GPS attribute input and interactive GPS relay.

GPS Attribute Input

To show how GPS can be an interactive methodology, image 1 in figure 5-8 represents a highway inventory operation. The GPS is linked to the GIS, providing time and location data. The computer's function keys have been set to record operator input, as shown.

As the car moves along the highway (speed is no concern) and a data notation is needed (say, notice of a section of road to be repaired), the operator presses the F1 key to enter a data point, noting the location and time. This "freezes" that location until the next time F1 is pressed, giving the operator time to input data while the car is still moving.

If the highway inventory data at this point is a repair notice, the operator presses F4 (Repair) and a submenu is presented, asking for the type of repair necessary: S stands for surface problem, P for pothole, and R for repaint. Other options are included in the illustration; for example, a notation for four-way stops (using F5, 4). Almost any type of data input scheme could be devised. This is a "custom" interactive input system linked to GPS.

Interactive GPS Relay

Not only can GPS operate with GIS interactively, data can be transferred to distant stations (image 2 of figure 5-8). From the field, data can be relayed directly to satellites by antenna or to cellular phone stations. In turn, receivers on the ground, or a phone system, can take the data and relay it directly to a central GIS, where either further processing occurs or the field data is stored. When the mission is finished (at the last data relay), the central GIS produces a map, possibly even before the field technicians have parked the car.

GPS processing and relay can provide very fast and highly accurate data (including maps) from a distant location (even half a world away). This type of technology is changing the way we view, collect, and manage data. As a result, our expectations of what we can do with data and information are also changing.

Further, the finished data can also be relayed by satellite to a customer at a distant third location, as noted. We can envision a final product from field to central GIS to customer, anywhere in the world, all before the field crew finishes packing the equipment at the end of field data collection. Location is becoming irrelevant. Marvelous possibilities exist. Not surprising, GPS has become a distinct field of its own, as a specialized technique and methodology, a profession, and a business, much like GIS, as discussed in Chapter 1.

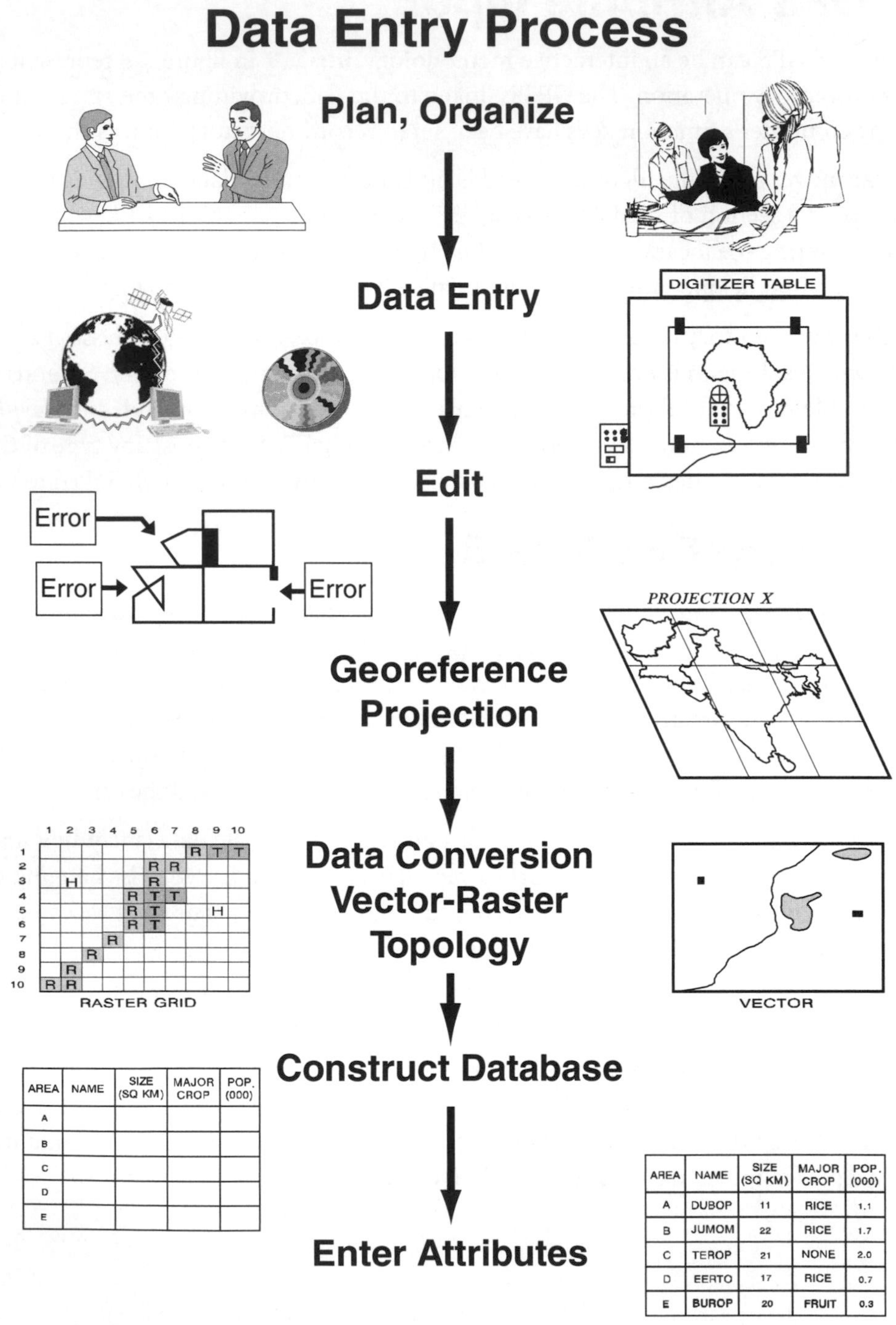

Data Entry Process

Plan, Organize

Data Entry

DIGITIZER TABLE

Edit

Error

Error

Error

PROJECTION X

Georeference
Projection

Data Conversion
Vector-Raster
Topology

	1	2	3	4	5	6	7	8	9	10
1								R	T	T
2						R	R			
3		H				R				
4					R	T	T			
5					R	T			H	
6					R	T				
7				R						
8			R							
9		R								
10	R	R								

RASTER GRID

VECTOR

Construct Database

AREA	NAME	SIZE (SQ KM)	MAJOR CROP	POP. (000)
A				
B				
C				
D				
E				

Enter Attributes

AREA	NAME	SIZE (SQ KM)	MAJOR CROP	POP. (000)
A	DUBOP	11	RICE	1.1
B	JUMOM	22	RICE	1.7
C	TEROP	21	NONE	2.0
D	EERTO	17	RICE	0.7
E	BUROP	20	FRUIT	0.3

Fig. 5-9: The GIS data entry process.

GIS Data Entry

As noted, data entry can be very time consuming, but it is the most important task of the GIS process. This section discusses the basic organization of entering data, manual digitizing, georeferencing data, projections, and other important aspects.

Data Entry Process

Data entry is a simple process that can take a great deal of time and effort, sometimes involving 75 percent of project's life span. Because data entry is perhaps the most critical stage, patience and care are needed. The following points, depicted in figure 5-9, describe the basic data entry process (discussed further in material to follow).

- *Planning and organization:* This is the most important part of the process, because without proper planning, subsequent steps will be inefficient or incorrect. The end product should be known and all steps in the procedure must lead to it effectively. Planning and organization must be given patience and focus, yet too often it is hurried or ignored.
- *Entering spatial data (digitizing):* Entering spatial data can be time consuming. Manual digitizing is the most common process, and it can be tedious, although semiautomatic scanning is rapidly advancing. Other techniques are used, such as typing, reading tape or CD-ROM data, downloading from the Internet, and using remote sensing imagery.
- *Edit and correct:* Regardless of entry procedures, the data must be checked for accuracy. Mistakes are normal, and must be corrected. This process can require much work, but it is necessary.
- *Georeference and projection:* The digitized data usually need to be referenced to the real world, typically using some established coordinate system, such as Latitude-Longitude. This step can occur before or after digitizing (depending on the GIS). In addition, a projection must be specified for translating world sphere data to a flat, 2D view.
- *Convert:* Digitized data must be given a specific data structure—either vector or raster. If vector is selected, topology may be assigned for those GIS programs that use it. These steps are fairly simple, often automatic, typically involving a few commands to select the desired operation. Some GISs work only with one data format type, but many are now able to accept either vector or raster, and selection may not be necessary. For example, many vector systems save in vector format automatically, but a raster gridding option is available. Perhaps this step may soon disappear from most GISs.

- *Database construction and entering attributes:* Once the geographic data set has been secured, the database must be developed. The first step is to create the database; that is, to construct fields (columns). Some databases require considerable work to develop, whereas others are very easy and flexible to produce. Many GIS establish initial databases at the digitizing stage, and records are entered as they are digitized, often along with some automatic data, such as item number, area, and perimeter where appropriate. Other fields can then be added to develop a complete database. There are various ways of entering attributes, from typing to loading tables. Data derived from outside sources usually have attached databases, although they can be modified or expanded as needed.

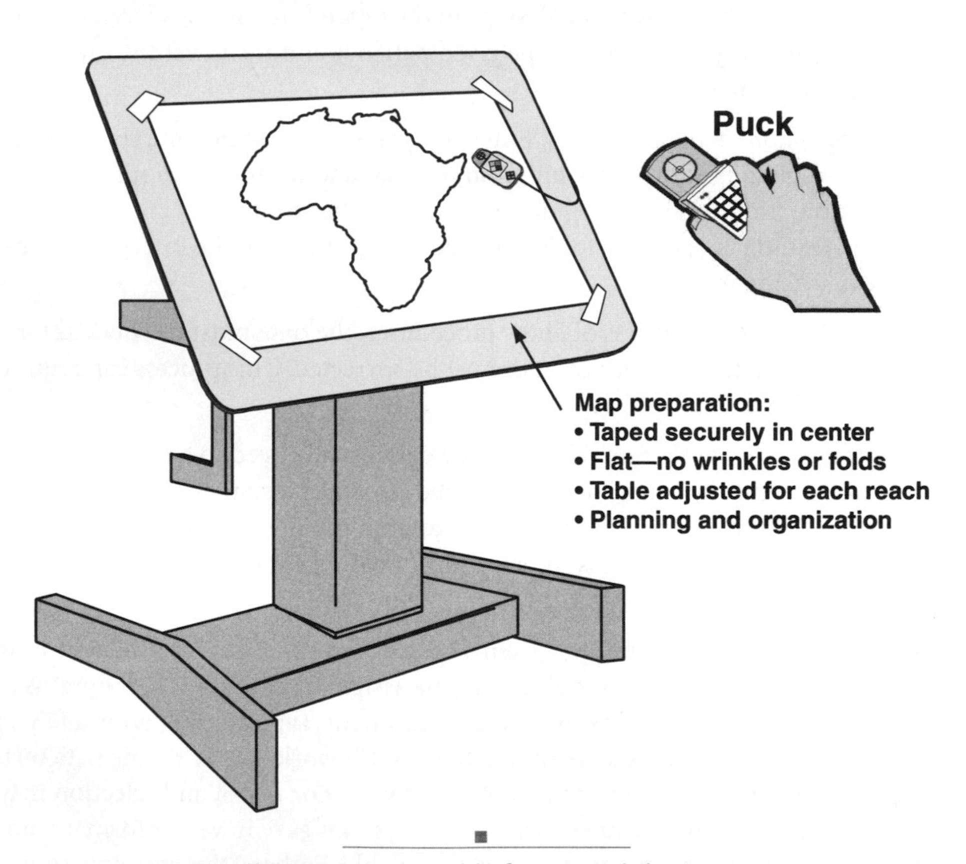

Fig. 5-10: Digitizing table for manual digitization.

Manual Digitizing Preparation

The most common method of GIS data entry is manual digitizing of maps. Although semiautomated scanning may be more important in the near future, manual digitizing will continue to be important for many users. Digitizing involves placing a map on an electronically sensitive table called a digitizer table (or just digitizer, shown in figure 5-10), and then tracing the map features with a hand device, called a puck or cursor. Number and letter buttons on the puck control operations, such as designating inputs as nodes, vertices, and labels that define features. A cross-hair aiming window helps with pointing.

Preparing a map for digitizing is not difficult, but care is needed to provide accurate data entry. First, the map is placed in the center of the digitizing table, taped securely, with no wrinkles or folds that create distortions (a flat map is best). The table must be adjusted for easy reach of all parts of the map; stretching can result in errors. As noted, planning and organization are critical if the process is to be efficient and effective. Development of a strategy is recommended—an organized procedure that establishes the region to be worked on, the features that are to be entered, how they will be labeled, and the order of features to be digitized. This last step is important in avoiding the omission of features.

Depending on the GIS used, the digitizing process can begin immediately or wait until several preparation steps are finished. In this example, the preparation steps are covered first before discussing manual digitizing in detail. Geographic data, by definition, represents the "real world" in terms of location, so digitized features must be linked to the world in some way. Otherwise, digitizing is no more than drawing pictures that have no connection or reference to the earth (although there are projects in which geographic reference is not important, such as making diagram maps of bus routes for the public). Two steps are needed: establishing Earth coordinates and defining a projection. These steps are discussed in material to follow, as preparation to actual digitizing.

Georeferencing is the procedure that establishes the necessary coordinates and projections that give the map the proper shape. Either step may be completed prior to or after digitizing. When georeferencing is the first step, world coordinates are established for features as they are digitized. When performed after digitizing, features are first entered according to digitizer table inches or centimeter coordinates, and later converted to a real geographic system.

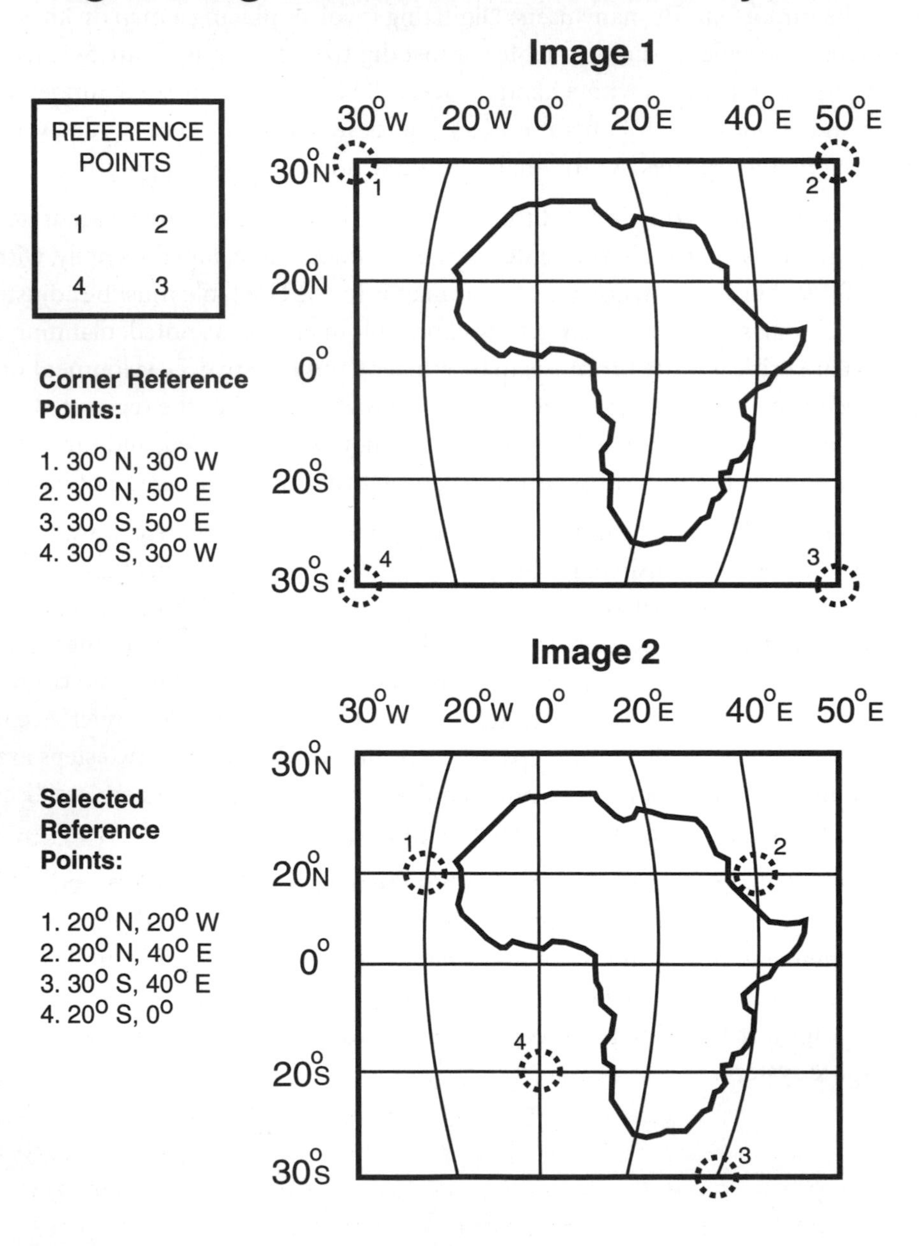

Fig. 5-11: Georeferencing is registering data to a fixed coordinate system.

Georeferencing

GIS data files usually must have a real-world coordinate system (such as Latitude-Longitude) if they are to be suitable geographic data. The process is termed *georeferencing*, defined as registering, or fixing, data to a standard coordinate system, thereby linking the map to the earth. The best method of establishing a proper georeference is to define at least four reference points (sometimes called tic points) around the area being digitized (close to the corners if possible), each with a precisely known real-world coordinate position that is typed into the program. With known reference points, digitized features can be properly located on the earth.

In image 1 of figure 5-11, the corners of the map are selected as reference point locations (dashed circles). It is best to choose coordinate points outside the mapped area when possible, because the reference points "frame" the digitizing. The corners make a nice pattern that is easily referred to when necessary (although the computer does not really care about patterns).

However, location is not rigid, and reference points can be selected inside the digitizing area if they are near the edges and preferably toward the corners (a square pattern is not necessary). Available known coordinate points are the best references, but are not always at ideal locations. Image 2 shows selected scattered coordinates, but they are still outside the digitized outline of Africa. Points inside Africa may be acceptable, but there is a chance for less precision.

Note that the coordinates specify latitude first (north or south), followed by the longitude (east or west). Negative latitudes refer to the southern hemisphere, and negative longitudes refer to western hemisphere.

Commonly, the upper left corner is the first point, although any corner can be used. Some GIS permit the points to be digitized first and the coordinates entered right away. Other systems have the points entered as digitizer table inches or centimeters first. After map features are digitized, these points are defined as world coordinates. Either way, the features ultimately become linked to the real world.

Georeferencing changes (transforms) the digitized file into the coordinate system of the map (in this case, Latitude-Longitude). This is where computers are necessary; translating one reference system to another is a number-crunching task. The program notes the reference point coordinates and establishes a proper georeferenced structure for the map. Good GISs render the chore rather easy by offering a selection of coordinate systems and then making the computational process invisible to the operator.

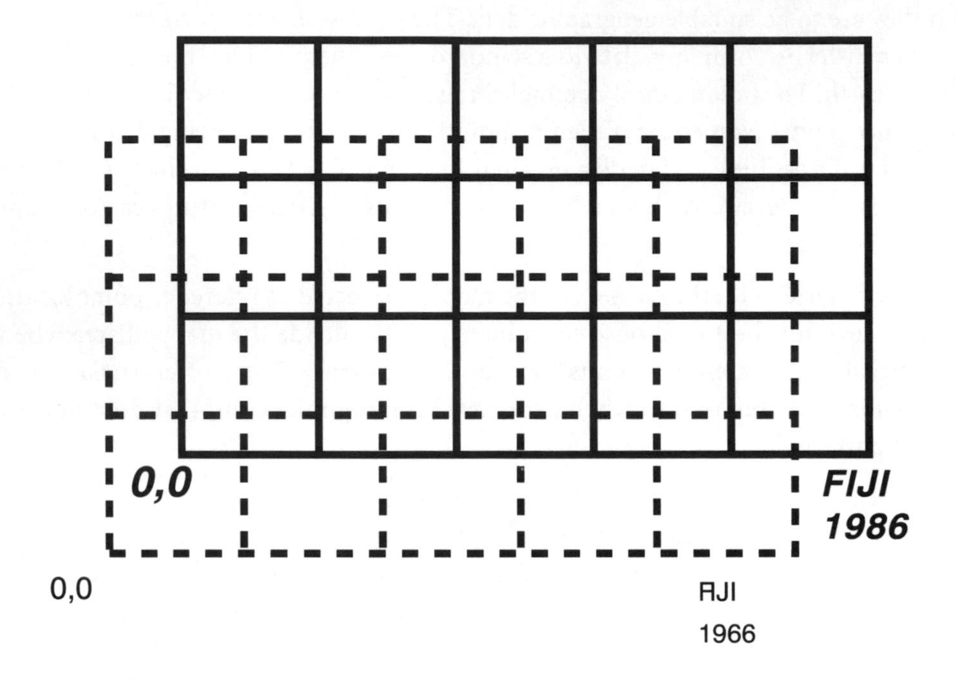

Fig. 5-12: A coordinate grid system may be changed.

Changing Coordinate Grid Systems

Latitude-Longitude is a world coordinate system, but smaller-scale systems exist for regional purposes and more accurate positions. Georeferencing systems are not necessarily stable; they may change over time, for a variety of reasons. Even the Latitude-Longitude system shifted slightly due to more accurate measurements, replacing the 1927 North American Datum (NAD) with a 1983 (NAD83) update. Some maps calculated with the old system show the new shift with a small cross so that both can be used.

As indicated in figure 5-12, Fiji's regional grid was moved in 1986 as an update from its original 1966 system. The 0,0 origin point at the lower left was shifted slightly in order to have the entire country within a convenient system.

The date of data must be considered in light of coordinate system adjustments. Otherwise, positions can be misplaced and uncoordinated. For example, if uncorrected, old map data will not properly overlay or combine with new data, making accurate change analysis difficult. Some GISs, or special programs, can make the appropriate changes (e.g., from the 1927 to 1983 NAD). Coordinate system management can be a problem, but it is a very important aspect of data reliability.

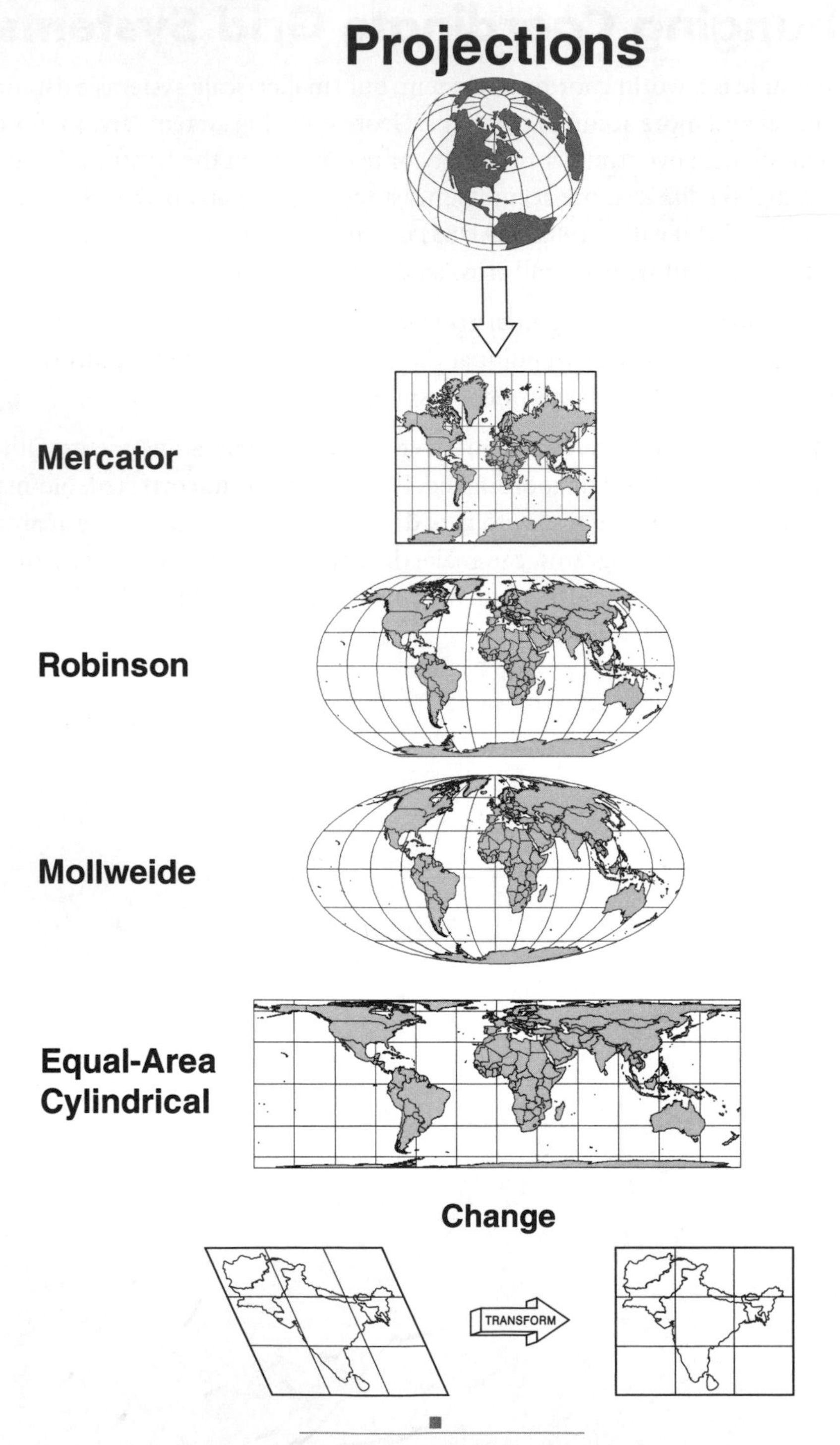

Fig. 5-13: Types of global projections.

Projections

Even when a map data file has been georeferenced to a specific coordinate system, the GIS still needs to know which map projection to use in order to give proper spatial characteristics to the features. Projections are spatial configurations used to fit a portion of the globe onto a flat view; that is, spherical data from the globe is converted into a flat, 2D presentation. It is impossible to make spherical maps into perfectly flat paper maps without some distortion.

There are many projections, each with its own advantages and disadvantages. No single projection can maintain all of the important spatial variables, such as shape, area, direction, and distance. A projection that keeps shape over the entire map is called "conformal" because it conforms to the real shape.

Projections must distort one or more of these variables to maintain a realistic depiction of one or more other variables. For example, a projection can retain realistic shape but distort direction. The familiar Mercator projection, used on many world maps, preserves direction but distorts distance and area, particularly near the polar regions. Many school children believe that Greenland is almost as big as South America, although on a globe, the truth is evident.

Projects and applications often have specific spatial needs that are met with particular projections. GIS can convert digitized map data into a selected projection, or change from one projection to another as desired. Establishing the map projection can be a part of the initial map setup process, along with georeferencing, or it can wait until after editing (depending on the GIS).

Figure 5-13 presents four types of global projections. The top projection shows the familiar Mercator. Originally made for mid-latitude navigation (with geometric latitude-longitude blocks), it greatly distorts the higher latitudes (north and south). Antarctica is greatly enlarged.

The Robinson and Mollweide projections are attempts at more true representations of the globe, even though both have distortions. Robinson may replace Mercator as the standard view. Cylindrical projections try to transform the globe onto a cylinder and then "unwrap" it to form a flat surface. The cylindrical projection shown maintains equal areas of the continents.

At the bottom of figure 5-13, an existing projection (X) is transformed to a more appropriate one (Y). These standard GIS data preparation operations are usually very easy to perform, typically from a software program menu selection.

Manual Digitizing

Image 1: Digitizing the Map

Hand tracing of lines to input shape point coordinates

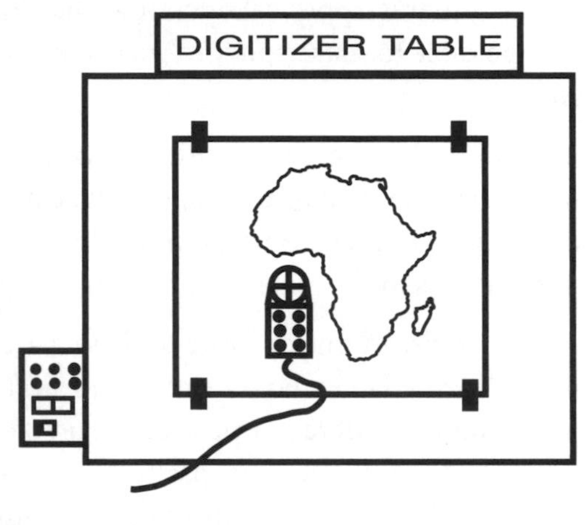

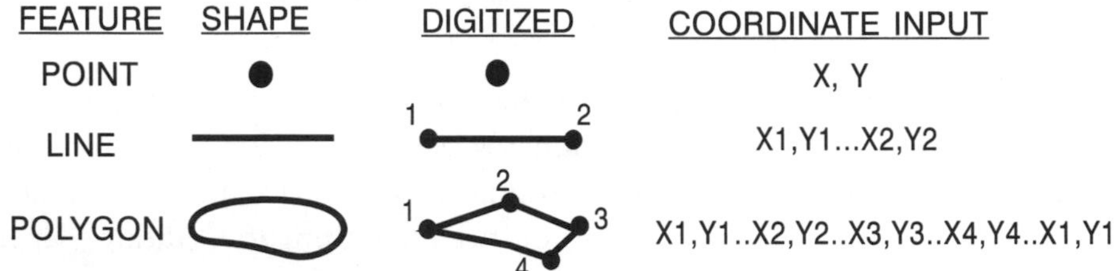

Image 2: Digitizing Database

Shape	ID	Area	Perimeter	Code
Polygon	1	122.6	35.9	1
Polygon	2	548.2	122.7	1
Polygon	3	444.9	111.8	2
Polygon	4	687.5	229.3	2
Polygon	5	774.1	151.8	3

Fig. 5-14: Manual digitizing related to features and coordinates.

Digitizing the Map

After georeferencing, the project is ready for digitizing, the process of tracing map features for conversion into a digital format. Normally, manual digitizing is in vector format and the digitized data are coordinate points that are later connected into chains (as discussed in Chapter 4). Features are recorded starting with a node, then connection to each vertex, and ending with a node (for lines and polygons, nodes only for points). Image 1 of figure 5-14 shows the shapes and coordinate inputs.

Because GIS data files are usually specific to point, line, or polygon features, each type must be digitized separately. For example, points cannot be digitized with line features. When digitizing from a general reference map, careful planning is needed to ensure that every feature of a given type is located and digitized. When possible, many users acquire *map separates* from map publishers. Each printed color layer, typically individual feature types, is a separate sheet that is combined with the others into a composite final color map. It is much easier to digitize selected features from a separate than from the final multi-color production map.

Each feature receives a label or identity of some type, either during digitizing or after. One method of labeling is to enter the classification code before tracing a given feature (a crop-type class number, for example). Most GISs automatically assign a sequence identification number (ID) to each feature, so that there is at least one way of distinguishing each point, line, or polygon. Every feature in the database will have an ID and a classification label. Image 2 shows an example of the database made during digitizing. Later, other attributes can be added to the database.

Normally, manual digitizing involves pressing a number on the puck at each node to start, then the vertices, and finally another number to signal the end of the feature. The person chooses where the inputs should occur.

Another system is called *stream mode*, which is automatic vertex input at set intervals. The puck is moved along the feature, without manual number entry except at the beginning and end, and the system inputs a vertex every 0.1 centimeters (or whatever increment is set). This can be faster and easier, but more prone to spatial errors because it is difficult to trace with consistent accuracy. The editing stage can make corrections, but that is a time-consuming task.

Simple digitizing jobs, with only a few features, can be relatively short, but overall, the task is time consuming. Some themes have many features, such as the very high number of polygons on soil maps. While digitizing, the project should be saved often in case of system failure; no one wants to repeat laborious digitizing work.

Digitizing Features

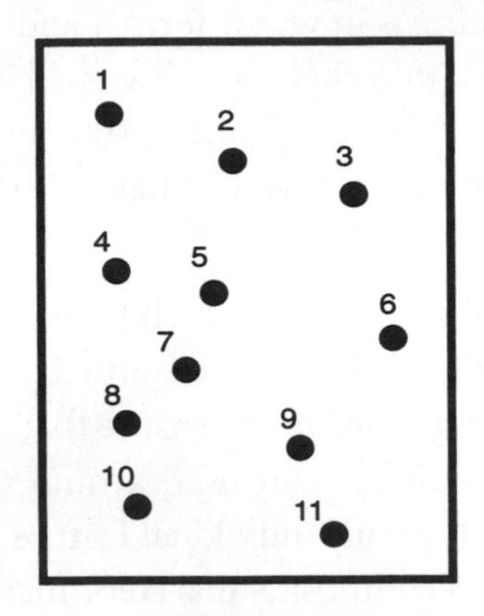

Image 1: Organized Digitizing

End node
Vertex 2
Start node
Vertex 1
chain 1
chain 4
chain 2
chain 3
Automatic node: Divides each line feature into two chains

Image 2: Lines

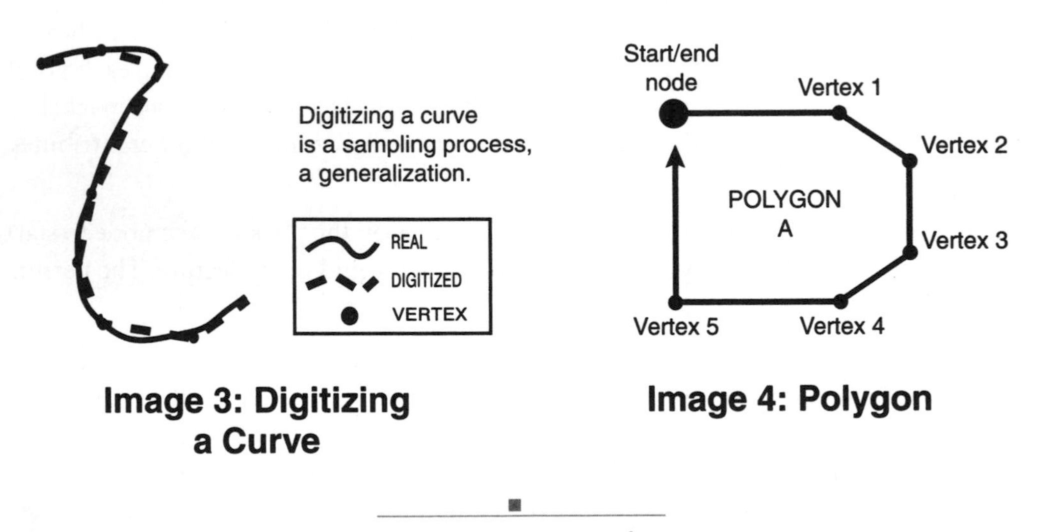

Image 3: Digitizing a Curve

Image 4: Polygon

Fig. 5-15: How to digitize features.

Digitizing Features

Digitizing is tracing of map features, which really means entering nodes and vertices, not just drawing lines. Tracing features with the puck is not difficult, but it can be tedious. Various tricks are suggested to help avoid errors, such as marking each feature on the map as it is completed (to avoid duplication and to help ensure that all selected features are digitized).

Organization of the order of input will help to avoid missing features. For example, the order can be from top to bottom of the map or all those features within a certain area. The order can be around a particular important feature (a central city and then its satellite towns) or along a road from north to south. It may be useful to number the input entries on the map first so that none are missed during digitizing. Image 1 of figure 5-15, for example, shows an order of points from north to south, making digitizing easy, fast, and complete.

Line features usually are not difficult, starting at one end, progressing to each vertex, and then to the other end. A typical puck sequence would be to start with the number 2 button to establish a node, use the 1 button for vertices (they will be sequenced automatically), and signal an end node with button 2 again (see image 2). A straight line would be created simply using button 2 twice (no vertices are needed). Note that crossing lines (intersections) automatically receive a node when topology is built; manual input of a node is not necessary.

Geometric features are fairly simple to enter, but curves and other complex lines have potential inaccuracies. Perfect tracing by hand is very difficult, and vertices input along the arc of the curve's path make straight-line connections to each other, thereby creating a generalized feature. In effect, the curved line is being sampled for representative points, not actually traced along every point on the curve.

Note the difference between the real and digitized lines in image 3. Stream mode could be used on curves, but as noted, this requires very steady hands to ensure accuracy—a rather difficult requirement for large or long features. The degree of detail needed in a project will help determine how close vertices should be spaced. A general map, or one that is to be displayed at small scale, does not require many vertices, and it would be inefficient to enter data at very small increments. Experience and tests can guide the process.

One possible solution to this problem of inaccuracy is the use of a curve generation routine (offered by some digitizing programs) that can draw a controlled curve between two vertices. This takes careful and slow supervision by the operator, but it may be necessary for high accuracy.

A square polygon is easily digitized by entering the corners. More complex polygons require patience and careful tracing; shape point locations must be decided. Automatic closure is a common digitizing capability: press button 2 after the last vertex and an automatic connection is made back to the first node, completing the polygon without having to manually redigitize on the same point accurately. The polygon in image 4 shows the starting node, vertices, and return to the starting node to enclose the area. Most GISs do not have a digitizing direction requirement, although it is probably a good idea to use one method consistently (e.g., clockwise) in order to minimize confusion.

Digitizing Polygons

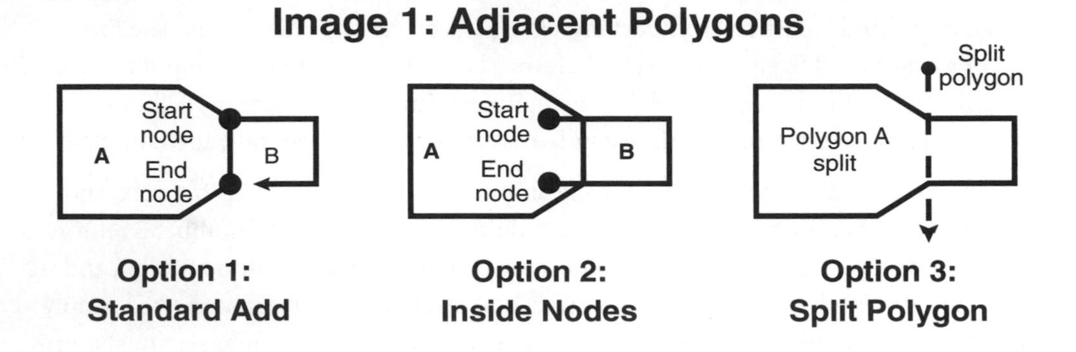

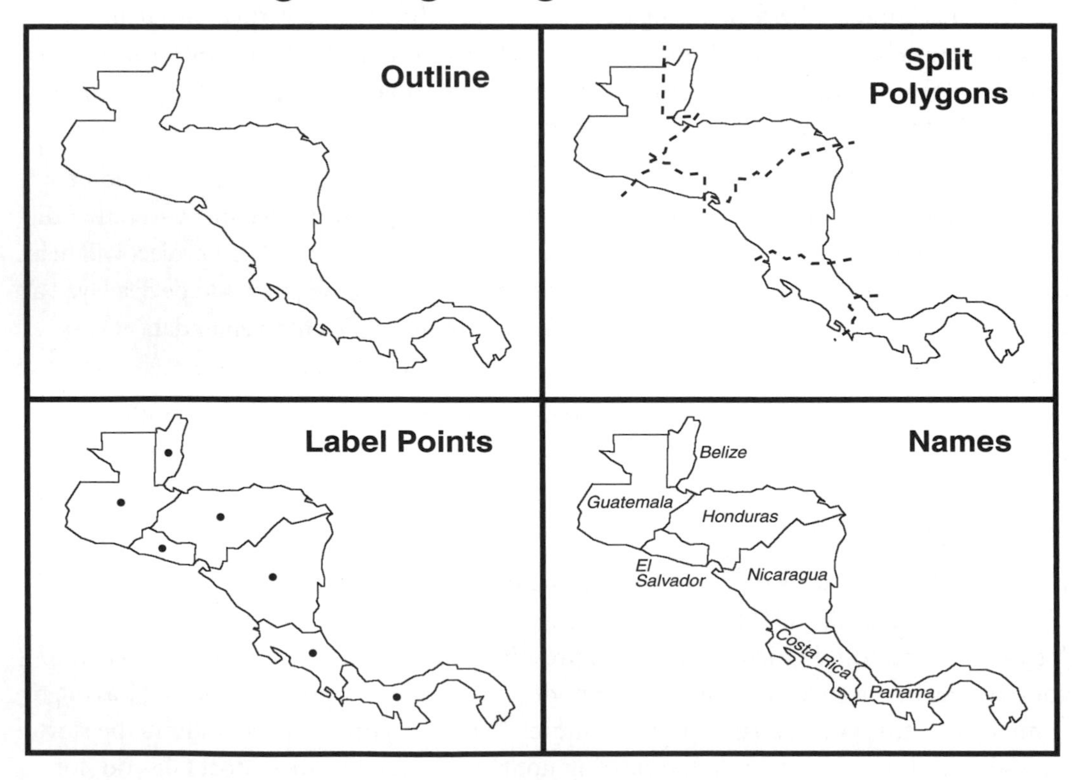

Fig. 5-16: Digitizing polygons.

Some GISs require that common borders be digitized twice—once for each polygon. Other GISs have adjacency options. Shown in image 1 of figure 5-16 are three ways of designating or entering adjacent polygons. Each example is chosen as a digitizing option. After polygon A is entered, the first method (option 1) to add polygon B is to begin at an existing vertex, node, or border position of polygon A, then draw polygon B, and end back at A. Nodes will be created at the start and end points on the polygon A outline if the starting position is a vertex or on the border.

Option 2, similar to option 1, is to make the additional polygon by entering the start node inside the first feature (A), drawing polygon B, and ending inside A. The software interprets the common border and establishes the correct chain/node structure. This option helps avoid accidental gaps between polygons.

Option 3 is to split a large polygon into two features. The entire A-B area is entered as polygon A, and then a split-polygon option is selected to divide them into two polygons. The split line starts outside the polygon, traces through the initial large polygon A along the intended border, and then ends outside. The software interprets the operation and builds two features with the appropriate nodes.

Splitting polygons is a relatively fast and easy way of digitizing some region of the world, such as Central America, demonstrated in image 2. The first step is to digitize the entire outline of the region (Outline map), and then apply split polygon lines to create the land borders (Split Polygons map).

Some GISs require *label points* to be entered inside each polygon in order to provide an "anchor point" for labels (e.g., name or any database attribute). Label points are usually placed at the polygon center, but the user can choose another location. These can be entered after digitizing the polygons. Other programs will automatically place the label in the polygon center, although the user can move it manually for better display. Shown in image 2 are the label points in the center of each country (which are not seen as part of the map except under a special editing view).

The user can select the country name to be displayed at the label point. When the names are first applied at the label points in the Names map, most run over the borders and are not easy to read. They have been edited by changing the font style and size and then moved to a new location for better visualization. Chapter 11, on GIS cartography, discusses appearances in more detail.

On-screen Digitizing
Puerto Rico

Original uncorrected Space Shuttle photograph

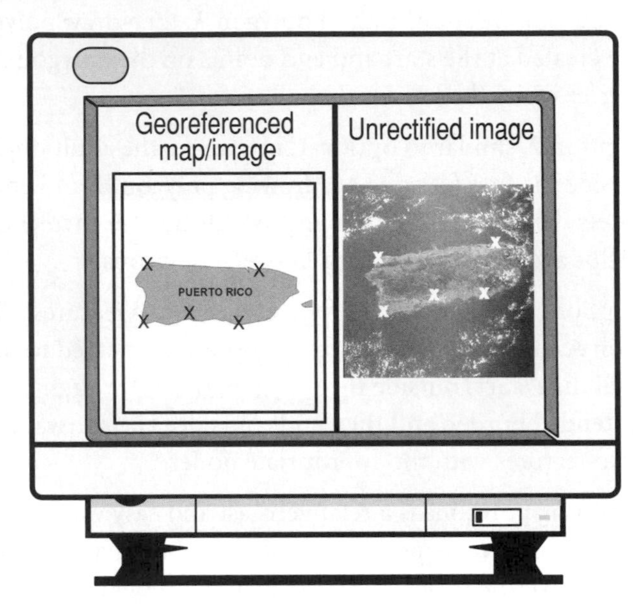

Image correction

Digitizing corrected image

Fig. 5-17: The on-screen digitizing process.

Digitizing at the Monitor

A digitizing table is not always needed for entering map data. Many GISs now offer digitizing capability on the monitor (also called *on-screen digitizing*, *heads-up digitizing*, and *desktop digitizing*), depicted in figure 5-17. Typically, a remote sensing image or a scanned map is displayed and traced with the mouse. As in standard digitizing, the feature type is set (point, line, or polygon), and then the selected features are traced using the mouse as a puck. A database is constructed automatically during digitizing, but it usually needs editing and attribute entry after the graphics have been completed.

One of the advantages of on-screen digitizing is the ease of zooming for detailed work. Magnification and moving the view are simple mouse click options. Also normally available are the full array of editing (delete, copy, and so on) and text feature (font and text editing) capabilities. In many ways, on-screen digitizing is easier than using a digitizing table. For education, it is good way of introducing manual data entry.

Because the scanned map or image is not usually "rectified" (corrected to a georeferenced location), it is only an "uncorrected" picture. On-screen digitizing is therefore not properly georeferenced. Some GISs offer an option for splitting the screen, with a georeferenced theme or map on one side (say, from standard digitizing or a scanned map), and the unreferenced image on the other. Using the cursor, features or specific locations that can be found on both views are selected, giving known coordinates to the image.

After selecting about a dozen or more common points (Xs in the illustration), the image is adjusted using an *affine transformation* program that stretches and deforms it according to the known coordinate points (a process often referred to as *rubbersheeting*, stretching and pulling the image to make it fit the map). In offices and agencies that do major GIS work, this can be a standard task.

Although the affine transformation sounds complicated, it is the computer program that transforms the digitizer table coordinates to the real-world coordinates for the digitized map, using the reference points and projection information. It is a highly useful and, thankfully, mostly invisible program in GIS.

On-screen digitizing is often a quick and inexpensive method for producing GIS data from images, although some work is needed to integrate the results into standard referenced and projected data. In some organizations, particularly in the developing world, this may be the only immediate means of acquiring useful data.

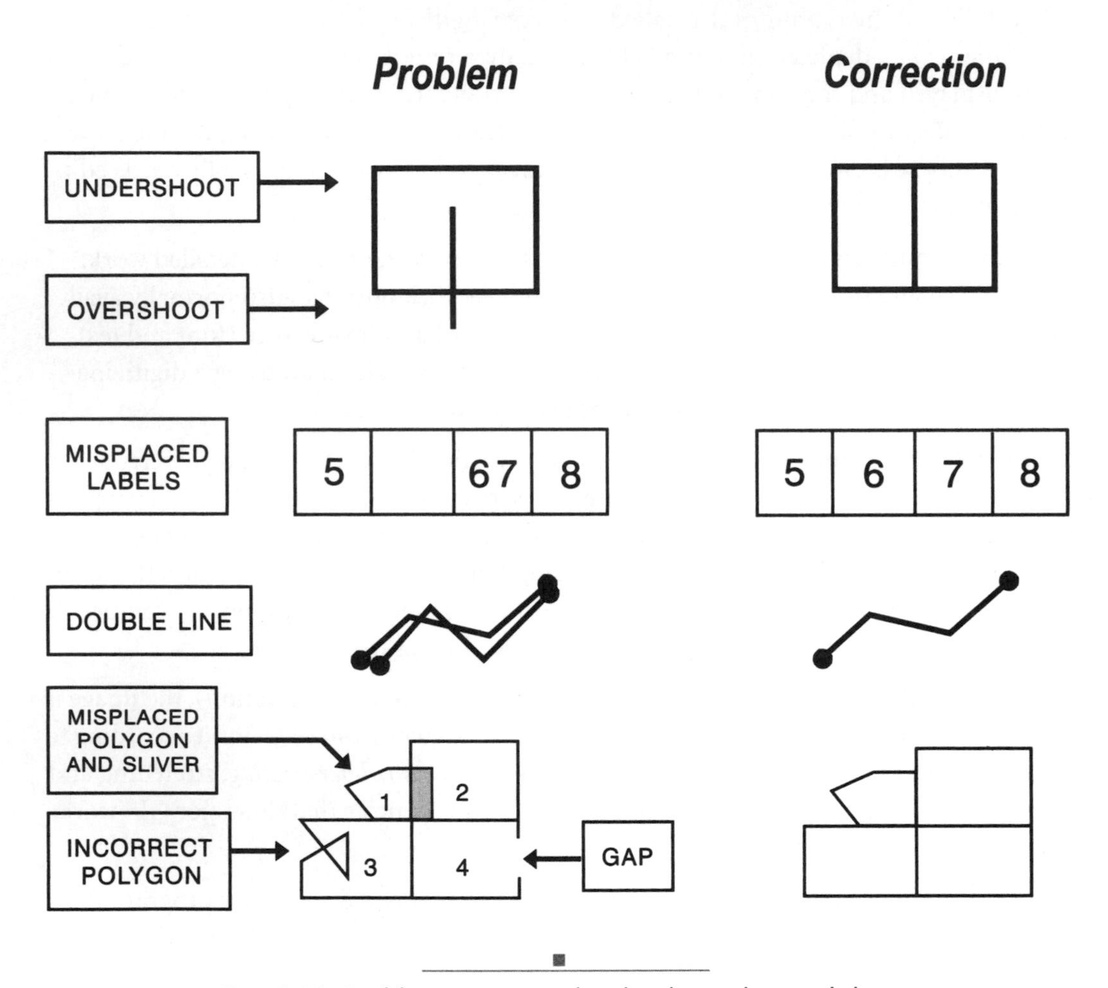

Fig. 5-18: Problems associated with editing digitized data.

Editing Digitized Data

It is easy to make mistakes in manual digitizing, and correcting them is a normal part of the data entry process. Editing is an essential task, although it takes time, a critical eye for detail, and a sense of perfection. Many GISs offer error detection options and easy operations for repairing problems. Some of the more common errors, discussed in the following, are shown in figure 5-18 (problem on the left and correction on the right).

- *Undershoot and overshoot:* End nodes that do not reach other chains as intended or that extend beyond the targeted chain must be corrected. This requires either extending or reducing the digitized chain length, usually just by pointing to the incorrect chain, then to the target position, and pressing a key to instruct the editing program to connect them properly. Most digitizing programs use a process of "snapping" (or automatically connecting) nodes and features if they occur within a specified sensitive distance (often called the *snap distance*; other names are possible). If the node is digitized within the snap distance of the target, it is automatically "snapped" to a connection. This helps prevent overshoots and undershoots.
- *Misplaced labels:* Some features may receive no label points, multiple labels, or incorrect labels. In the illustration, the second polygon's label point was accidentally placed into the third polygon. It is easily moved by pointing to the incorrect label and placing it where it belongs.
- *Double line:* It is easy to enter an original data feature twice, particularly from complex maps. In the illustration, one line has been repeated. The least accurate line is removed by pointing to it and pressing the Delete key.
- *Inaccurate digitizing:* Manual digitizing is not always accurate. Imprecise hand motion, slips, poor vision, distraction, occasional electronic "glitches" (small faults), and other momentary problems can create imperfections in tracing features. Shown are several common problems. A bullet-shaped polygon (number 1) has been digitized out of place, too far to the right, creating a sliver in polygon 2. It can be grabbed and moved to the correct location. Repair can also be accomplished by moving the vertices to a correct position, usually a simple matter of placing the mouse or puck cursor over the vertex and manually moving it to the correct location. Polygon 3 contains several additional and out-of-place vertices, perhaps due to hand control problems. The two inside vertices can be deleted by pointing to them and pressing the Delete key, thereby making a straight-line border. A gap, or incomplete chain, is evident on polygon 4. Closure is made by pointing to the open end of each chain and instructing a connect operation, similar to the undershoot repair.

These types of corrections are usually easy to make, often by simply pointing to the area to be corrected and making a single correction command. However, editing takes time, and although is the temptation is to move to the next step, editing must be completed with care and patience.

Scanning and Vectorization

Image 1: Scanners

Image 2: Vectorizing Software

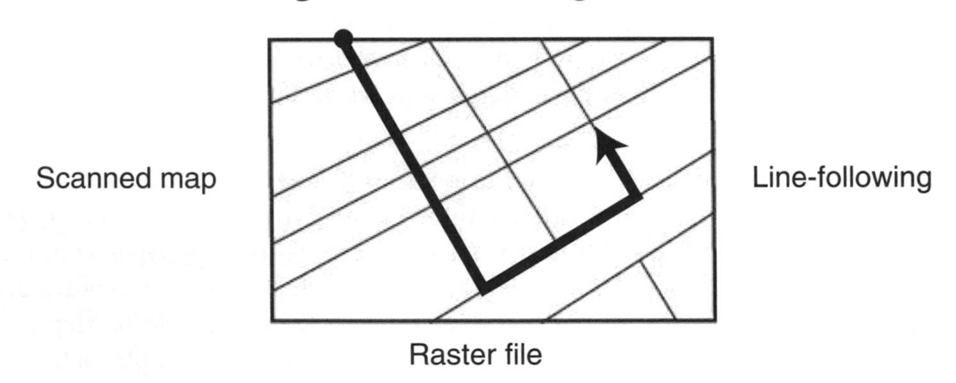

Image 3: Humans in the Loop

Interactive decisions

Fig. 5-19: The scanning and vectorization process.

Automatic Digitizing: Scanning and Vectorization

Automatic digitizing is a misnomer (misleading name) of the non-manual process of digitizing. Although hardware and software offer some automation to data entry, it is really a semiautomated, human-assisted alternative. Maps can be scanned and vectorized. The following sections discuss the nature of automatic digitizing, the major steps in the process, and the potential problems.

What Is Automatic Digitizing?

Digitizing would be less trouble if hardware or software could do much of the tedious work. There has been considerable progress in some areas of digitizing technology, and too little in others. There is no real "automatic digitizing" because human intervention is necessary during the process. Although there are some systems that can run alone ("batch mode"), they need substantial editing after. "Automatic digitizing" is really semiautomatic. Perhaps the near future will see substantial advances, but it may be a while before humans are no longer needed for guiding automatic devices and editing the products.

The basic process, sometimes called *scanning data entry*, involves initial raster scanning for either on-screen digitizing or a raster-to-vector conversion (*vectorizing*), followed by editing. As indicated in figure 5-19, there are three major components: scanners, vectorizing software, and humans. The material that follows describes the process in detail.

Scanners are important peripherals (accompanying equipment) for computer systems today (image 1). They copy documents, images, and maps for use in various programs. They are fast (much faster than manual digitizing), accurate, and easy to use. Scanners range from the very affordable desktop units for home and office to large, expensive specialized equipment for scientific and commercial applications.

The desktop scanner may be satisfactory for background images in on-screen digitizing and elementary maps, but more sophisticated equipment is needed for controlled, precision work. Shown are a standard table model and two larger, mid-priced commercial scanners (names in parentheses) that use sheet feeding for loading maps (working somewhat like a plotter in reverse).

Scanners produce raster files, which can be used as background images for on-screen digitizing, as raster images to be combined with remote sensing data, as images for display in the project, or as raster data in the GIS. However, scanned maps in raster format are not always acceptable for GIS, primarily because feature intelligence is not produced. That is, each raster cell has only a color or tone value and nothing else; there is no connection with

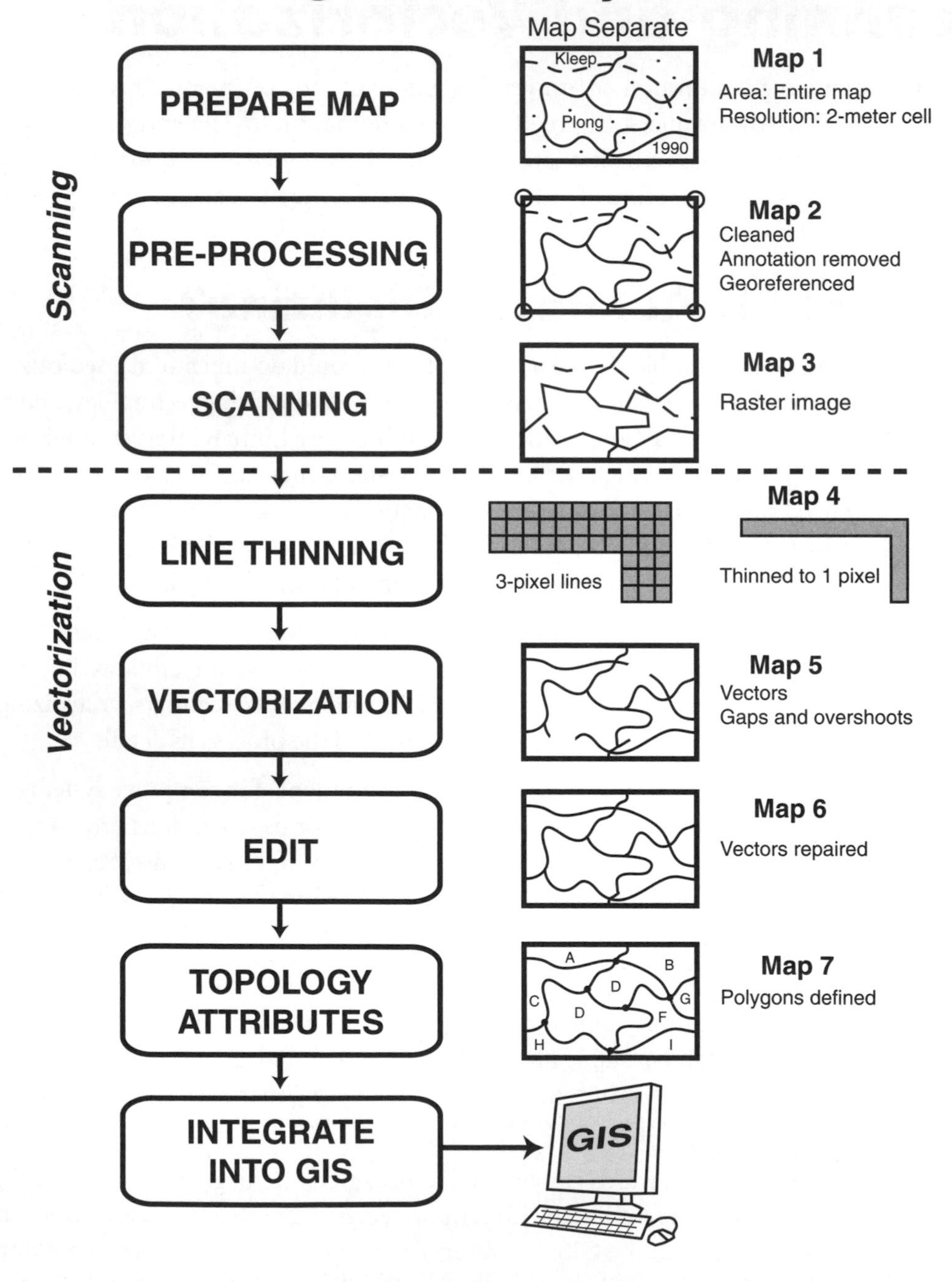

Fig. 5-20: The scanning data entry process.

surrounding cells and no recognition of features. Attribute databases are not part of the results, and topology does not exist. Even though very high-resolution raster data can be generated, many GIS projects prefer vector data with feature definition when possible.

Vectorizing software converts the raster data into a vector format. There are advanced systems that can run alone but need substantial editing. Image 2 of figure 5-19 shows a line-following operation, where each line is traced with high accuracy. Line-following software uses a person to guide progress and to make decisions (which line to follow and which way to turn at intersections). It can connect the cell centers, as shown in figure 3-11, or follow the edge of common value cells to provide polygon outlines.

Vectorization does not automatically produce features, because there is no feature data to read. Humans are needed in the loop to provide attribute information (image 3). Other problems, discussed in material to follow, deal with completing the vectorizing process.

Scanning Data Entry Process

The steps for entering maps, air photos, and other hardcopy data into GIS are no more complicated than in standard digitizing. The basics of each step, depicted in figure 5-20, are discussed here, although additional considerations may be needed for application.

The first step is to prepare the map. Although scanners will copy any data, map scans produce best results using bi-tonal data (black-and-white lines and features, not gray or color). Lines produced in the vectorization process that follows will be more accurate. For published color maps, it is possible to purchase the individual sheets, called separates, which contain black-and-white theme features. Some advanced vectorization systems can accept grayscale and color maps.

Other preparation steps include selecting the area to be scanned, reviewing procedures, and selecting the desired resolution (map 1). Resolution can be defined using either grid structure arrangement (number of rows and columns) or real-world cell size (e.g., 30 meters on a side). High resolution gives better detail but makes much larger data files and takes more time than lower resolutions.

Map pre-processing involves cleaning the map of stains, specks, and other useless marks that make "data noise." A clean map reduces the editing process. When possible, "clutter" should be removed, such as annotation text, map symbols, and other elements that do not contribute to the features. Finding real-world coordinate control points for later georeferencing is also necessary. Map 2 has map 1's dot speckles and annotation text removed, and the corners selected as georeference control points (circles).

The next step is to scan the map. Each scanner has particular controls and procedures, but it is usually a standard process of either positioning the map in the scanner or feeding it into the system. A raster image is produced, which can be edited and used as a raster GIS

Problems of Automatic Digitizing

Image 1: Line Problems

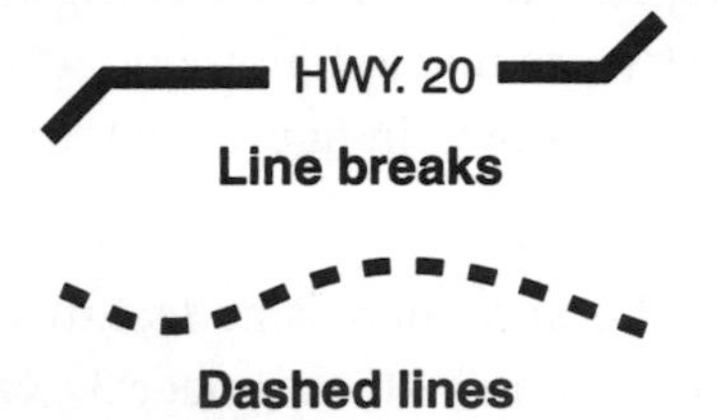

Image 2: Annotation

S O I L

Annotation
or features?

Image 3: Guidance Needed

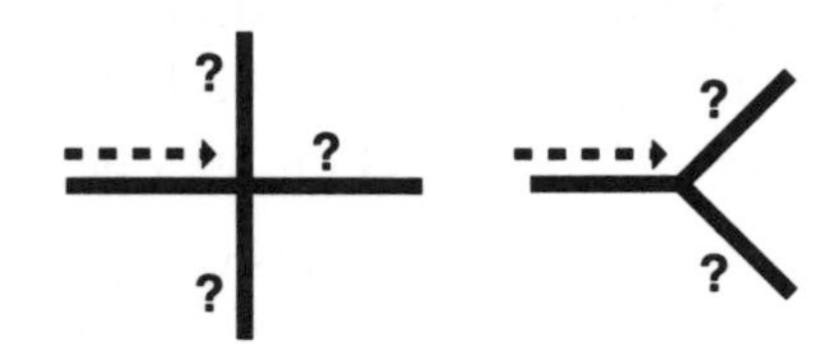

Which direction?

Image 4: Converging Lines

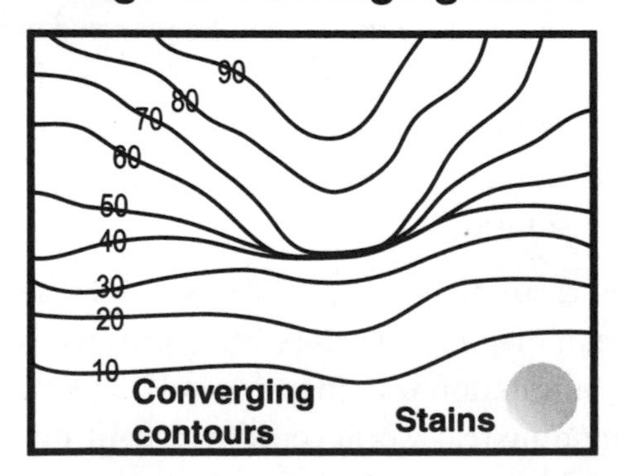

Fig. 5-21: Problems associated with automatic digitizing.

theme. For some applications, this is the last step, but many GIS projects prefer vector data. Note how map 3 of figure 5-20 is somewhat generalized and degraded from the original vector (map 1).

Scanning records what it sees, and some features may have thick lines and filled polygons. Lines may be several cells wide, making the line-following process (next) less than precise. A line-thinning procedure can be applied to narrow the features to a single row, thereby helping the vectorization process. Some human control may be needed. Map 4 shows how a wide line is thinned to a single row of cells.

Vectorization is the application of line-following software to convert rasters to vectors. This is an interactive procedure, normally requiring human intervention to make decisions, as discussed. Map 5 demonstrates some of the problems of an imperfect scan. Despite the problems and need for a human director, the process is much faster and usually more accurate than manual digitizing; therefore, the high quality and consistency of data can be worth the effort.

After vectorization, the new map usually needs editing, which includes deleting unneeded lines, correcting overshoots and gaps, and other steps identical to editing standard digitizing (map 6). This process can require more time than all of the other steps, although the results are highly beneficial and necessary.

After editing, topology and georeferencing can be applied. Some advanced software can build topology on the edited vector data. Enclosing lines become polygons, intersecting lines become networks, and spatial relationships are recognized. Attributes can also be assigned to individual features, such as contour elevations, polygon identification, and any other information the user wishes to attach (map 7). As in standard digitizing, the database may need further development and additional items.

The last basic step is to incorporate the new data into the GIS project. This process typically includes simple data management steps, such as copying the data into the correct location, renaming, and confirming satisfactory integration with existing data.

Problems with Automatic Digitizing

Line-following software cannot make decisions about features. Human monitoring and intervention (in effect, "baby-sitting") are necessary, both during digitizing and in editing, thereby slowing down the process. Sometimes as much time can be spent editing the automatic data as would be required for manual digitizing editing. The following, depicted in figure 5-21, discuss problems associated with line-following digitizing.

- *Line breaks:* Although the human eye dismisses small and logical breaks in a line (e.g., a highway number or an intermittent stream), a line-following program stops when an interruption occurs (image 1). The operator can manually continue

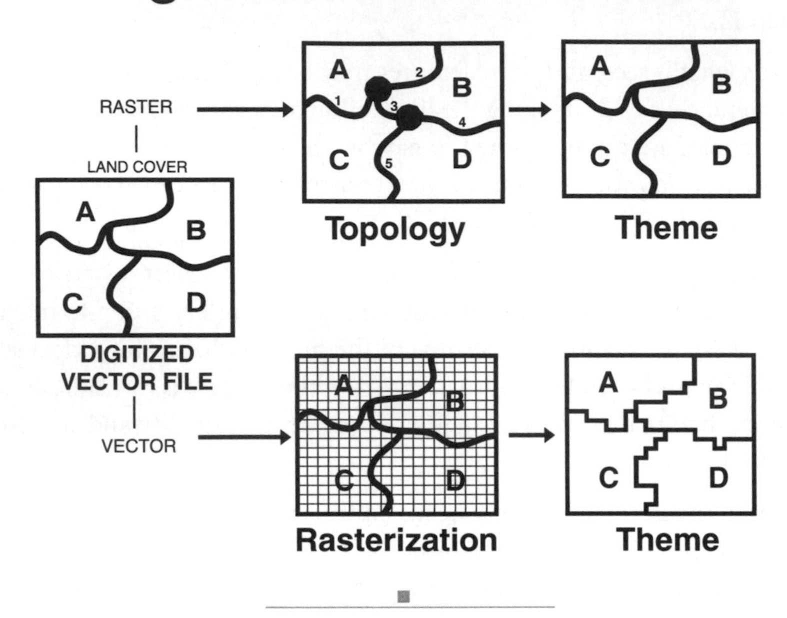

Fig. 5-22: The digitized-file conversion process.

the line over the text and have the scanner continue on the other side. If the system does not permit interruptions, the human editor must ultimately erase (or ignore) the text and manually draw the line over the gap—a tedious process.

- *Dashed lines*: Dashed lines are common symbols on maps, used for road classification, types of boundaries, and other features. Although software is advancing and will soon be able to recognize broken lines, at present dashed lines still require human guidance.
- *Annotation or features?*: Scanners cannot read, so they may recognize text as features (image 2). The "O" in *SOIL* might be seen as a polygon, and the other letters may be interpreted as lines. Again, the operator must make corrections.
- *Which direction?:* If a laser following a horizontal line from left to right comes to an intersection, it has no way of knowing which direction is correct. It may stop, awaiting a human decision before continuing, or it may take a wrong direction. In the intersections shown in image 3, are the vertical lines separate features, sides of polygons, or continuation of the current polygon? These are important questions when topology is to be used.

- *Converging lines*: When topographic lines converge at very steep features (e.g., a cliff), the knowledgeable interpreter understands the symbology and landscape, but the scanner does not (image 4). Coming from the left, the laser follows a single contour line, but when it meets the diverging lines in the middle, it must stop and "ask directions" from the human. It cannot understand which line is the correct one (a problem for the human, also, who must pause to read the map and search for annotation, count the number of lines, or use some other information).
- *Stains, wrinkles, and "extras"*: Because scanners of all types are just machines, there is no intelligence to separate extraneous (irrelevant) visual features from the important information. Stains are recorded, as are the innocent pencil marks, rips, wrinkles, and other items the eye automatically dismisses. The odd dark circle in the lower right of the contours in image 4 may be a stain, a printed logo, or some other artifact that does not contribute to the map, but it probably will be digitized anyway. These features may be minor inconveniences (easy to delete), or they may interfere with feature recognition. Human intervention and editing are necessary, which unfortunately are time-consuming tasks.

Truly operational automatic digitizing may be a major GIS innovation of the near future, but today it is not readily available for most GIS operations.

Data Entry Final Steps

The final phase of the data entry process involve data conversion to vector or raster and development of the database. The following topics discuss how digitized data is converted into the working format for GIS and how databases are completed.

Digitized File Conversion

The next step in the data entry process (see figure 5-9) is to convert the digitized data into either vector or raster format, depending on project needs and system requirements. Most of the concepts and considerations have been discussed and this section is primarily a review. Figure 5-22 illustrates the options. When a digitized data file (in spaghetti vector format) has been edited, georeferenced, and projected, it is ready to be converted into a GIS theme file.

Vector systems maintain the data in the original digitized form and there is no appreciable change in the visual appearance. However, topology adds significant power and value to the data. In most systems, after applying topology, nothing else is needed for the program to read the data as a theme or set of features. Note that the original polygon values have been retained, awaiting database work to add labels and meaning to them (next step).

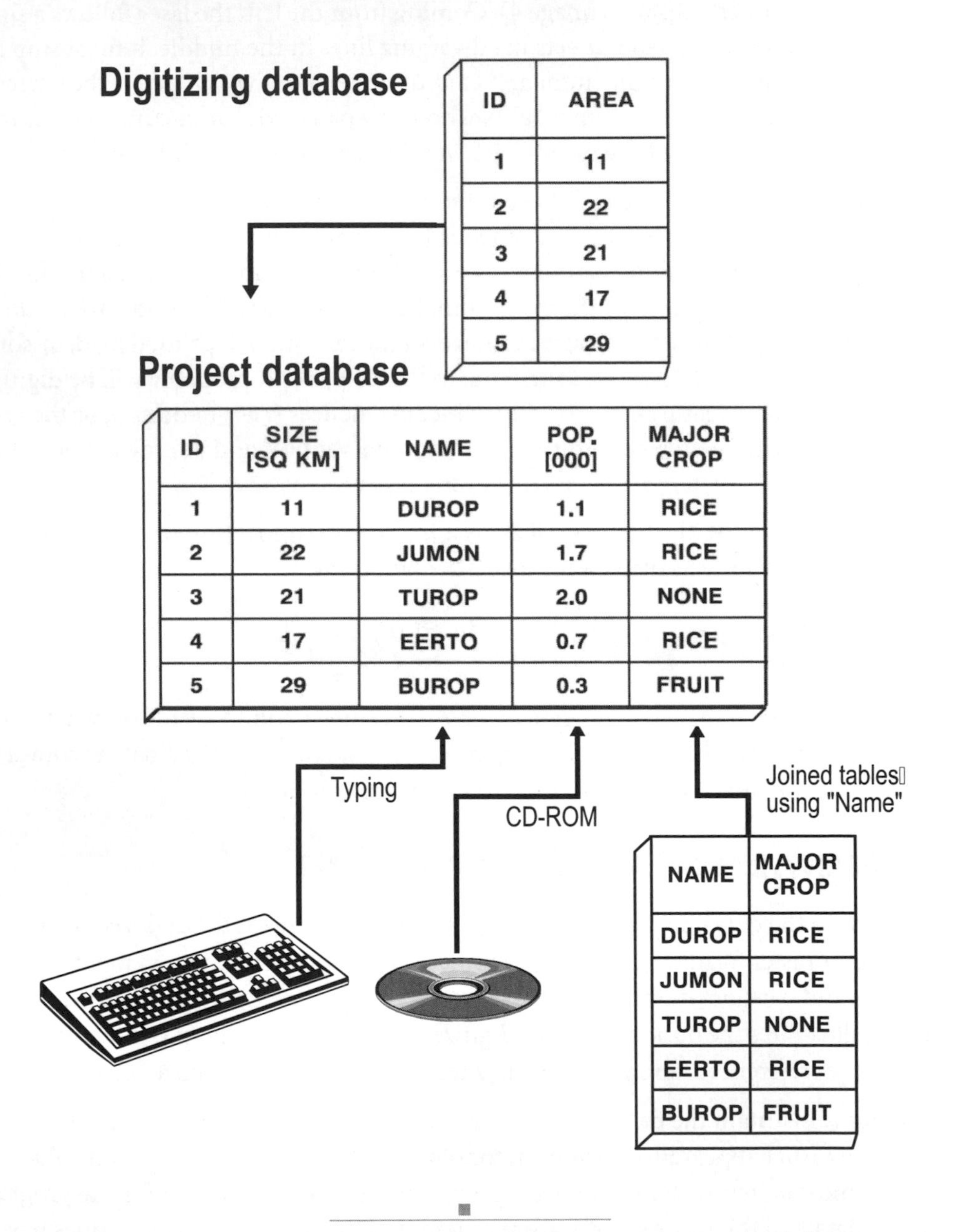

Fig. 5-23: Databases can be constructed from many sources of data.

Raster systems require gridding for data conversion. The size of the grid may be expressed in the number of rows and columns, or in real-world sizes for the cells. Higher resolution (say, 100 meters) will require considerably more cells than a 1-kilometer grid (there are 100 100-meter cells for each square kilometer).

There is much better detail on higher resolutions, but the file sizes are much larger, slowing down calculations and display times on the screen (although these problems are less of a concern today because of the fast computers). The effect on the original digitized vector data can be pronounced or unnoticeable, depending on the data, resolution, data presentation (monitor or maps), and project needs. The operation is usually fairly simple, requiring little more than filling in needed details in an on-screen form. Conversion is a fairly easy process.

Database Construction

Databases are central to GIS applications, and their flexibility gives strength to operations. They should be constructed carefully and logically. Some GISs automatically make an initial database from the digitized data, usually with only a few attributes, such as feature ID and perhaps a spatial measure (area and perimeter), as discussed. From these tables, attributes can be added to make a complete database. New records can also be included.

Other GISs require the operator to set up the database, such as establishing the anticipated size of fields and other structural items. As database technology progresses, however, development is becoming easier and less demanding for the user to define specific structural formats.

Figure 5-23 shows a database automatically constructed in digitizing or after topology has been established. Each digitized polygon is a record with a unique ID number and area (the units are normally defined in the digitizing preparation step and in the metadata). Other fields (attributes) are added by the user at any time.

Data input can be from a variety of sources. In this example, the Name is entered by keyboard typing and the Population data (abbreviated for visual convenience) comes from a census CD-ROM. Another project table with the crop types is joined to this database, using Name as the common item to make the connection. That is, a simple command joins the old table to the new one, using Name as the linking field, thereby making a final combined table with all information. A complete database can be constructed, but it remains open for editing and additions. From here, the "real" GIS work can begin.

CHAPTER 6

DATA QUALITY AND MANAGEMENT

Introduction

WE TEND TO BELIEVE WHAT WE SEE ON MAPS, that everything is true and accurate. However, data is not always so clean and authentic. Just because a map appears exact and perfect does not mean it is necessarily correct. There are a number of issues concerning data that the user should recognize. Because data is the most important part of GIS, it must be understood to be used effectively. This chapter looks at some of the more important data issues in GIS management, operations, and applications.

Data Quality: Error, Accuracy, and Precision

The concept of data quality deals with various aspects that affect the utility and credibility of geographic data in GIS. Data quality relates to several aspects, including completeness, consistency, accuracy, precision, timeliness, and effectiveness. These terms are often used rather loosely, but they do have significance for GIS. The sections that follow examine issues related to data error, accuracy, and precision, which are three related fundamental qualities relevant to almost every GIS project.

Error, Accuracy, Precision

Image 1: Classification

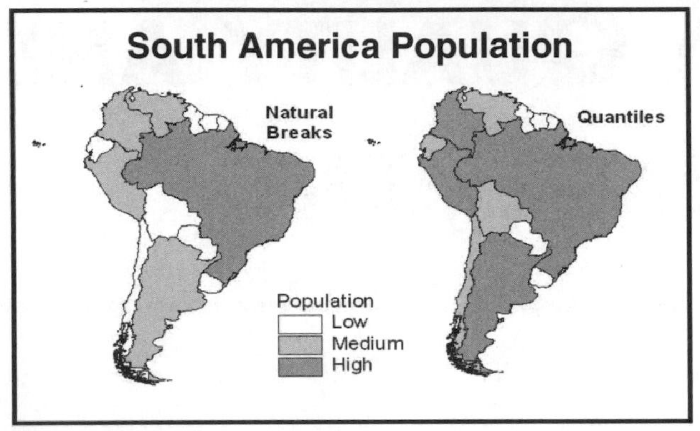

Which classification is best for the project purpose?

Image 2: Inaccurate Data

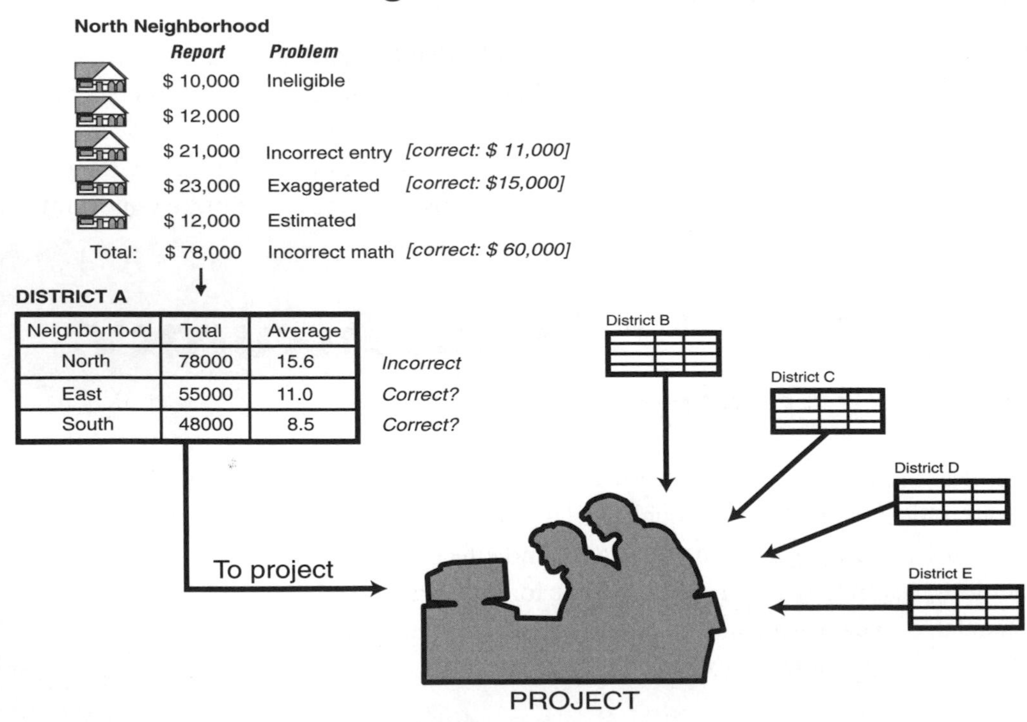

Fig. 6-1: Error, accuracy, and precision affect the reliability of a GIS project.

Error

The concept of error goes beyond mere mistake and includes technical issues in the nature of GIS operations, processing algorithms (e.g., mistakes in topology can create serious error), misuse of statistics, field collector or operator bias, equipment quality, and others. Each of these concerns could be discussed at great length, but the point is that the GIS user should have some awareness of the many influences on data. It is easy to appreciate that error can come from numerous sources, many beyond the detection and correction capability of most users. Several types of error are discussed in this chapter, such as time and dates, consistency, and scale; others exist.

A great deal of trust is given to data, whether acquired from the outside (maps, other data sets, digitizing), obtained by the project (field data, surveys, tables), or derived within GIS processing (recoding, calculations, classification, overlays). Errors that go undetected and uncorrected will be perpetuated or continued in the project, and possibly made worse. Some errors may be small and irrelevant, but others can be dangerous. An incorrect classification, for example, could lead to improper policies or decisions.

Countries that depend on accurate statistics for development assistance can be misclassed as medium population rather than high, simply due to poor arithmetic calculation or incorrect statistical measures. Image 1 of figure 6-1 shows two maps of South American population, both using the same data but presenting two different statistical methods for classification. Six of the countries included fall into a different classification, and stand to gain or lose based on which mapping classification system is used.

Accuracy

Data accuracy is defined as the truth or validity according to reality, within acceptable limits. Data must be expressed correctly according to what actually exists. Correct values, names, spatial extent, and other properties are expected. The South American population figures are assumed to be correct, and the statistical methods used for classification are presumed to be acceptable.

Consider a simple example of collecting and processing income figures for districts. Image 2 shows reports for individual houses, with five houses in the North neighborhood. There are a number of potential inaccuracies, such as the ineligibility of the house 1 owner (perhaps he actually resides in another district), an incorrect entry for the third house (mistaken entry from $11,000 to $21,000), an exaggerated income from $15,000 to $23,000 (human variability and credibility problems), and only an estimate for the final house (answer was "I don't know; about $12,000"). In addition, the total was miscalculated (careless work).

The incorrect total is entered in the database as part of the district data. Then the averages are calculated, though the average for the North neighborhood is incorrect because of the

Precision Issues

Image 1: Fieldwork Problems

Image 2: Digitizing Generalization

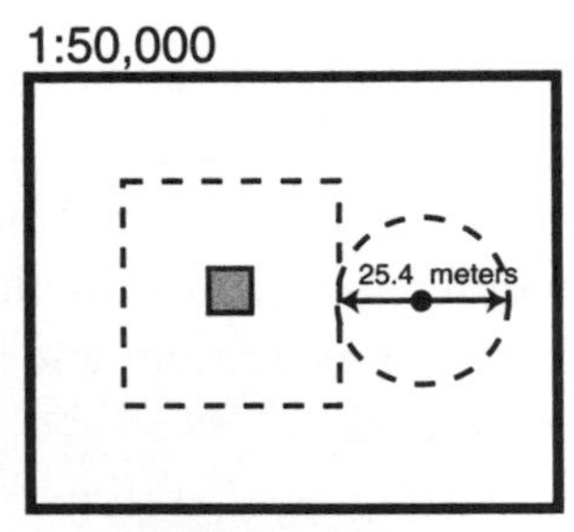

Image 3: Map Imprecision

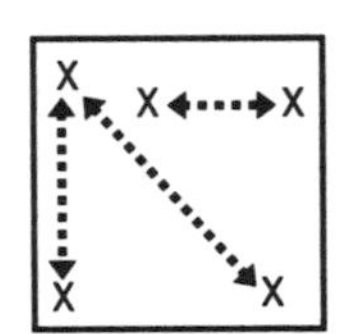

Image 4: Raster Cell Imprecision

Image 5: Project Questions and Complications

Fig. 6-2: Numerous factors influence data precision.

data problems. Perhaps other neighborhoods have similar mistakes. However, all of these inaccuracies occur at the initial stage and are passed to the project.

Project personnel, too far removed from the initial data collection, have no way of verifying accuracy and therefore must trust the reported data. When the database is incorrect, the final effects can be dangerous. Inaccuracy is a major source of data error. These types of problems support the premise that GIS users should have knowledge of the project's topic, area, and potential influences on the data. Computer proficiency alone is not sufficient for GIS work. The *geography* is more important than the systems.

Precision

Data precision deals with spatial accuracy. Field data is usually precise, meaning that a person can return to the exact spot if directions are recorded properly (GPS is particularly useful in this regard). However, careless fieldwork is both easy to do (intentionally or accidentally) and difficult to verify. Image 1 of figure 6-2 shows lack of confidence by the field crew. Sometimes there is too much reliance on the accuracy of field data.

When maps are digitized, the precision of features is generalized because there is some loss of spatial accuracy when copying data, particularly in manual digitizing (image 2). Recall how curves, for example, cannot be duplicated precisely without difficulty. Data generalization results in decreased precision.

On the ground, a person is working at a 1:1 scale, but when the data is transferred to a map, precision declines because any scale smaller than 1:1 is a generalization of reality. Spatial accuracy is defined according to scale.

In the United States, for maps larger than 1:20,000, at least 90 percent of the points tested should be within 1/30 inch (0.08 centimeters) of their true position; for scales 1:20,000 or smaller, 1/50 inch (0.05 cm). On a 1:50,000 map, this is 25.4 meters (83.3 feet). This means that any point on the map is not necessarily precisely where it is depicted, but only within 25.4 meters of its true location. The vertical accuracy standard requires that the elevation of 90 percent of all points tested must be correct within half the contour interval. On a map with a contour interval of 10 meters, the map must correctly show 90 percent of all points tested within 5 meters of the actual elevation.

These are demanding standards that cannot always be met. Image 3 shows that two features are located within the map's spatial tolerance of 25.4 meters (dashed lines are precision ranges). Are they really adjacent to each other or separated by up to 50.8 meters? Recall the similar issue of location within a raster cell discussed in Chapter 3 (image 4).

Problems occur when the original scale is not known; the precision of location cannot be determined, and features do not have known spatial accuracy. (Other factors of scale, accuracy, and precision are discussed in material to follow.) The end result is that questions may

Generational and Improved Data

Image 1: Second-generation Data

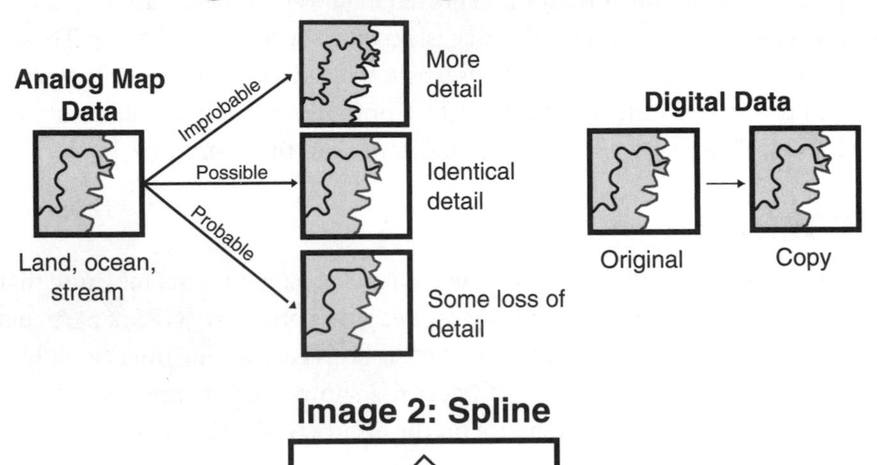

Image 2: Spline

Individual Segments

Spline Curve

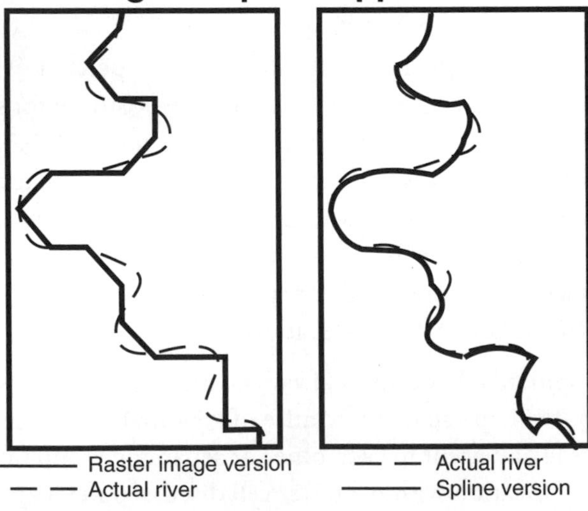

Fig. 6-3: The concepts of generational and improved (derived) data.

arise in the project and some data may have suspected imprecision, but few options exist other than checking original data. The project personnel in image 5 may suspend progress until these problems can be resolved.

Generational and Derived Data

Collecting data, from the ground or by remote sensing, is by nature generalization of the real world. The first data set collected is considered the *first generation*, and each copy or reworked version thereafter is a succeeding generation. That is, the second copy is the second generation.

Generational Data

Copies normally have loss of data quality, principally due to imperfect reproduction (a photocopier, for example, does not make exact copies). For analog data (such as maps, photographs, and photocopies), there is almost always some loss of quality from the original. Without special processing, the accuracy of the second generation can be no better than the first generation, and probably will be less.

However, duplicating digital data normally results in exact reproduction, with no loss of quality; each succeeding generation of data is exactly like the original, primarily because the process is simply copying numbers. In either case, the second generation can be no more accurate than the first. Accuracy and precision cannot be improved unless some new information is obtained and used. Image 1 of figure 6-3 diagrams the effects of generation data.

There are various computer routines (programs) that help data to appear better, but the emphasis is on *appearance*. For example, photographs can be visually improved by adding contrast, which helps interpretation, but it is a second-generation photo. Spline operations, discussed in Chapter 5 in regard to digitizing curves on maps, can be used on digital data to smooth out curves, adding vertices or mathematical shapes to give a graceful arc to a set of angled lines (image 2). However, the result is a mathematical curve rather than an observed one.

Perhaps the curve is more representative of the true ground feature, or perhaps it is not. The user must be careful in inferring a shape that is not necessarily genuine. Because the receiver of the data usually does not know that an artificial improvement has been applied, the data will be used as truth.

Image 3, on the left, shows a stretch of river with its raster version (from a satellite image) and the actual river. Recall that the raster image is a series of straight lines connecting the raster cell centers. The actual river, from a map, is naturally more sinuous. On the right is an attempt to spline the raster data into a more realistic shape. Although it is an improvement over the raster version, is still does not match the actual river. Applying a spline operation to raster data, whether successful or not, is an artificial modification that may present better appearance, but not necessarily better data.

Derived Data
Image 1: New Data
A
B
C
D
E
F
G
Districts
District | Avg Income
A 5,888
B 6,022
C 7,400
D 5,500
E 6,000
F 6,600
G 7,100
District income
District | Avg Income | +- Mean | Class
A 5,888 -471 Low
B 6,022 -337 Low
C 7,400 +1341 High
D 5,500 -859 Low
E 6,000 -359 Low
F 6,600 +241 High
G 7,100 +741 High
Derived data
Class map
High
Low
Image 2: Resampling
40 50 60
20 10 20
Original data
40 45 50 55 60 65
30 30 30 35 40 45
20 15 10 15 20 25
Derived data
Original visual
Derived visual
Image 3: Interpolation
Missing scan line data
Interpolated repair

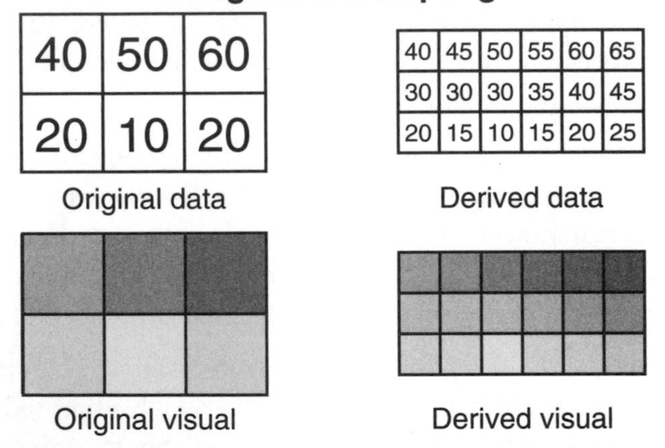

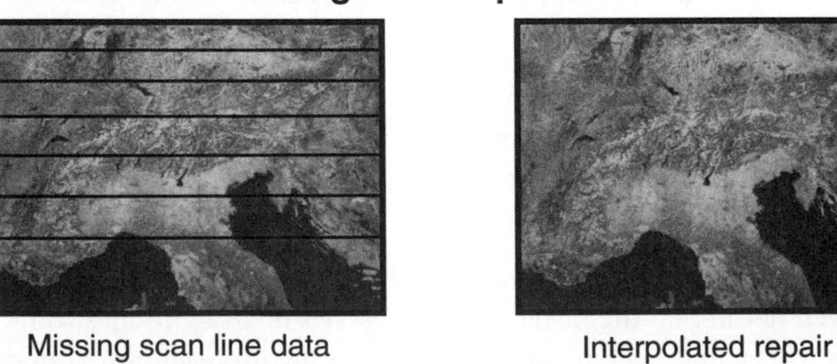

Fig. 6-4: Derived data, resampling, and interpolation.

Derived Data

Another option for improving data is to generate derived data; for example, using procedures to create new values from existing ones. Derived data comes from many sources in GIS, from simple mathematical calculations to sophisticated operations (discussed in chapters 7 through 10).

Image 1 of figure 6-4 shows a database of seven districts and their average income. (Actually, the average income figures are derived data from the original individual household income data.) Spatial analysis of the incomes is needed, and therefore new figures were calculated in terms of above and below the mean. Districts were classed accordingly (Low to High) and then mapped, with the results presenting a definite spatial pattern. The derived data is very useful in understanding the income patterns in the districts.

One method of improving a raster image is to increase the number of raster cells (for better resolution and spatial detail) and give them derived, calculated values. Image 2 shows two versions of original data acquired by the satellite scanner from measurements 100 meters apart on the ground. The cell values are in the top line of the illustration, and the visual versions are in the bottom (gray tones correspond to the number values; higher numbers are darker).

A process of *resampling* is applied to the original data, which reconfigures the grid and generates estimated values. The number of cells is doubled in this example, and new values are derived. The new values are calculated to be intermediate between existing ones, both along rows and columns. For example, the value 45 is between 40 and 50, and 30 is between 10 and 50. This provides a smooth and logical transition, but it is easy to see how they are artificially derived, not actually observed. The new values represent the best estimate. Three resampled rows of derived values are shown, but the new bottom row is not presented because the new values depend on the third, unseen, original row.

Resampling is used in satellite imagery for restoring lost data. Scanners sometimes "drop out" or miss lines of data when the instruments fail to record a line or set of pixels (image raster cells). Although there is no way of "restoring" data that was never recorded in the first place, resampling reads surrounding pixels and makes the best estimates of new values for the missing ones (a process termed *interpolation*). Image 3 presents the Alps in Europe, with several lines missing. Resampling is applied to fill in the no-value lines, effectively constructing a complete image.

It may be safe to assume that ground values should be a linear transition between collected measurements (the 45 between 40 and 50), but it is still an assumption. The user who receives the new data set will accept the level of spatial resolution (50 meters versus the original 100 meters) and values to be correct, whereas there is an extra magnitude of detail that was added. Artificial additions may be sound, but they are still presumptions of what *probably* exists.

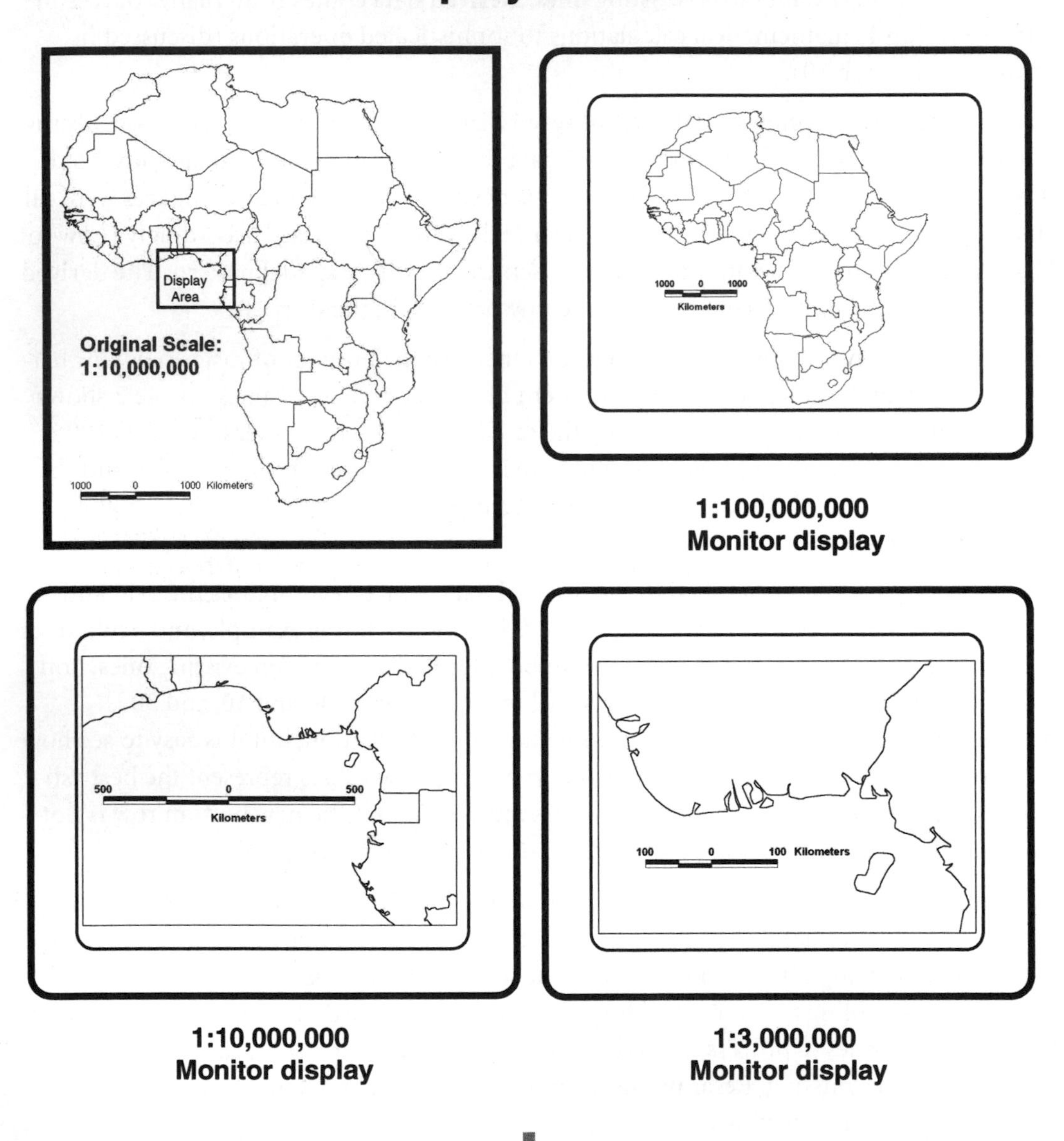

Fig. 6-5: Africa shown at various scales.

Scale

The sections that follow discuss issues related to GIS and scale. Scale and precision, scale differences, and scale incompatibility are important factors that are not always considered. They are discussed in the following as they relate to GIS.

Scale and Precision

It is often said that in GIS there is no scale. That is, GIS can enlarge or reduce data to any output scale (on the monitor or a paper map), regardless of how sensible the results may be. GIS *can* enlarge a small scale well beyond the precision limits of the data, even though it *should not.*

The scale of original data can be enlarged, but precision is still defined by the original scale and cannot be improved simply by zooming in. (Recall the scale limits discussed in regard to precision and depicted in figure 6-1.) Figure 6-5 contains a map of Africa with an original scale of 1:10,000,000 and various scale views on computer monitors. The map's precision limit would be 5 kilometers under the exacting official U.S. standards, but perhaps it was originally developed with a 20-kilometer precision. As displayed on the monitor at 1:100,000,000, it is well within margins of precision. The visual scale is very small and appears to be a low-resolution theme because the view is restricted by computer presentation limitations.

The 1:10,000,000 display effectively duplicates the original data and presents some of the coastal and border detail. The 1:3,000,000 view is too much of an enlargement because the features are no longer credible. For example, the shapes of the bays are too general and certainly there are more details that could be seen on data at this scale (there are too few vertices for the scale). Further, the scale infers a precision that cannot be supported by the original scale; any measurement would be too exact to be authentic. In effect, the enlargement exceeds its authority.

Again, knowledge of the original data is very helpful to the GIS user but it is often unavailable. However, the situation is improving as the concept of *metadata,* documentation of data, becomes more accepted. Metadata is discussed later in this chapter.

Scale Differences

Image 1: Enlarging Scales

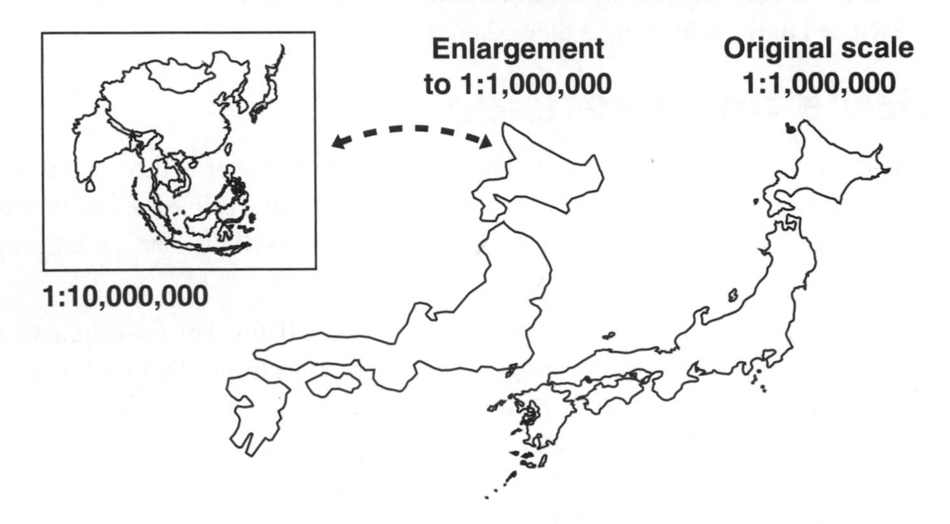

Image 2: Scale Incompatibility

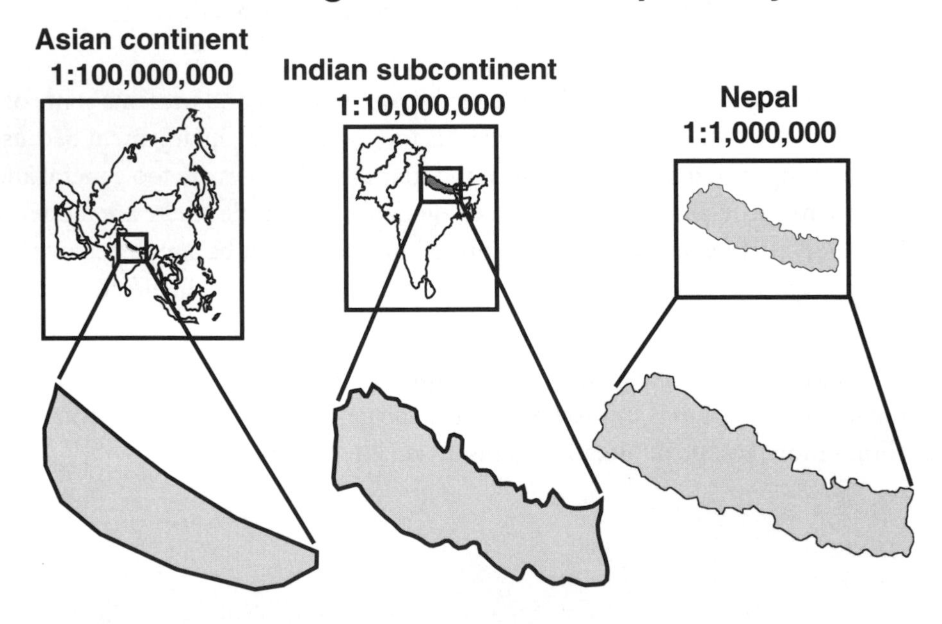

Enlarged to 1:50,000

Fig. 6-6: The concepts of scale difference and scale incompatibility.

Scale Differences

The precision issue can be important when changing or using various scales. A strength of GIS is its ability to integrate features that come from different scales, so there can be confusion concerning spatial accuracy. Image 1 of figure 6-6 presents two images of Japan: one is a 1:1,000,000 enlargement from a 1:10,000,000 scale map, and the other is from original 1:1,000,000 data. The differences in detail and data quality are obvious. As noted in the previous section, simple magnification of a feature or theme does not improve data accuracy.

Scale Incompatibility

Image 2 of figure 6-6 shows three original scale maps or themes (boxes), with Nepal clipped (cut and removed) and enlarged to a common scale. The resulting scale is the same, but the details are considerably different because the original map scales are different. Precision differs, of course, but other types of accuracy may be equally variable. The purpose of the project should guide which original scale is used.

Obviously, the largest original scale (1,00,000) has the best resolution and data quality for depicting Nepal, but if the regional view is needed, perhaps one of the smaller scales is more appropriate. Scales may be incompatible when incorporating data from various maps and care must be taken. A feature from the 1:10,000,000 data will not have the precision and accuracy of a similar feature from the 1:1,000,000 data.

To repeat an important concept: In GIS, there is no scale; data can be enlarged or reduced beyond the spatial accuracy limits of the original data. The user must be careful in making decisions based on presented data. This is one major reason why data sources, including original scale, should be documented, as will be discussed under the section on metadata in this chapter.

Area and Scale Coverage

Image 1: Incomplete Area Coverage

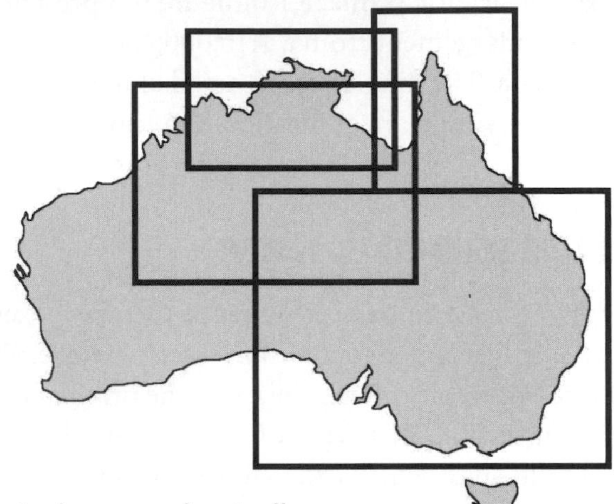

Original study area: Australia
New study area: Exclude Tasmania and Western Australia

Image 2: Smallest Scale Rule

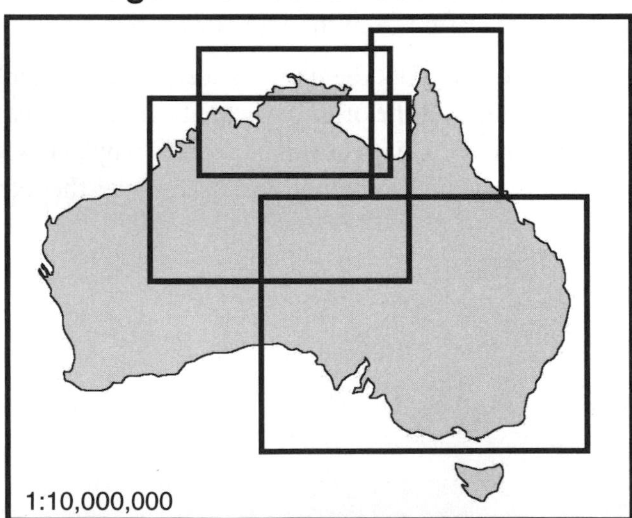

Smallest scale map: 1:10,000,000
Project base map: 1:10,000,000
Database precision: 1:10,000,000

Fig. 6-7: Incomplete area coverage and the "smallest scale" rule.

Area and Scale Coverage

GIS projects attempt to obtain complete data for a study area. The accuracy of a project is determined by the smallest scale used to examine the study area. The sections that follow explore these concepts.

Incomplete Area Coverage

A major practical problem in GIS is getting complete data of the study area. Typically, there are gaps and missing coverage, especially for large regions, which forces either extra work locating the missing data or a redesign of the project to a more restricted area. Image 1 of figure 6-7 shows map and data coverage of Australia for a national project, but there are missing parts. The project must be redesigned to exclude Western Australia and Tasmania (the island to the south of Australia) or to include the additional time, money, and effort necessary to find and incorporate the missing areas.

Smallest Scale Rule

Regardless of area covered, the smallest applied scale determines the accuracy of the project. In image 2 of figure 6-7, all of Australia is covered now because a 1:10,000,000 map was found to correct the data shortage of image 1. If all of Australia is used and all of the maps are incorporated into the project, accuracy limitation will be determined by the smallest original map scale of 1:10,000,000.

Higher accuracies from larger scales will not improve the lowest-accuracy data. This is a GIS equivalent of the principle that a chain is only as strong as its weakest link; a data set is only as accurate as its least accurate data. Therefore, the base scale and data quality of this GIS project are from the 1:10,000,000 map, although subsets and selected parts can use the best scale available.

Continuous Data Interpretation

Image 1: Earthquake Intensity Zones

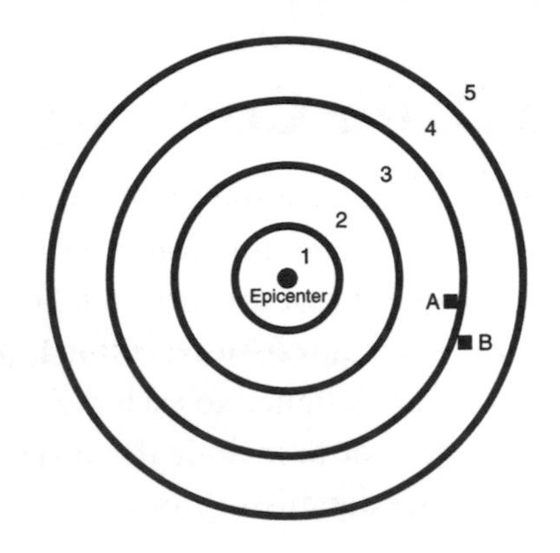

Image 2: Gravity Models

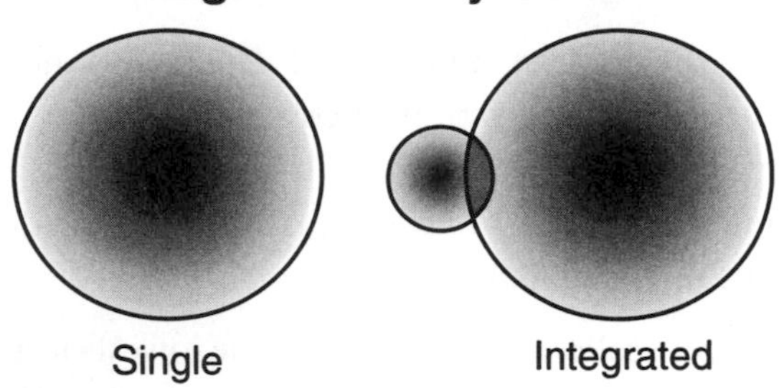

Image 3: Buffer Zones

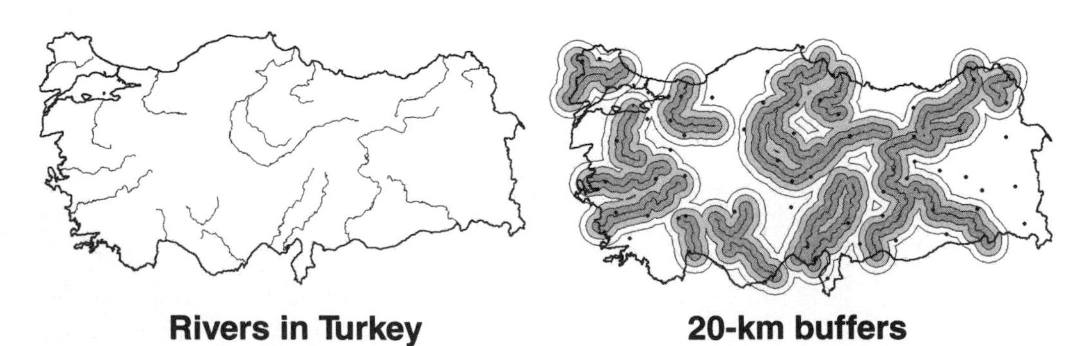

Fig. 6-8: Examples of continuous data interpretation.

Data Problems

Because data is the heart of GIS, it should receive the greatest attention and care, yet too often it is taken for granted and given insufficient consideration. Numerous issues concerning GIS data tend to confuse users, but they must be taken into account if satisfactory progress is to be made. Explored in the following sections are a few of the problems, including the use of continuous data, and considerations of incomplete and inconsistent data sets.

Continuous Data Interpretation

Some data and information can seem to be more accurate or important than they really are (too much credibility simply because they are published). By nature, radiating phenomena are continuous, transitional data, but they are difficult to depict on a map or display and are therefore usually presented in artificial discrete zones for convenience. An example of this, using temperature gradients, was presented in Chapter 2. Precision or discrete classes of data should not be inferred, however. It is important to understand the nature of the real-world features being mapped and analyzed.

Earthquake impact is often presented in zones radiating from the epicenter, as shown in image 1 of figure 6-8. Sites A and B are in different earthquake intensity zones, but because their distance from the epicenter is almost identical, they probably experienced little difference in impact. Any variance in damage is more likely caused by local geology and building conditions than by distance.

Earthquakes are inconsistent in strength, epicenter location, duration, and effects; the next quake probably will generate different zones anyway. This means that the illustrated map cannot be used for prediction but only as a very generalized indicator or model of earthquake impacts. Unfortunately, such maps are too often employed as specific planning data rather than general guides.

There are other types of gradational phenomena, such as gravity models. Shown on the left of image 2 is a single city with radiating lighter shades indicating decreasing influence or attraction ("gravity") from the center. This can be used in planning and marketing, as in evaluating the impact of distance from the central business district. The "integrated" model on the right shows a smaller city's effective area merging with the outskirts of the larger city's area. The integrated zone is a bit darker than the surrounding vicinity because it is influenced by both cities in the gravity model.

Use of continuous data is important in this type of application, although it is difficult to display properly. Pollution zones are typically gradational and could be depicted in continuous data, but resource managers prefer discrete zones as planning and management units.

The figure 6-8 image 3 maps show Turkey, with the rivers on the left and 20-kilometer pollution zones on the right. Although perhaps a bit generalized, the zones are clearly recog-

Incomplete and Inconsistent Data

Map 1

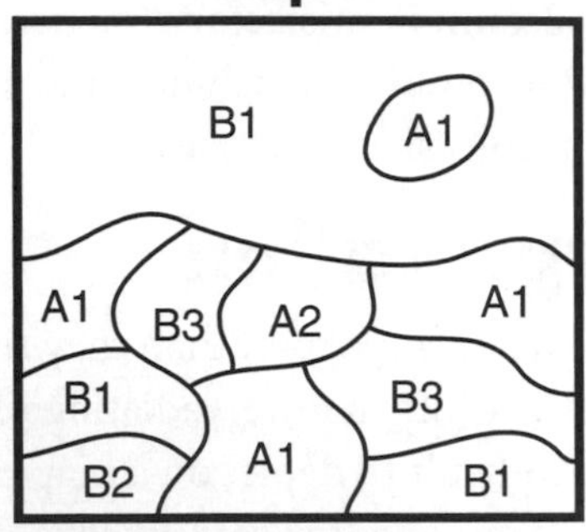

Complete mapping?

Map 2

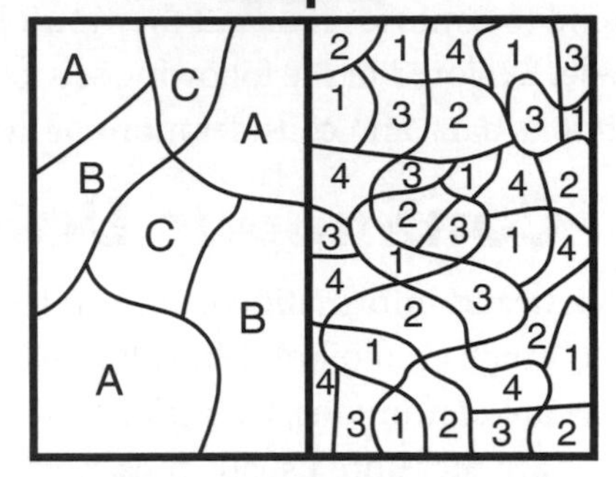

Inconsistent mapping and names

Map 3

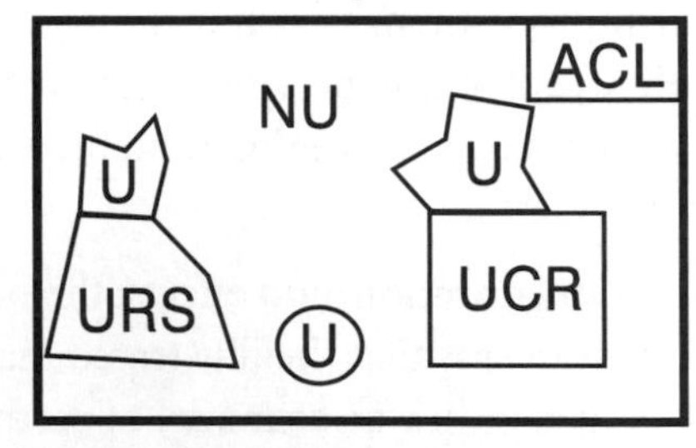

Inappropriate classes

U - Urban
NU - Non-urban

URS - Urban, residential, single family
UCR - Urban, commercial, retail
ACL - Agriculture, citrus, limes

Map 4

13
A1-X
A2-X
A1-X
A1

Incorrect attributes

Are A1 and 13 valid?

Fig. 6-9: Examples of incomplete and inconsistent data.

nized and can be combined with other features without difficulty. For example, analysis of the cities (the small dots on the map) in relation to the pollution zones can assist management planning. These types of zones are called buffers (discussed in detail in Chapter 8).

Complete and Consistent Data

At times, it can be difficult for users to know if data files and maps are complete in terms of theme coverage and use of attributes. In addition, there can be questions of consistent quality. Some obvious problems may be easily spotted, but others require careful examination and possibly time-consuming investigation. Map 1 in figure 6-9 shows a high level of soil mapping on the southern half, but the north half seems very weak or incomplete. Is the map correct and consistent, or is there an unusual landscape situation (different geologies)?

Another possibility is that data collection could be careless and incomplete for the northern part (or some data was lost in the process between field mapping and cartography). These ideas are worth investigation if data is suspect. It is important for the GIS user to have some knowledge and understanding of the study region.

The soils data display in map 2 is from two nations (west and east), for which data collection methods differ. Even the naming conventions are different (letters versus numbers), and work will be needed to make them comparable. Each nation has satisfactory soil detail for itself, and the data is accurate according to local systems and criteria. However, regional analysis will be uneven and perhaps even misleading because of the contrasts in data. Either the differences between nations will be noted or the more detailed classification will have to be generalized to match the level of the more general data (equivalent to the "smallest scale" rule, which says that the weakest or least accurate data is the standard data quality level). International mapping presents problems.

Choosing an accurate and efficient classification scheme can be difficult. It is easy to use incorrect or unsuitable classifications, such as too many or too few categories. Sometimes there is a tendency to give every feature a separate class or value instead of reducing the data to manageable categories for better understanding. On the other hand, too few classes can be used, leaving the user confused.

Map 3 is a poor land-use display due to an inconsistent scheme. Basically, these are two different levels of classifications. The use of a general Urban/Non-urban system could be acceptable, or exclusive use of the more detailed scheme would be preferable (URS, and so on), but mixing is not good design. Some very good data sets may be organized and written in a confusing manner, making the data useless. Fortunately, GIS permits easy reclassification and rectification of the problem, providing there is available information to make good decisions.

We normally assume that imported data is correct, but there can be mistakes in data collection, reporting, copying, and transfer. Undetected and uncorrected, these mistakes can result in permanent error. Sometimes careful examination of attributes reveals inconsistencies or concerns. Notice in map 4 of figure 6-9 that the attribute labels are inconsistent. Perhaps they are correct names, but they are worthy of investigation to make sure that an error is not continued. This means more work for the GIS user.

Accessibility and Data Sharing Distributed Geographic Information 1

Image 1: Global Issues

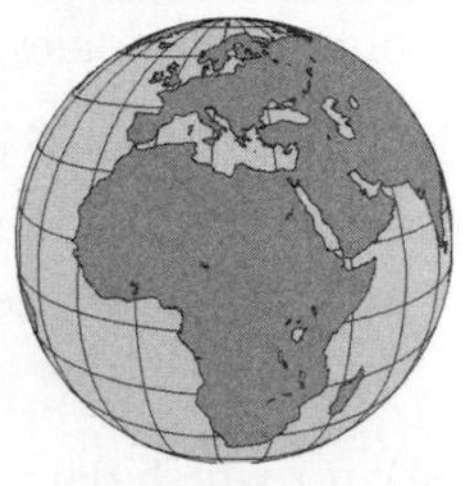

- Global distribution
- Third World access
- Haves vs Have-nots

Image 2: Ownership Issues

- Large investment
- Fear of misuse
- Legal issues
- Protection

Access negotiations

Image 3: Data Copyright Issues

Unauthorized use = theft; crime

Authorized use = purchase/permission

Agreement required

Fig. 6-10: Issues related to data acquisition and distribution.

Acquiring and Distributing Data

Project data can come from many sources, such as entering original data or incorporating outside data. There are numerous issues in acquiring data, which therefore also relate to distributing it. Data sharing, whether getting or giving, must consider various issues. The sections that follow discuss data accessibility, data costs, data standards, and metadata.

Data Accessibility

Obtaining and distributing data (maps, remote sensing, or digital data sets) is becoming easier, global, and standard, yet problems and considerations remain. The terms *data sharing* and *distributed geographic information* (DGI) are often used. Figure 6-10 illustrates some of the issues involved. Although today's communications technology supports global distribution of data, there are now global issues.

Access to data can be a major problem for many users, especially those in the developing world and regions out of the data mainstream (the data Haves versus the Have-Nots, image 1). Most users know what is needed, but often it may be out of reach for various reasons, including ownership, distance, bad format, and cost.

Many organizations are reluctant to distribute their data freely because of strong feelings of ownership. "Ownership" is a sensitive issue for several reasons: (1) the typical large investment associated with creating the data (work, time, resources), (2) fear of misuse, (3) legal concerns, and (4) illogical protection reasons (fear of losing control of the data or "guarding turf" attitudes). Sometimes, potentially valuable data can be found locked away in an inaccessible filing cabinet, unavailable for use. Negotiations may be needed to make it accessible (image 2).

Data can be copyrighted (protected by law), meaning that authorization is required for its use (image 3). Sometimes permission is merely a formality, perhaps involving no more than verbal agreement, but it can also involve significant expense and time. A major global concern today is recognition of "intellectual property," the protection of data and information. Illegal copying is considered theft, though some countries do not abide by international law. Purchases and agreements may be required.

There are legitimate legal concerns for protecting and restricting data, such as misuse for purposes unacceptable to the owners, use in matters that put the owners at risk (e.g., lawsuits), or questionable practices for commercial purposes. The concept of "value added" improvements is tricky, where a buyer of inexpensive (or even free) public data adds something to make the data more valuable, and subsequently makes the data available at a higher price. That is, distributor A sells or gives data to B, who in turn adds value to the data and sells it to customer C at a profit.

Accessibility and Data Sharing
Distributed Geographic Information 2

Image 1: Value-added Marketing

District	Area
North	2065
East	1546
South	3321
West	2367
Central	1789

Original data

District	Area	Pop.	Crop	Value
North	2065	22567	Rice	133567
East	1546	11453	Rice	144327
South	3321	16529	Taro	223761
West	2367	12548	Taro	168322
Central	1789	10442	Rice	116542

Value-added data

Image 2: Global DGI

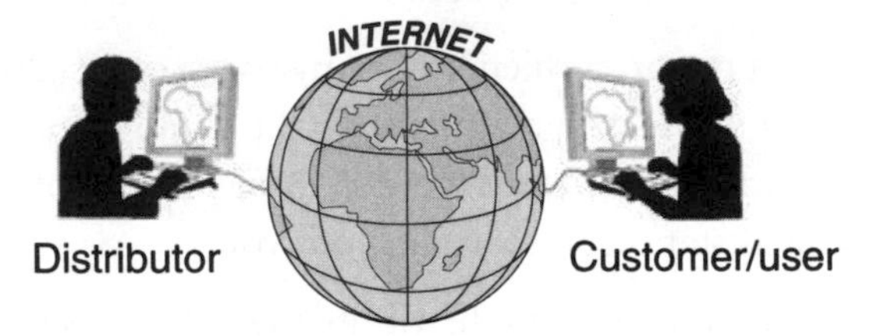

Fig. 6-11: Further issues related to data acquisition and distribution.

The nature of the "added value" can create conflict. Often, clearly useful additions are made, such as integration with other data, new classifications, and extra, derived data. Image 1 of figure 6-11 is a data set that has been enhanced with population and agricultural data. However, sometimes dishonest commercialization occurs and the value added is no more than labels or very simple changes. These are ethical and legal issues (which add to the ownership concerns).

There is much free and "public domain" data around the world, but finding and acquiring specific data can a problem. Today, the Internet and World Wide Web (WWW) offer global network capabilities and a convenient medium for finding and delivering data. Many organizations on the Net offer access to data by the global community. In fact, the Net has revolutionized GIS for both distributors and customers. The concept of "data mining" deals with sophisticated techniques of locating and acquiring specialized data from sources around the world. The ability to use the Internet is a valuable modern GIS skill, and becoming "web literate" is easy and fun. GIS data is a true global commodity, and distributed geographic information is becoming part of the "data globalization process" (image 2).

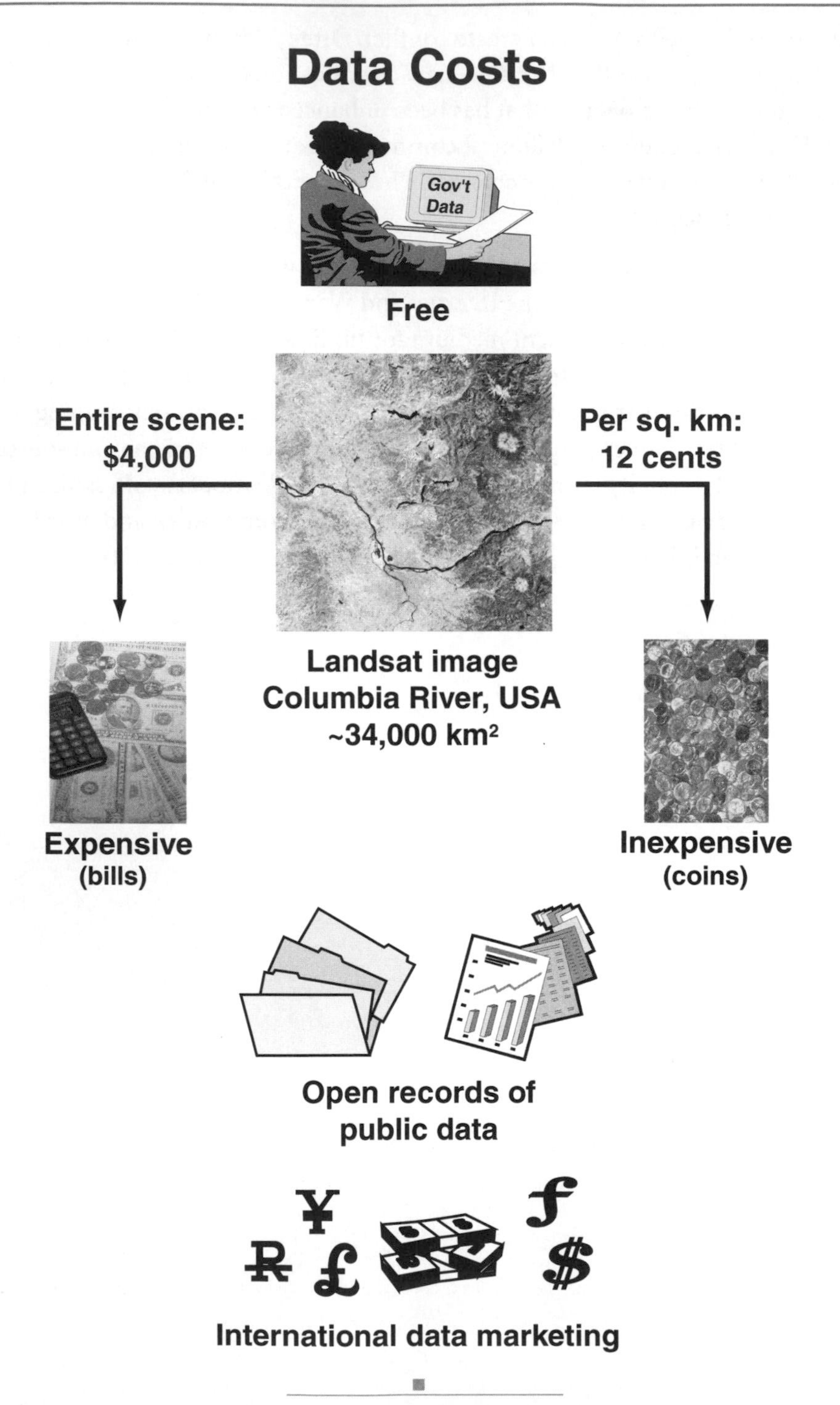

Fig. 6-12: Issues related to the cost of data.

Data Cost

There are no established price lists for geographic data. Market demands drive some pricing, whereas other influences make for widely varying pricing. Many public agencies offer low-cost data, whereas commercial ventures charge much more. Some data sets are free, especially government data, whereas others cost too much for all but the wealthy (figure 6-12).

For example, satellite data can be expensive by nature, but because of the large area coverage, images can be cost efficient per unit area. A Landsat scene, for example, can cost up to $4,000, but when calculated for the 34,000 square kilometers it covers, the price is a very reasonable 12 cents per square kilometer. However, when the rate is U.S. $4,000 per scene, pricing is beyond many users, regardless of the spatial economics.

A great deal of work is usually involved in developing data sets, and therefore some compensation is expected by many organizations. Even government agencies that are not allowed by law to profit from their work are often permitted to charge a fee to help meet the cost of development and distribution. "Open records" laws in some countries (such as the Freedom of Information Act in the United States) determine what is accessible public data and information. That is, if the public paid for the data, it is entitled to have access to it. Public agencies, therefore, may not hide or restrict data except under special circumstances (e.g., personal privacy, national security, interference with conflicting private-sector rights, and so on).

Numerous nations now have land-imaging satellites and market the data globally. These countries include the United States, Canada, Japan, Russia, France, Sweden, and others. Many countries and international agencies are selling or buying data, making the acquisition and distribution even more complicated. Shifting currency exchange rates, for example, make costs variable and unpredictable.

Data is a commodity still seeking an economic level and the cost issues continue to get more complicated. In a free-market system, the price will find it own level eventually, but in the meantime, there can be "casualties" along the way, such as the poor nations.

Sometimes clever arrangements are made between distributors or provider agencies and various levels of customers. For example, some poor nations, with low volume needs, may form a consortium to purchase a data set at standard commercial rates, but with permission to share among themselves. With cost sharing, each country pays only a fraction of the original price. Sometimes a donor agency, such as the United Nations, will make the initial high-charge purchase and then share among a group of nations under a development program scheme.

Standards

Image 1: Distributed Geographic Information

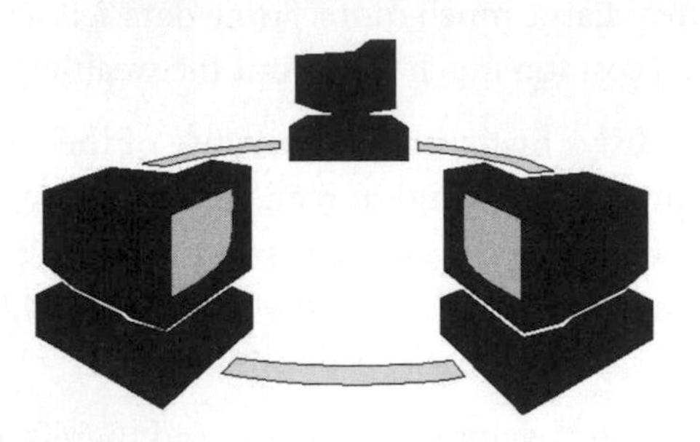

Data Should Be
- Credible
- High quality
- In readable format
- Consistent:
  - — Accuracy
  - — Precision
  - — Definitions/terminology
  - — Classifications
  - — Names and symbols

Image 2: Development of Data Standards

Difficult, time-consuming, many meetings, many people, technicalities

Various Standards:
Data
Networking
Display
Database
Exchange

Image 3: GIS as a Common Language

"Debemos hablar el mismo idioma en GIS."

"Bhu suchana pranilima hami autai bhasa bolchhaun."

(Spanish)

(Nepali)

"In GIS, we talk the same language."

Fig. 6-13: Issues related to GIS standards.

Data Standards

Data quality issues have become major topics for GIS users, particularly regarding sharing and exchange. Today, national and global data interchange needs some conformity to give users confidence that what is being distributed is credible, understandable, acceptable, and useful. Several major GIS principles have been developed in recent years to facilitate distribution and acquisition of data and to ensure satisfactory data quality. Standards, metadata, and open GIS are three of the more prominent developments. Standards, depicted in figure 6-13, are discussed in this section.

The criteria for common definitions, an established nomenclature (set of names), established accuracy, and other data factors are called "standards." Most projects need and want to be compatible with other similar projects and applications, particularly those in their area of interest. Perhaps uniformity is not technically necessary, but it makes good sense for purposes of regional and time comparisons, future partnerships, and data stability.

For distributed geographic information, GIS users need reliable data. They prefer consistent formats and classifications, dependable accuracy, common names and definitions, familiar symbols, and compatible media for transferring data. For example, a user should expect a uniform land-use scheme, with a common interpretation of features and standard classification definitions, from a government agency data set (image 1).

The development of universal standards can be difficult because there is little agreement what should be included and what measures should apply. In addition, there is little agreement as to which projects, which organization, and which data any standard should apply to. Even the idea of universal standards is controversial, primarily because there are too many levels of users, from individual consultants to national governments, with too many individual requirements and applications.

One set of acceptable centralized definitions will be impossible to develop and too demanding for too many users. The basic idea of easy compatibility is good, yet forcing everyone into a single system is both unworkable and unacceptable. Standards, then, is a non-universal concept, meant for organizations or administrative units, not every GIS project or data set in the nation or world.

Many types of standards exist in GIS, dealing with numerous aspects of data (quality, content, classification, and others), networking, operating systems, display, database queries, exchange, and others. Determination of standards can be difficult, time-consuming, technically demanding, and involve numerous meetings and many people (image 2). In the United States, standardization efforts were begun in the 1980s, and in 1992 some GIS standards for data distribution were adopted by the federal government.

The Spatial Data Transfer Standard (SDTS) established systems for data exchange, addressing accuracy, terminology, definitions for features and data structures, completeness, and

Metadata Example

TITLE: South Asia
AREA: India and surrounding nations
DATA DATE: 1995
PUBLICATION: 1998
ORGANIZATION: UNESCO
ORIGINAL SCALE: 1:1,000,000

Legend: Map metadata

Identification: Title, area, dates, owners, organization, etc.

Data quality: Attribute accuracy and spatial precision, consistency, sources of information, and methods of data production.

Spatial data organization: Raster- vector format, location guides (street addresses, mailing codes, districts), and organization of features in the data set.

Spatial reference: Map projections, grid systems, datums, and coordinate system information.

Features and attributes: Database content, feature types and des cription, database attribute organization and definitions, and naming system.

Distribution: How to obtain the data, contacts, available formats, fees, and copyright.

Fig. 6-14: Metadata can include many types of information on data.

other issues. Related efforts in the early 1990s began establishing standards for data collection, content, classification, and management. This means that government data should conform to the established standards and that users can have confidence in distributed geographic data quality. Many states, agencies, and companies have adopted the SDTS (or a modification) and it serves as a model for GIS standards. Similar programs are underway internationally as well. Perhaps global standards will be achieved for easy exchange of global data sets in the near future.

Although the SDTS is aimed at standards for exchange of data, it also helps to define content. Although each GIS project is free to interpret and characterize data as it needs, conforming to uniform terminology and definitions for content is a good idea. If the GIS community speaks the same "language" for its data and methods of data management, everyone benefits (image 3 of figure 6-13).

Metadata

One primary means of establishing confidence in data quality for both the producer and user is to develop proper documentation so that everyone knows and understands the form, usability, accuracy, precision, content, and other aspects of the data. Documentation is a significant process today, given the large amount of data exchange and prominence of the Internet. Standards in documentation are also evolving, and the concept of *metadata* is important. Essentially, metadata is information about data. It considers quality, condition, distribution, and other attributes that may be helpful to the user.

Documentation is valuable, even critical, for users who invest considerable time, effort, and expense in the production, exchange, analysis, and application of geographic data. Examples of metadata are presented in figure 6-14. Actually, a map legend is a form of metadata in that it documents the data in a readable format. A separate metadata file is an extensive "legend" for a data set. Metadata provides a common, uniform set of terminology and definitions for the geographic data, although there are many forms the documentation can take in conveying that information. Metadata typically includes the following.

- *Identification:* Title, area, dates, owners, organization, and other elements
- *Data quality:* Attribute accuracy and spatial precision, consistency, sources of information, and methods of data production
- *Spatial data organization:* Raster-vector format, location guides (street addresses, mailing codes, districts), and organization of features in the data set
- *Spatial reference:* Map projections, grid systems, datums, and coordinate system information
- *Features and attributes:* Database content, feature types and description, database attribute organization and definitions, and naming system
- *Distribution:* How to obtain the data, contacts, available formats, fees, and other useful information

Figure 6-14 shows a map of South Asia and a sample metadata set. Metadata typically includes much more than this, but the main idea is to give the user sufficient information about the quality, lineage, definitions, and other factors listed. There are software tools (special programs) to help define and document data for distribution and content purposes. Most of these programs are typically free or low cost and most assist by using some type of fill-in form. Obviously, the software is convenient and helps to ensure conformity to existing standards. Metadata is a very important item to include in GIS projects.

Distributed GIS Advantages

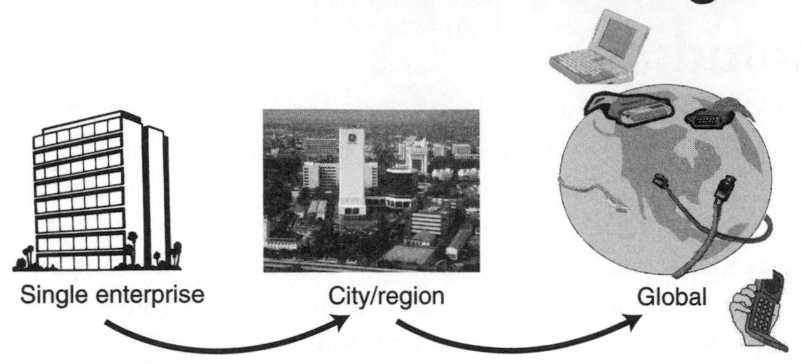

Image 1: Distributed Operations and Applications

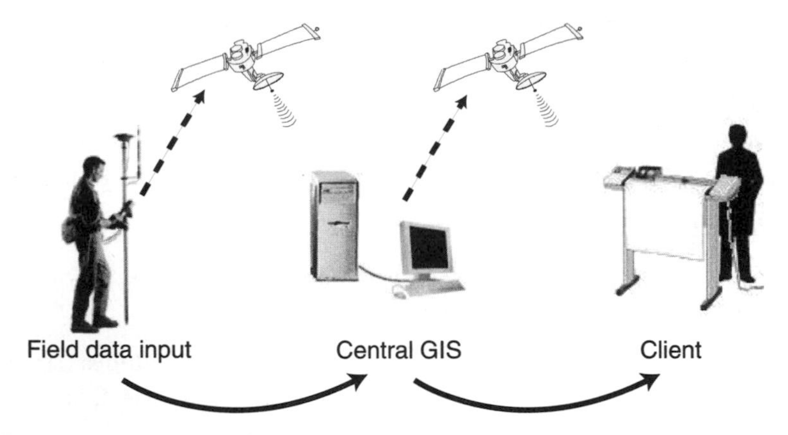

Image 2: Decentralized GIS Infrastructure

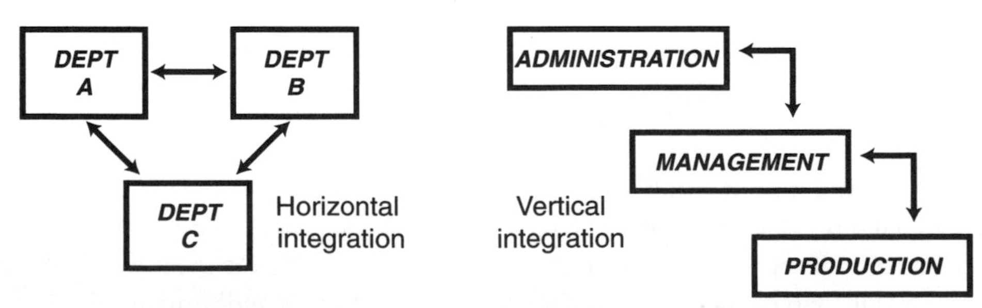

Image 3: GIS as Organizational Integrator

Fig. 6-15: The nature of distributed, decentralized, integrated GIS.

Distributed GIS

In the "old days" of GIS, projects were largely independent and private, with in-house data and locally distributed results (paper maps, reports, perhaps a disk of data, and maybe a cable or local network connecting computers). Today, GIS is a *distributed* technology, meaning that a project no longer needs to be confined to the work area where the computers are located, and can be a regional or global operation. Those are far-reaching concepts, but certainly not exaggerated.

Advantages of Distributed GIS

Networking technology, especially the Internet and World Wide Web, basically has removed location as a restriction for GIS work. Data and input can come from anywhere and can be distributed to anywhere. Image 1 in figure 6-15 depicts the evolution of GIS from a single enterprise activity to an entire city (perhaps through networking) and eventually to global distribution. Some of the principles and issues of distributed data were discussed in previous sections, and it is clear that useful data sources can be found in many places.

Distributed GIS also pertains to operations as well as data. It is no longer necessary to be in front of a fully loaded computer to perform GIS work. With the advent of cell phone technology, the Internet, and other communications technologies, a GIS infrastructure can be decentralized and distributed. As noted in the GPS section, a field crew can collect detailed data and relay processed information to a second location (e.g., a central GIS lab), which in turn can relay final results to other locations (image 2). As will be seen, interactive GIS is now available on the Internet, permitting data and operations to reside in one part of the world while giving access to anyone from anywhere.

Another aspect of decentralization, largely unplanned but highly beneficial, is that the spread of GIS throughout an organization has helped make better connections between departments and between the various hierarchical levels. That is, within large and small organizations (government, education, and industry and commerce), offices and departments tend to be somewhat independent because each has distinct responsibilities, budgets, management, and facilities. Unproductive competition, isolation, and redundancy often result. However, because GIS uses data, projects, methods, and personnel from all over the organization, it tends to be an integrative force.

Recall the advantages of a central database. People and departments begin cooperating, speaking the same data and technical language, sharing the GIS, and working together. GIS becomes a horizontal integrator (image 3). Similarly, the various authoritative levels within an organization find common ground. For example, technical data workers must work closely with middle management and with the administration. There is better cooperation and understanding among the once separated hierarchical levels; GIS becomes a vertical integrator.

Distributed GIS Issues

Image 1: Different Audiences

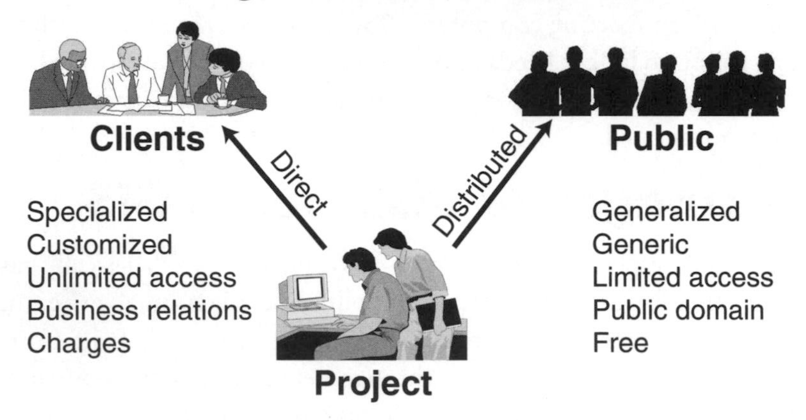

Image 2: Legal Issues

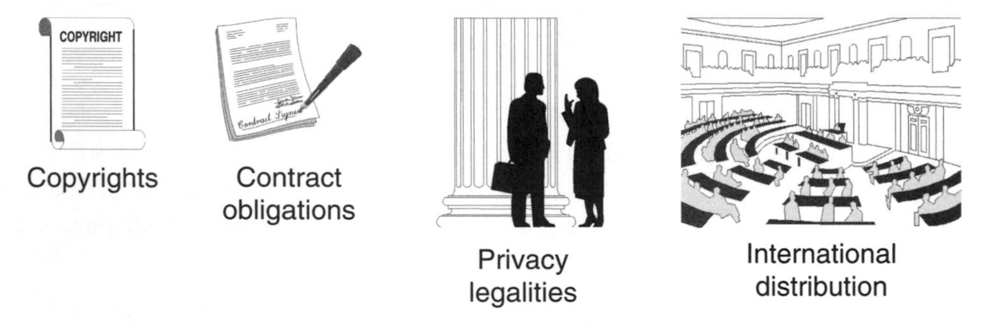

Image 3: Access Restriction Issues

District	Area	Pop.	Crop	Value
North	2065	22567	Rice	133567
East	1546	11453	Rice	144327
South	3321	16529	Taro	223761
West	2367	12548	Taro	168322
Central	1789	10442	Rice	116542

Read only

Read

Download

Fig. 6-16: Issues related to distributed GIS.

Issues Related to Distributed GIS

Distributed GIS has inherent problematic issues as well as benefits. First, the project must consider a wider and usually less specialized audience than the immediate clients; for example, use of accepted terms and definitions ("standards"), well-designed map and data presentation, and the depth of knowledge the potential audience may have regarding the project application (image 1 of figure 6-16). Moreover, there may be a need to protect some of the information and to limit access. Legal issues also may be important, such as copyright protection, contract obligations to the client, charges for data, privacy concerns, and international distribution restrictions that involve legislative determinations and oversight (image 2).

As the Internet changes the established ideas about freedom and access of information, a project must decide if the distributed GIS will be read-only of presented products (maps and tables that cannot be changed), or if it will permit access into the database and retrieval of data (image 3). This means that either the audience can see only what is designed and presented or that there is operational functionality attached for the user, allowing use of the data, such as for queries or GIS operations. The project will have to decide if data can be downloaded into other GIS and reworked. These are major concerns by modern GIS projects, particularly those using on-line GIS, discussed in the following section. Distributed GIS may be highly beneficial for many users, but there are cautions to be observed.

On-line GIS 1

Image 1: On-line Generation

Technology-oriented
Global expectations
New communications paradigms

Image 2: On-line Atlas

Univ. of the South Pacific
South Pacific GIS Atlas

Fig. 6-17: Web page for an on-line atlas.

On-line GIS: The Internet

A new and major part of DGI is the Internet, especially the World Wide Web (WWW), but also including other Net communications, such as FTP for data exchange, e-mail, and possibly the multiple connection discussion forums and interactive "chat" lines. (The terms *Internet, Net, WWW, the Web,* and *on-line* refer to the Internet generically, though there are subtle and technical differences between them. *Web site* and *web page* often mean the same, although a page is really one part of a larger site. It seems that common usage of Internet terms and definitions is rather flexible, which may be an indicator of evolving paradigms.)

Because the Net is global, instantaneous, friendly, essentially free, and widely available, it has created revolutions in business, science, publishing, education, and now GIS. The sections that follow discuss issues regarding the Internet and GIS.

GIS and the Internet

It is said that society and knowledge are being revolutionized (see the material on the Information Age in Chapter 1), and that a new "on-line" generation is in the making, with global expectations and new communications paradigms (image 1 of figure 6-17). GIS is taking advantage of Internet technologies and is rapidly developing a web-based DGI.

The Internet is a wonderful medium that can support practically all types of GIS products, such as maps (in various formats), images, data sets (tables and lists), text (reports, papers, articles), and links to other web sites. GIS is well represented on the Net by all of its many components, from services to data to education. There are probably several thousand GIS-related web sites now, with more added every day.

There are numerous geographic data sources on the Internet, some as dedicated resource sites and others as subsets of larger projects. Public domain (free) and commercial data are available on-line. In fact, in many parts of the world, almost any substantial business that does not have a web site is out of sync with the commercial and technological worlds.

GIS atlases are a relatively new on-line product. They may be by-products of a project that offers views of its work, or they may be dedicated atlases. Image 2 shows the home page of the University of the South Pacific's South Pacific atlas, which displays thematic maps selected by topic from the list on the left. Interactive atlases are one of the latest on-line developments, offering topics and maps that can be controlled by the user.

On-line GIS 2

Image 1: ArcView Internet Mapping Server National Geographic Map Machine

Image 2: Distributed Processing

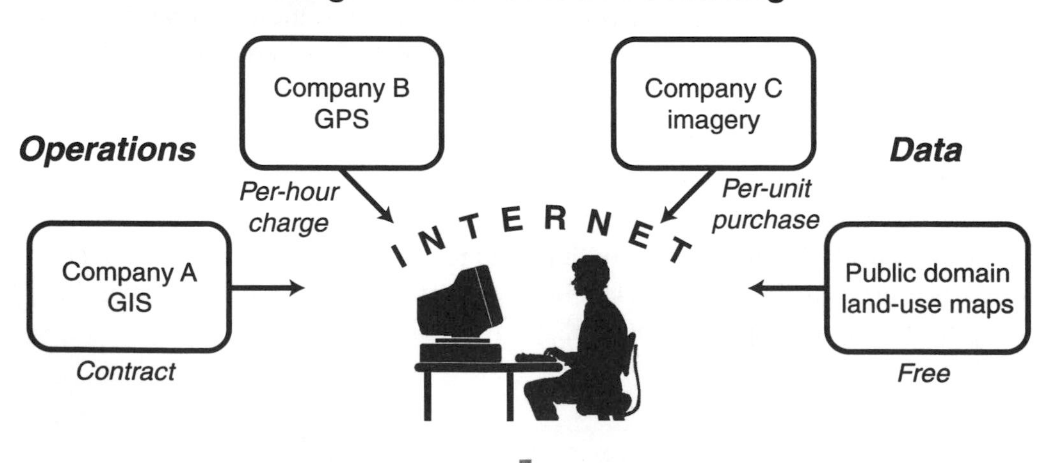

Fig. 6-18: The ArcView Internet mapping server and the concept of distributed processing. (Courtesy of National Geographic Maps)

GIS Internet Trends

Image 1 of figure 6-18 shows the home page of the National Geographic Map Machine, which uses ESRI's Internet Mapping Server (one of several commercial products). The Map Machine is a rich and dynamic atlas full of maps, pictures, text, and on-line information sources (though its effective color and interactivity are not readily apparent from this single monochrome view). Although the Web is in its infancy, the near future will see even more exciting multimedia sites with advanced interactivity. Chapter 13 examines these probabilities and possibilities in regard to GIS.

One of the major software technologies under development is distributed operations (distributed processing), meaning that software will reside on the Internet, possibly at a major vendor's site, and users (customers) will operate through the Net, renting the centralized software. Depicted in image 2 is a user operating rented GIS software from company A (on a long-term contract basis), incorporating GPS operations from company B (at a per-hour rate), imagery data from company C (credit card purchase per unit of data), and free public domain land-use maps and data from a government agency.

Although controversial and still problematic, this type of infrastructure may be the wave of the near future for many major software applications, including GIS, probably as an option at first and later becoming the standard method for commercial GIS. GIS users will be able to operate their GIS and projects from anywhere, no longer confined to specific machines or labs or to specific data residing at project sites.

A field worker in Malaysia can stop by a cyber-café in Kuching, for example, and perform useful GIS work to evaluate on-site data and its compatibility with the overall project. Sophisticated networking already permits this type of connection and interaction, although it is the user who maintains the server and network, rather than a commercial provider. Nonetheless, the Internet is moving GIS in new directions and toward new capabilities.

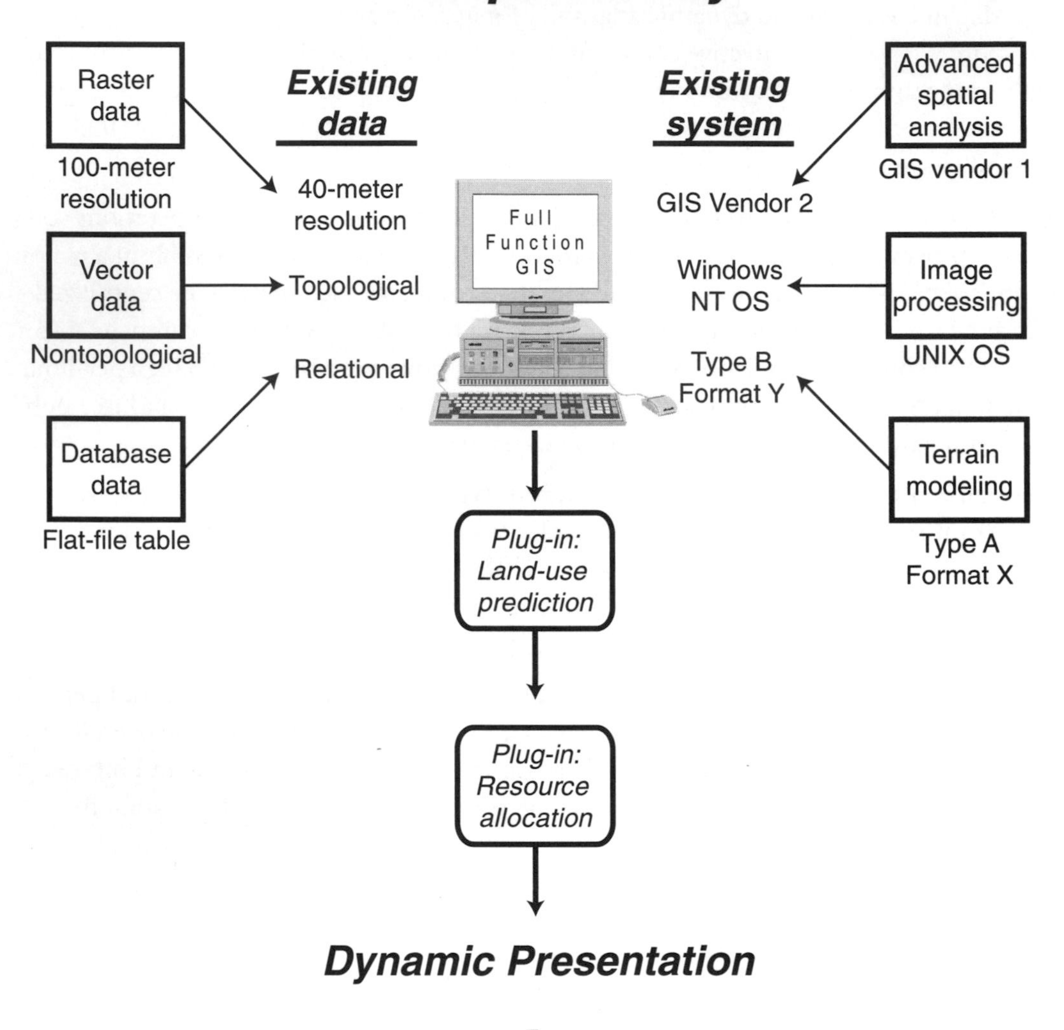

Fig. 6-19: The new paradigm of open GIS.

Open GIS

The concepts previously discussed are leading the field of GIS into a new paradigm, from the old ways of individual, private, and independent (but disconnected) GIS projects to distributed, shared, holistic, and integrated GIS. (Read that important sentence again and take special note of the major words.) A major new direction is underway that relates to a truly open structure for operations and data. Termed appropriately *open GIS*, the concept means true freedom from difficult data format conversions and a truly shared and connected infrastructure.

For example, although significant progress has been made in transforming data from one format to another, the user is still faced with limitations on which type of data, and therefore which data sets, can be integrated into the application at hand. Simply being able to read other data does not ensure the capability to use it in the software's advanced analysis (much the same as one word processing program translating another's text, but with loss of formatting, such as underlines and selected fonts). It will be nice to incorporate any data into a project conveniently, without the difficulty of reformatting and without questions of functional operation.

Open GIS leads to *interoperability*, which makes a project capable of using whatever data, data structure, tools, operations, or GIS systems are necessary and available for the application. The application will drive the project, and will not be limited by restrictions on data or systems. This is true data sharing and flexible operations; that is, it is a "seamless" integration of data from (potentially) around the world, combined with the power of various GIS. It is the ultimate in distributed data and distributed processing.

Combining data from various sources into a single-source interactive GIS web site is one of the next major steps in DGI. This integrates the concepts discussed previously and presents a revolution in GIS. Users, customers, and applications will be unconstrained by location and systems limitations. They will have a global infrastructure and global data resource as a standard paradigm for GIS, and they will have advanced GIS functions available from several sources, such as statistical analysis, modeling, and dynamic presentations.

Faced with a specific problem or application, a user (either an applications specialist or GIS professional) goes to the Internet, enters the problem or need, and obtains access to the necessary data and to the suite of GIS capabilities to work the problem. The capabilities can consist of standard geoprocessing tools (such as overlay, database functions, image processing, and cartography), as well as "plug-ins" (separate programs) for specific applications. These separate programs are written for uses such as hydrologic models, resource allocation processes, land-use predictions, decision support systems, and other specialized operations and processes.

Figure 6-19 depicts a project that uses additional resources to complete the project. Raster data at 100-meter resolution is incorporated with existing 40-meter data without difficult and time-consuming transformations. Nontopological data is integrated into existing topological data, and flat-file tables are brought into the relational database. An advanced spatial analysis module from another vendor is combined with the existing vendor's capabilities for analytical synergy. Special remote sensing image processing programs that work

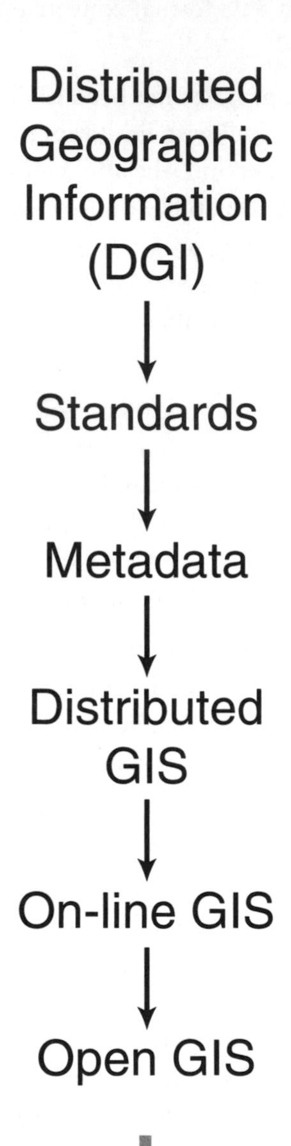

Fig. 6-20: A summary of advances in the field of GIS.

only with the UNIX operating system (OS) are incorporated into the current platform's Windows NT system without major reconfigurations.

Terrain modeling is needed but exists only with type A algorithms (specialized numeric processing programs), which produce format X output. Interoperability permits the algorithm to work with the existing type B algorithms and to produce format Y output. After the data analysis, a plug-in module to predict land use is applied, followed by another plug-in to translate the land use into a resource allocation structure. The final results can be presented in standard reports or as a dynamic presentation such as a DVD multimedia movie.

Open GIS and interoperability could have been discussed under Chapter 13, on the future of GIS. The concept has started, and parts of it are working, with others under development.

A New Paradigm

As discussed, perhaps the cost of the external, web-based capabilities and data will be annual subscription fees or charges per use. For the customer, high-priced hardware and software systems and their maintenance will no longer be required. Occasional users, the poor, and the disadvantaged (particularly in the developing world) will not have to purchase and maintain large, expensive, and complicated technical and data support systems. Open GIS, when fully functional, will be a truly comprehensive, flexible system and a new, powerful paradigm for GIS. However, although promising advances are being made, this "pie in the sky" vision may not be a reality for a long time.

Figure 6-20 serves as an overview of some of the advances presented in this chapter, all of which lead to a comprehensive open GIS infrastructure. DGI offers a data sharing process, whereas standards can help to ensure the consistency of and confidence in the data (within the confines discussed under standards). Metadata provides documentation to support the distributed data. Distributed GIS uses shared functions as well as data, particularly through on-line GIS capabilities. Global open GIS is a dream that is already under development and continued evolution is encouraging.

Each of these developments and their impacts offer exciting potential and opportunity. When combined, they constitute a synergistic new paradigm for GIS. The future may bring a GIS we only dream about today.

CHAPTER 7

INVENTORY OPERATIONS

Introduction

THERE ARE MANY OPERATIONS IN GIS, from simple data inspection to advanced analytical processes. To help learn and to appreciate what GIS can do, these functions are grouped into four sequential chapters, beginning with this chapter, Inventory Operations; proceeding to Chapter 8, Basic Analysis; then to Chapter 9, Advanced Analysis; and ending with Chapter 10, Site Suitability and Modeling. Only a sampling of the variety of GIS capabilities can be offered here.

One of the simplest yet most useful sets of GIS operations is termed *inventory,* the first steps in preparing and starting a project. They include procedures to extract and use data and information, basic work on themes, and a few preliminary steps for initial analysis. Discussed in this chapter are some of the more common and important inventory options available on all GISs.

Each brand (company) of GIS uses different and various types of commands, procedures, names, and display formats. Some GISs operate by command-line entry (typing the operation and needed details), some by pull-down menus, and others by mouse point-and-click techniques. Most offer a combination of these options. The ESRI ArcView system is used here; many GISs operate in a similar way.

Viewing GIS

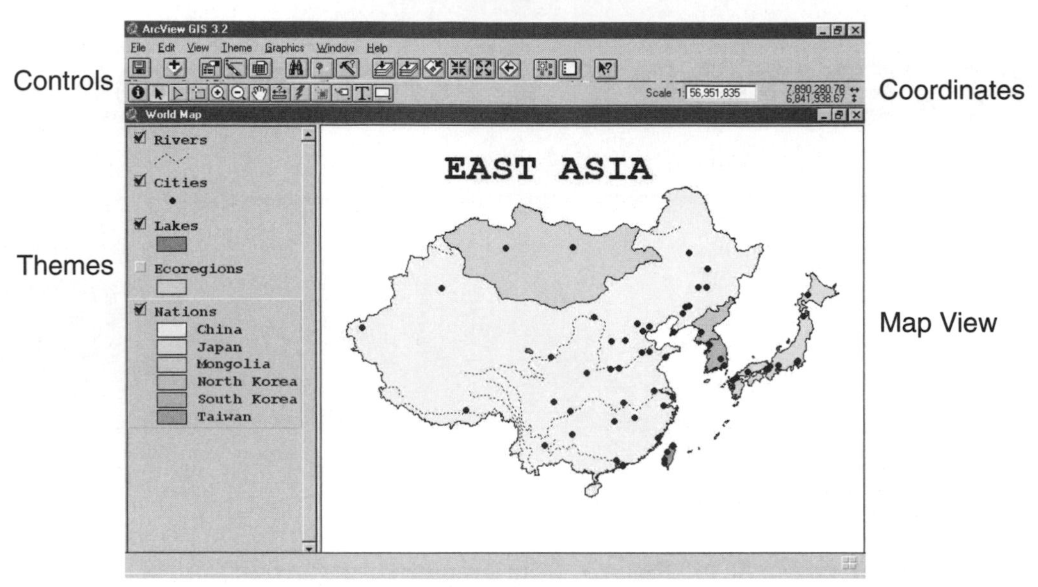

Image 1: Project View

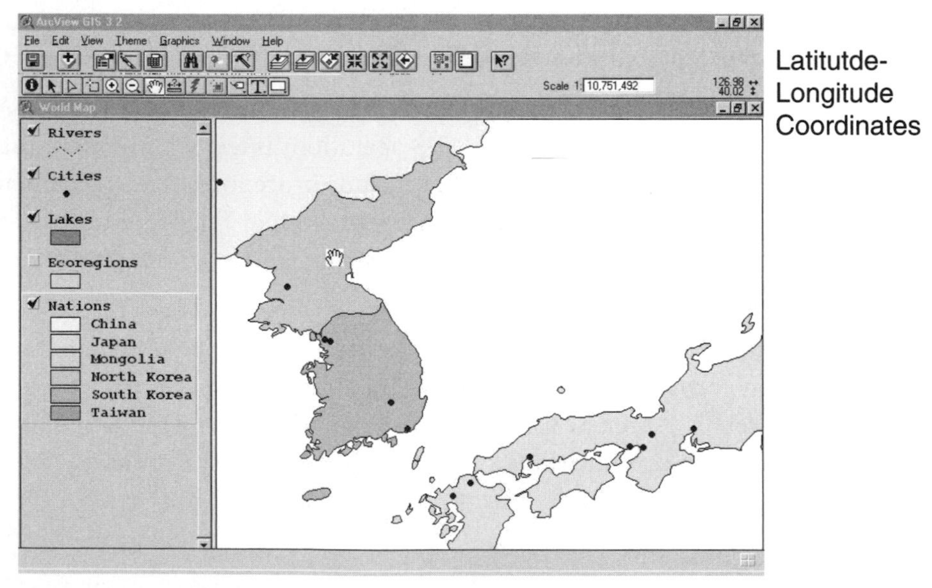

Image 2: Magnified View

Fig. 7-1: Viewing a GIS project.

Viewing GIS

After data has been entered into the GIS, one of the first GIS operations is to view the project themes. Image 1 of figure 7-1 shows an East Asia mapping project display in ArcView GIS, with the themes in the legend window at left (rivers, cities, lakes, ecoregions, and nations). Each theme has one or more features of the same type (for example, cities are points, rivers are lines, and nations are polygons).

Clicking on the small box next to a theme name displays the theme in the map view. Note that all of the themes except Ecoregions are turned on. Viewing the themes is a very important step, being the first use of the geographic data and starting point of project activity.

Display of the features and themes together in map form shows spatial relationships, an initial step toward spatial analysis. The eye can recognize patterns, landscape morphologies (appearances), and other meaningful information. It can discriminate details and subtle differences between features and between themes; for example, sizes, relative locations, and so on. A great amount of information and interpretation is available simply by observation, much of which is beyond the computer's capacity.

Navigating and reading the map data is easy, typically by moving the cursor (screen pointer) and using the control menus and icons. For example, the initial view displays the full extent of the mapped area, but zooming in for a magnified view is possible by clicking on either the magnifying glass icon or arrows pointing inward or by using pull-down menu selections. Image 2 shows a view zoomed in to Korea.

Other viewing options are available on most GIS displays, such as panning (moving the map around by using the hand icon). A small window in the upper right displays the coordinates of the cursor. The coordinates in image 1 are in meters, as defined by the type of projection used in the view. However, using a pull-down menu control, the projection can be changed, which can convert the coordinate system. Latitude and longitude are used in image 2 because the projection of the magnified view was modified. Are you able to see in the two projections the subtle differences between the shapes of Korea and Japan?

Using the Database

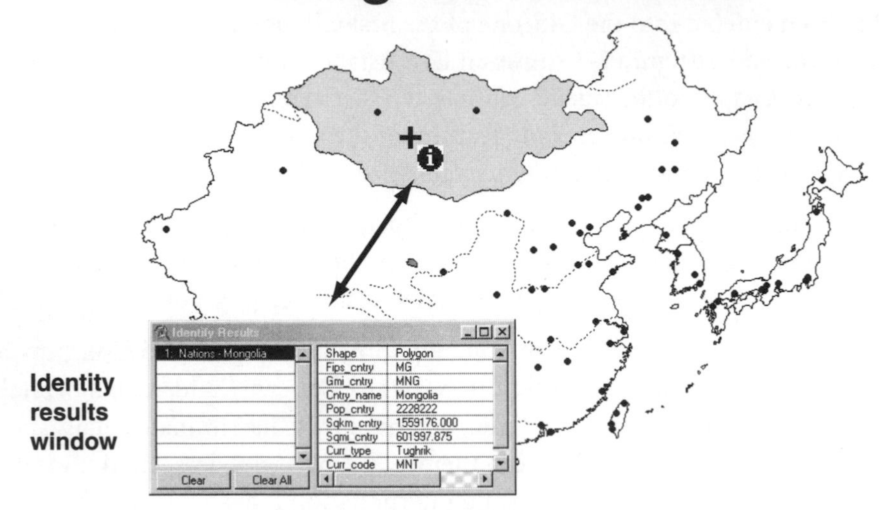

Image 1: Single Record Graphic Select

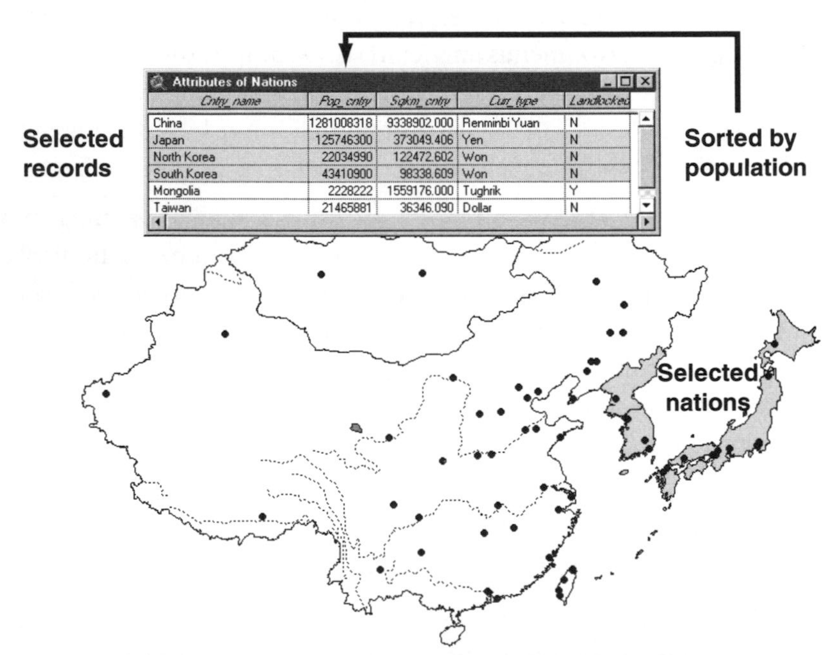

Cntry_name	Pop_cntry	Sqkm_cntry	Curr_type	Landlocked
China	1281008318	9338902.000	Renminbi Yuan	N
Japan	125746300	373049.406	Yen	N
North Korea	22034990	122472.602	Won	N
South Korea	43410900	98338.609	Won	N
Mongolia	2228222	1559176.000	Tughrik	Y
Taiwan	21465881	36346.090	Dollar	N

Image 2: Database View, Multiple Select

Fig. 7-2: Views of maps and associated databases.

The Database

Viewing and using the database can be as important as viewing the themes on the map display. The sections that follow discuss reading databases, database queries and summaries, relational database queries, Boolean queries, and graphic selection query.

Reading the Database

Databases (sometimes called tables) are central to GIS; they give "intelligence" to maps by attaching information to make geographic data useful. GIS performs much of its work at the database, and understanding what the database offers and how it functions is important. Basic principles of GIS databases were discussed in Chapter 2. This section presents some of the common options for using them.

One easy way of examining a specific feature's attributes (its record in the database) is by clicking on the Identity icon (small, circled *i*) and moving the pointer to the feature. Illustrated in image 1 of figure 7-2 is the icon graphic on Mongolia and the country's data displayed in the pop-up Identity Results window. The pointer can continue to select other nations for quick views of feature records. This is a very useful initial interactive link between the map and the database, one of the major strengths of GIS.

By clicking on a feature with an icon pointer or using a keyboard "shortcut" (set of keys), the entire database for a selected theme can be displayed to show all of the records and their attributes in a table format (image 2). There are various ways of examining individual records, often by a simple point-click on a specific row or set of rows. The records are highlighted across the database for visual convenience, such as South Korea, North Korea, and Japan in this illustration. Useful questions can be answered, such as: Which nation is the smallest in area? Smallest in population?

The highlight also "selects" the record or records for further work. When database analysis is made (e.g., to obtain descriptive statistics), the analysis will be applied only to the selected features (as will be demonstrated). If no selections exist, operations apply to all records. Notice in the map view that the three selected nations now have the same color (gray tone here) as the selection highlight in the database. When selections are made in the database, they are also selected and highlighted on the map, and vice versa, as will be seen.

Attributes in a field can be reordered for easy interpretation or analysis; for example, in ascending or descending order. Reorganizing the population numbers from low to high presents a useful hierarchy of nations in a region (as in image 2). This can be the beginning of investigating spatial relationships; for example, selecting the top two nations and seeing where they occur on the map.

Database Queries

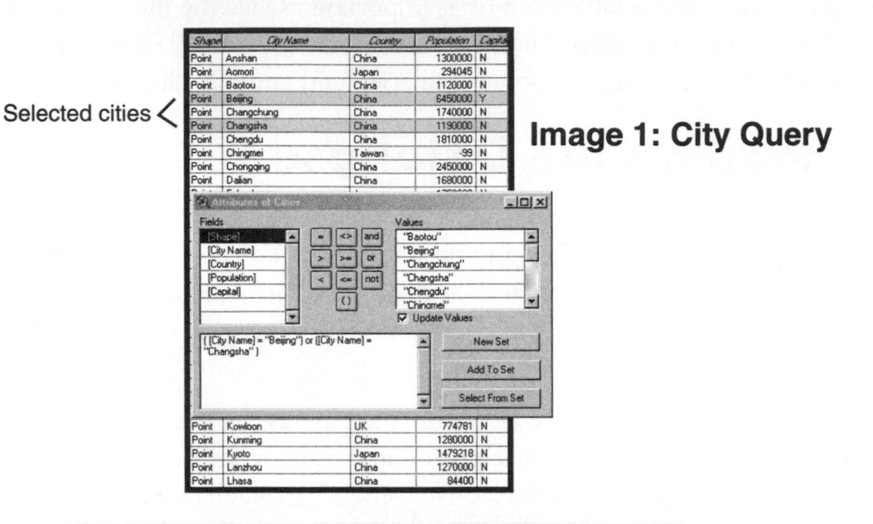

Selected provinces

PROVINCE	NATION	REGION	POP	AREA KM2	TYPE
Sichuan	China	Eastern Asia	118811800	563529.125	Province
Henan	China	Eastern Asia	94755800	166048.406	Province
Shandong	China	Eastern Asia	93518330	155603.000	Province
Jiangsu	China	Eastern Asia	74307460	99678.992	Province
Guangdong	China	Eastern Asia	69623060	176831.594	Province
Hebei	China	Eastern Asia	67687330	186638.594	Province
Hunan	China	Eastern Asia	67218970	211504.203	Province
Anhui	China	Eastern Asia	62255730	140206.406	Province
Hubei	China	Eastern Asia	59804980	186188.000	Province
Guangxi	China	Eastern Asia	46813840	237006.000	Autonomous Region
Zhejiang	China			.398	Province
Liaoning	China			.500	Province
Jiangxi	China			.406	Province
Yunnan	China			.906	Province
Heilongjiang	China			.812	Province
Kanto	Japar			.301	Region
Shaanxi	China			.703	Province
Guizhou	China			.297	Province
Fujian	China			.500	Province
Shanxi	China			.703	Province
Jilin	China			.094	Province
Gansu	China			.594	Province
Nei Mongol	China			.000	Autonomous Region
Kinki	Japar			.648	Region
Xinjiang	China	Eastern Asia	16794560	1604712.000	Autonomous Region
Shanghai	China	Eastern Asia	14784580	5994.313	Municipality
Kyushu	Japan	Eastern Asia	13296000	41852.480	Region
Beijing	China	Eastern Asia	11989330	16599.369	Municipality
Seoul-jikhalsi	South Korea	Eastern Asia	10612580	579.648	Special City

Statistics for POP_field

Sum: 451016450
Count: 5
Mean: 90203290
Maximum: 1`8811800
Minimum: 69623060
Range: 49188740
Variance: 3E1596240040900
Standard Deviation: 19534488

OK

Image 2: Statistics

Nation	Count	Total Pop.	Avg. Pop.	Total Area Km2	Avg. Area Km2
China	31	1258529096	40597713	9338902	301255
Japan	9	108543031	12060337	373049	41450
Mongolia	20	2891822	144591	1559176	77959
North Korea	11	21219431	1929039	122473	11134
South Korea	14	37583927	2684566	98339	7024

Number of provinces in each nation, with total and average province population and total and average provincial area.

Image 3: Summary Table

Fig. 7-3: Examples of database queries and a query summary.

Reordering can also begin the process of interpreting associations between attributes, such as population and area. For example, does area descend in order along with descending populations? A quick review of the numbers shows the answer to be no. Perhaps many answers are not obvious at this point, but interesting questions are raised to warrant further research.

Even at this early stage, valuable work can be accomplished. Useful data can be examined, extracted, and translated into information. Many GIS tasks are performed at the database, without having to view themes. Additional options are discussed in the sections that follow.

Database Queries and Summaries

A *database query* is a question or task operation, instructing the database to find or calculate specific data or information. The database can be searched to present records that meet certain criteria. Queries may be typed, or fields and operations can be clicked in a query box. The query *City_Name = Beijing* will highlight the appropriate city record on the database, and *City_Name = Beijing or City_Name =Changsha* will highlight both cities, as shown in image 1 of figure 7-3. Sometimes these tasks can be accomplished by hand, but using query operations is more accurate, safe, and efficient, particularly in large databases.

Some GISs offer descriptive statistics of the attributes, either for the selected records or the entire field. For example, image 2 shows the five highest populated provinces of China (selected at top), with the population field selected for descriptive statistics. The pop-up report presents the total population of the five provinces (Sum), number of provinces considered (Count), and other useful demographic statistics. Fast and easy selection of the records and fields allows the user to explore various analytical combinations.

Summary tables condense selected data to present a compact or reduced version of the database. Image 3 shows a summary table of East Asian provinces, ordered by nation and presenting the total and average provincial population and area within each nation. Summary tables are easy to construct and are very useful analytical devices.

Relational Queries

Image 1: Best Property

Best Property:
Area = >40,000 meters ^{2}
Owner: Not Silima
Tax Code: B
Soil Quality: High

PROPERTY NUMBER	AREA (SQ M)	OWNER	TAX CODE	SOIL QUALITY
1	100,000	TULATU	B	HIGH
2	50,100	BRAUDO	A	MEDIUM
3	90,900	BRAUDO	B	MEDIUM
4	40,800	ANUNKU	A	LOW
5	30,200	ANUNKU	A	LOW
6	120,200	SILIMA	B	HIGH

Image 2: Conditional Query

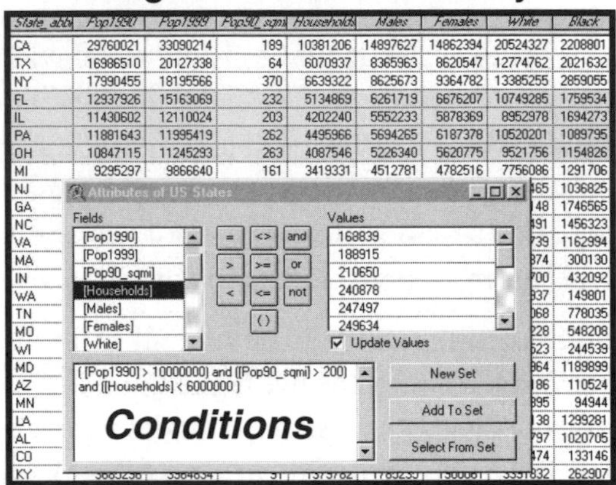

States meeting all conditions

Image 3: Additional Query

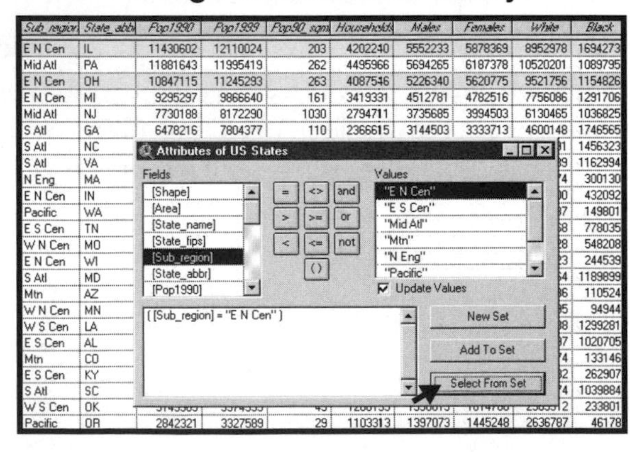

Fig. 7-4: Examples of relational database queries.

Relational Database Queries

The relational database offers significant advantages over "flat-file" databases, which are little more than simple tables. Because the data can be related and linked between fields, it is possible to make queries that connect multiple attributes (referred to as *conditional queries*). The query in image 1 of figure 7-4 addresses location of the best property. The program finds the requested set of conditions in each attribute column, selects the records satisfying these conditions (or sifts out the unneeded ones), and eventually arrives at the records meeting all of the expressed conditions.

It is easy to see that complicated queries can be made from a relational database. It appears that the imagination is the only limit regarding what information can be extracted from the available data. This is a true data-to-information operation. Image 2 shows the query language and results of finding U.S. states having populations over 10,000,000, population density less than 200 per square mile, and number of households under 6,000,000. Only four states meet all of these conditions, and they are selected (light gray above the query box).

The next step is either to map those states or use the highlighted list in the database for further work. The image 3 database is an additional query on the states selected in image 2. Of the four selected states, only those in the East North Central subregion are needed. The query expresses *subregion = E N Central*, but the Select From Set button is used to choose from only the previously selected four states. Two states meet the East North Central condition. Multiple sets of queries can be performed, with each query reducing the number of records meeting the sequence of conditions.

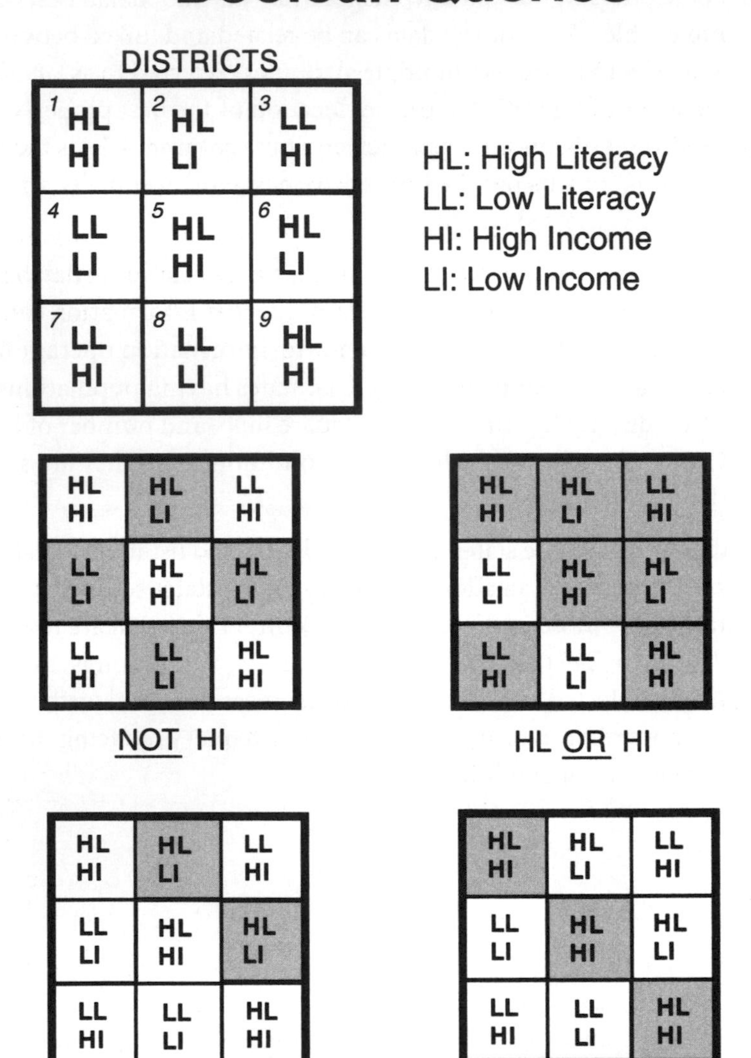

Fig. 7-5: The concept of Boolean queries.

Boolean Queries

George Boole was a mid-1800s mathematician who attempted to combine properties of logic with mathematics. Known as Boolean logic, this method basically determines the binary properties of features (such as true-false, yes-no, or presence-absence). That is, rather than using numbers and magnitudes of features, as in the relational database approach, Boolean logic reduces measures to some binary condition. The operation determines if one of two choices exists, which is a fairly simple calculation for a computer.

Boole's methods include comparison between two or more sets of binary properties in a given feature, using *and*, *or*, *and/or*, *not*, and other combining language. For example: "Does feature X have trees [Yes-No] *and* a river [Yes-No]?" GIS is very good at Boolean logic. Figure 7-5 shows examples of testing the relationship between literacy and income as nine districts. In this example, a district is said to have either high or low literacy (HL or LL), and either high or low income (HI or LI).

The larger map in the upper left describes each district's level of literacy and income. The four shaded maps are illustrations of Boolean query results; they show particular conditions or sets of literacy-income properties. The first theme looks at each district and asks the GIS to show all districts that do *not* have high income (NOT HI). Literacy is not considered in this query. The four darkened districts meet the Not High Income condition.

The second theme queries districts having high literacy *or* high income; that is, one or the other (HL OR HI). All but two districts have at least one high of either literacy or income.

The lower-left theme shows districts having high literacy but *not* high income (HL NOT HI), with only two meeting these conditions. This could be stated as HL AND LI, but the user wanted to note the absence of high income (an efficient approach when more than two properties are used). The last theme asks for districts having both high literacy *and* high income (HL AND HI); districts must have both conditions.

Boolean logic queries are very useful in GIS analysis. This concept could be extended to more complicated sets of Boolean properties, such as High-Medium-Low. Some GISs use Boolean logic to simplify the results of overlay operations, a topic explored later in this chapter.

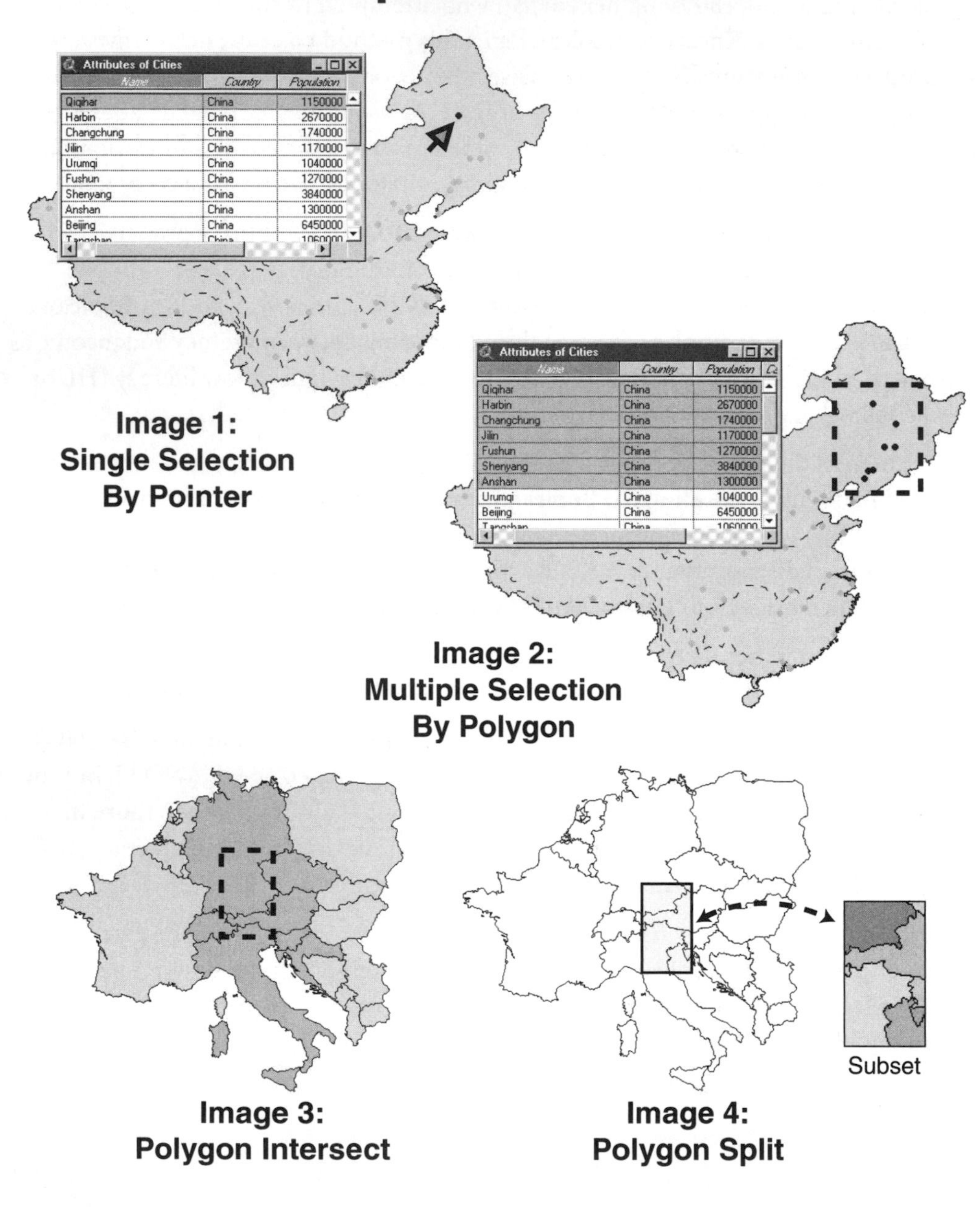

Fig. 7-6: Selection and query of graphic features.

Graphic Selection Query

Selecting features within graphics is convenient and efficient for many purposes. This can be done by using a special cursor (a selection icon) to point to or draw around the needed feature or features. Image 1 of figure 7-6 shows a map of China, with the selection of the city of Quqihar using a single selection pointer (and its record highlighted in the database). Image 2 groups all of the northeastern cities, selected by drawing a polygon around them with the selection pointer (although each could be pointed to individually while holding down the Shift key). The selections are automatically registered in the database (darker records). As noted earlier in this chapter, database selections can then be analyzed; for example, for descriptive statistics.

When selecting line or polygon features, those that intersect (touch) the selection polygon will be selected. The Western Europe map in image 3 shows a relatively small selection box, but all of the intersecting nations are activated, even if only a small part touches (e.g., Switzerland and the Czech Republic).

Some powerful GISs (particularly those using topology) have options for selecting only parts of features. For example, image 4 is the same selection box but the nations have been split rather than grouped by intersection. The database will report the same attributes from the original database, but the areas will be recalculated. They can be saved as an individual theme (shown in the associated inset map). This type of operation is difficult and requires a sophisticated GIS.

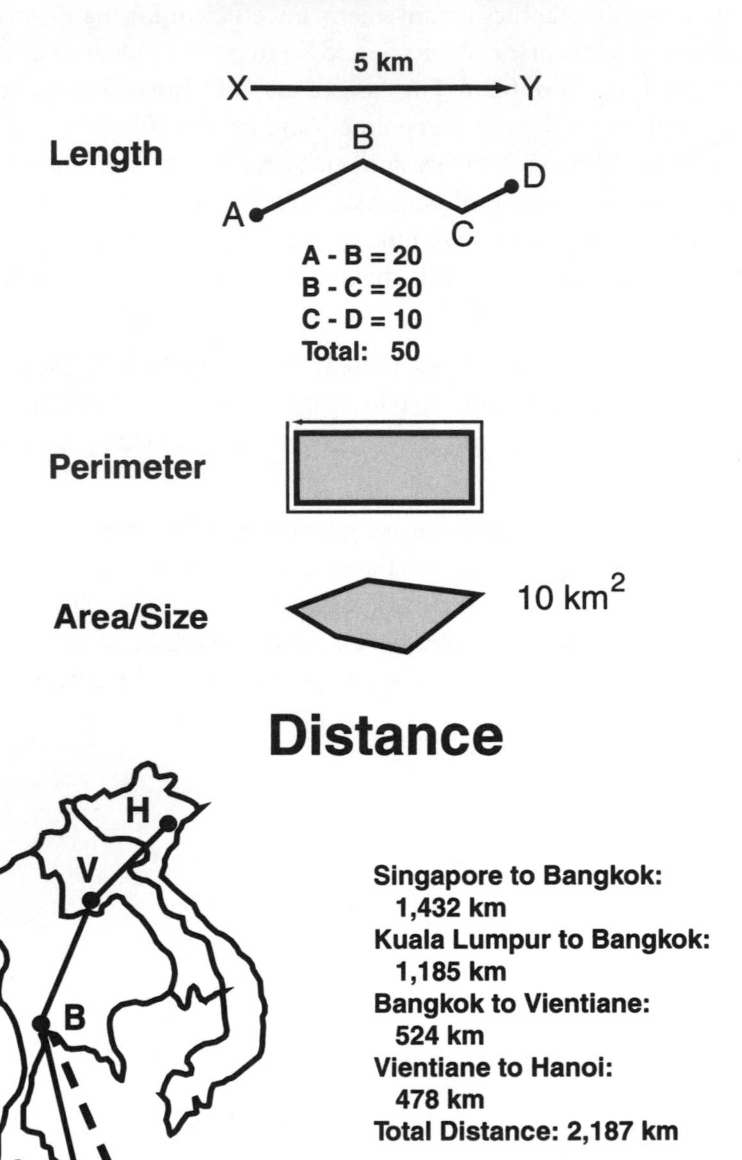

Fig. 7-7: Types of measurement.

Measurement

The sections that follow discuss types of measurement, the concept of distance, and GIS reports. All of these constitute ways of "measuring" GIS data.

Types of Measurement

Measurement applications are common and very useful in GIS. Most GISs have measurement tools, such as an icon that can be clicked to activate a measuring pointer, which is used between locations to determine length and distance. A single line is simply the length between the start and end points, as shown in figure 7-7. Complex lines measure the length between points and give the accumulated total length (identical to topology that measures chains in a line or polygon feature).

Some GISs offer sophisticated spatial analysis that can calculate distances between many points automatically, rather than depending on individual manual operations. Chapter 4 showed the example of determining the shortest route between cities, which requires the user to designate the cities (two or more) either by pointing, typing, or database selection. The program then rapidly performs all possible routes to connect the cities, presenting the shortest one in the end, along with distance between points.

In building topology, chains must be connected to complete line and polygon features. Because chain lengths and polygon perimeters and areas are calculated in the process, they can be included in the database automatically as feature attributes. Raster systems compute length, perimeter, and area in terms of the number of cells, and then convert to real-world figures.

Distance Application

The image at the bottom of figure 7-7 shows a simple application of city distances in Southeast Asia (map letters correspond to cities in the text). City-to-city distances were requested and a total trip length is displayed. Because databases are powerful and flexible, the important cost-benefit items for each type of aircraft can be entered and analyzed, such as fixed cost per distance, varying landing and ground charges at each city, passenger capacity, and so on.

Computers are very good at playing "What-If" games to test various air travel situations. Different passenger (income) scenarios can be tried in order to estimate cost-benefit trips, such as projected number of passengers boarding and exiting at each city. Perhaps a Singapore-to-Hanoi route may be more cost efficient with only a stop at Bangkok, thereby excluding Kuala Lumpur and Vientiane. GIS can be very useful in these types of applications.

Reports

Western Europe

CNTRY_NAME	POP_CNTRY	SQKM_CNTRY	CURR_TYPE	LANDLOCKED
Albania	3,416,945.00	28,754.50	Lek	N
Austria	7,755,406.00	83,738.85	Schilling	Y
Belgium	10,032,460.00	30,479.61	Franc	N
Bosnia and Herzegovina	2,656,240.00	51,403.38		N
Czech Republic	10,321,120.00	78,495.16	Koruna	Y
France	57,757,060.00	546,728.88	Franc	N
Germany	81,436,300.00	356,108.81	Mark	N
Croatia	5,004,112.00	56,287.79	Kuna	N
Hungary	10,310,410.00	92,782.20	Forint	Y
Italy	57,908,880.00	300,979.50	Lira	N
Slovakia	5,374,362.00	48,648.31	Koruna	Y
Liechtenstein	29,342.00	164.80	Franc	Y
Luxembourg	387,064.00	2,594.12	Luxembourg Franc	Y
Macedonia	2,104,035.00	25,321.29	Denar	Y
Monaco	27,409.00	11.99	Franc	N
Montenegro	635,442.00	13,743.05		N
Netherlands	15,447,470.00	35,492.69	Guilder	N
Poland	37,911,870.00	310,715.09	Zloty	N
Slovenia	1,951,443.00	20,245.69	Tolar	N
San Marino	23,758.00	63.17	Lira	Y
Serbia	9,979,116.00	88,201.76		Y
Switzerland	6,713,839.00	41,178.40	Franc	Y

Image 1: Selected Data

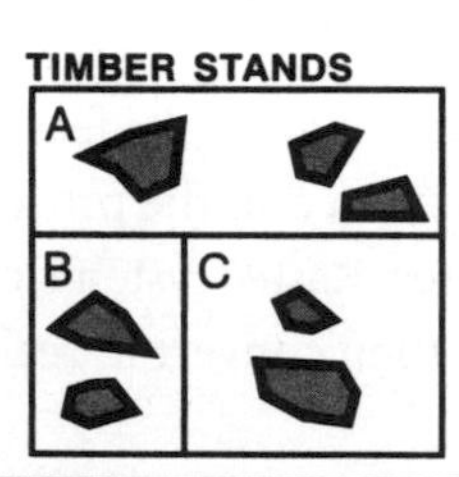

DISTRICT	STAND	STAND AREA (ha)	DISTRICT PERCENT	AREA PERCENT
A	1	40	40	20
	2	30	30	15
	3	30	30	15
B	1	30	60	15
	2	20	40	10
C	1	10	20	5
	2	40	80	20
TOTAL	7	200	--	100

Image 2: Basic Statistics

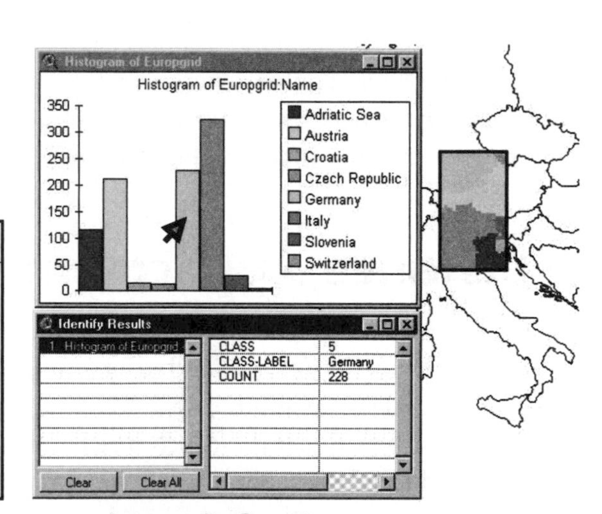

Image 3: Graph

Fig. 7-8: Examples of GIS reports.

Reports

Reports are very useful in GIS inventory operations, and they usually accompany graphic presentations. There are many types of reports that address various aspects of the project, from basic data display to graphs demonstrating summary statistics. Statistics and summary tables produced in database queries were discussed previously. Most GISs use standard reporting options, and some have sophisticated reporting modules.

Image 1 in figure 7-8 shows a conventional report that reproduces selected parts of a Western Europe database. Because databases can be very large and contain many cryptic and confusing fields, a summary report of selected data can be useful.

Generating a descriptive statistics report is not difficult. These can provide measurements, including total number of all or selected features on a theme. The statistics in the image 2 report give the area of each timber stand from the map above it, its percentage area in the district, and its percentage in the region. Each data field is useful for various timber management activities. This type of report can provide the beginning of spatial analysis for understanding spatial relationships between timber stands and districts.

Image 3 shows a graph of the selected area of Western Europe used earlier. The map has been gridded into a raster format (visible at close viewing). The graph presents the number of cells in each nation. Of course, the total area can be presented if the cells are converted to real-world units of measure. The pointer selects the bar for Germany and an individual report is presented below, giving the class number, name, and number of cells (Count).

There are other types of reports, some with high-quality presentation formats for professional output. In-house projects usually need only the basics, but clients prefer visual art as well as content. Reports and graphs are particularly useful in the analytical phases of GIS, such as displaying results and statistics of derived data.

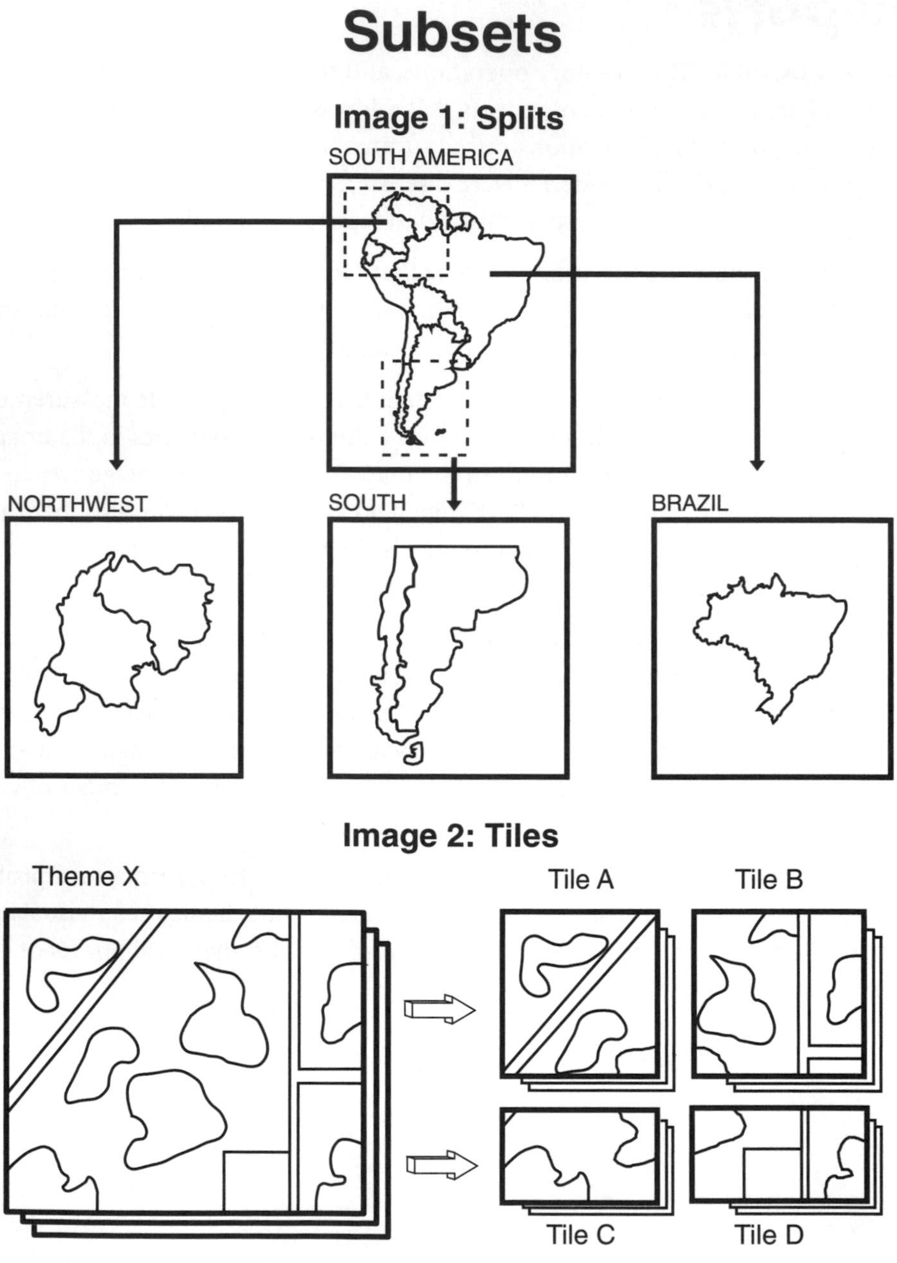

Subsets

Image 1: Splits

SOUTH AMERICA

NORTHWEST

SOUTH

BRAZIL

Image 2: Tiles

Theme X

Tile A

Tile B

Tile C

Tile D

Fig. 7-9: The concepts of themes and tiles.

Theme Modification

Modification of themes can be an important part of preparing for additional inventory operations and the beginning of analysis. Several types of modifications are presented in the following sections, though there are many others that could be used.

Subsets

It is normal to have a large area as the project base map or base theme, but often it may be unnecessary to apply GIS operations to the entire area when only a part is under investigation. All of South America, for example, does not need to be used if the current project deals only with Brazil. Reducing the theme supports time, effort, computer processing, project storage demands, and cost.

Splitting the theme into subsets for a particular task is a common operation, similar to some of the steps already seen in this chapter. Drawing the appropriate lines and instructing a split may be all that is required (though many GISs use more steps). Image 1 in figure 7-9 shows three versions of subsets. On the right, Brazil has been selected, typically just by pointing to it as a single feature and making the GIS operation command. The database is automatically reconstructed for Brazil as a separate theme, and work is performed for a separate Brazil project. If needed, any derived data can be incorporated back into the larger South America project.

On the left, the three northwest countries are selected. Again, these nations are set up as an operational subset and are disconnected from the larger South America theme until reconnected.

The theme at the center shows Argentina and Chile split by the interactive box, a process that can be difficult for the some GISs. That is, it may be a problem to split features and recalculate the records to reflect only the selected area. This is similar to the Western Europe selections previously discussed.

Tiles

Image 2 in figure 7-9 shows the concept of tiles. When a large base area needs to be split into separate areas of responsibility or work, it can be divided into selected major areas or into sub-themes or tiles. Perhaps the entire Theme X is too much work for one operator, so reduction to four sections becomes a more appropriate project resource allocation.

Tiles are considered subsets of a larger area, but they can be treated separately, as individual GIS projects, later to be reunited into the original large area if needed. A good GIS automatically constructs a database for each tile so that work can continue normally.

Spatial Delete
TIMBER STANDS
Image 1: Delete Stands < 10 Ha
ROADS
Image 2: Delete Roads < 5 Km
VILLAGES
X
Image 3: Delete Villages Within 5 Km of Point X

Fig. 7-10: Examples of spatial deletes.

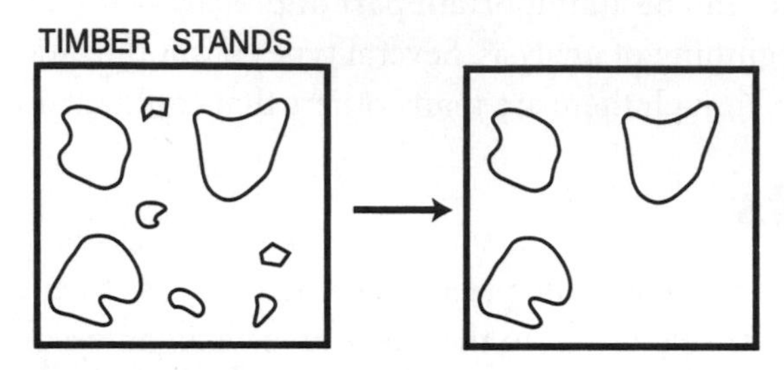

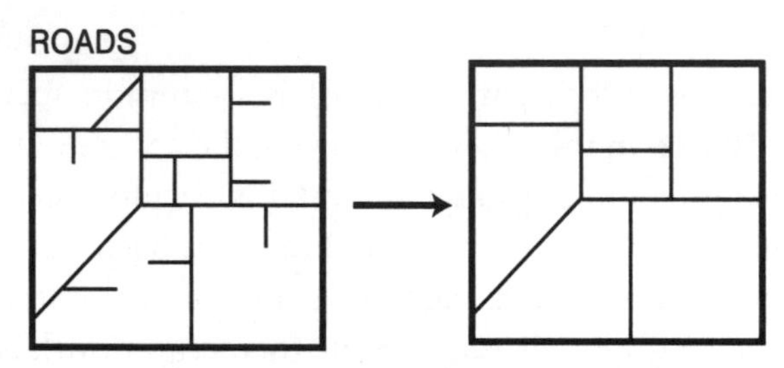

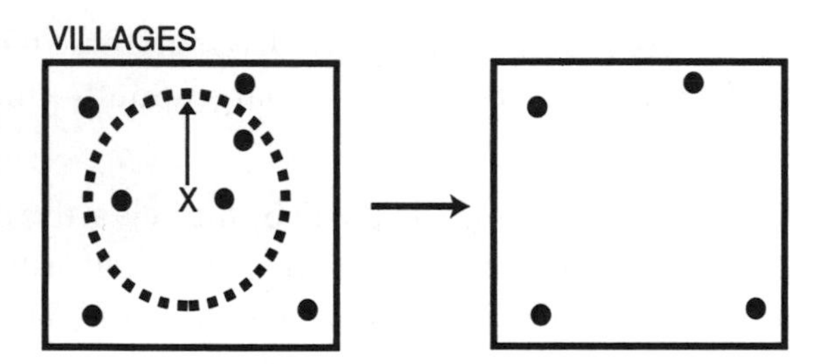

Spatial Deletes

Deleting unimportant or confusing features can be very useful, especially for projects that have many themes and features. One of the basic operations was noted at the beginning of this chapter, concerning turning themes on and off simply by clicking. *Spatial deletes* are procedures for designating certain types or measures of features no longer needed for immediate work. (When making changes, the original data should never be modified by the delete or other operations. Changes are saved under a new data file name.)

Three spatial deletes are shown in figure 7-10. Image 1 keeps only the large timber stands. Those less than 10 hectares are too small for the management process under study. The command to erase them may be simply stated as "Delete stands <10 ha." This can be done directly as a command in some GISs, or by making a query to select the small stands and then deleting them. Of course, they can also be selected visually and deleted.

Image 2 is similar: deleting roads that are less than 5 kilometers in length. Some GISs allow point-and-click selection for deleting. In a Delete mode, the user points to each offending road and clicks. Again, a common method is to make a query for selecting the short roads and then delete them.

Image 3 shows the use of a designated area around a selected point, typically called a *buffer zone.* A 5-kilometer circle is made around the X position and all villages within that area are deleted from consideration, leaving only those outside. The process is simple: point to the center spot, ask for a circular 5-kilometer zone, and then give the command to delete the villages within that zone. Buffers are discussed in Chapter 8.

Dissolve and Merge

Image 1: Dissolve/Merge Process

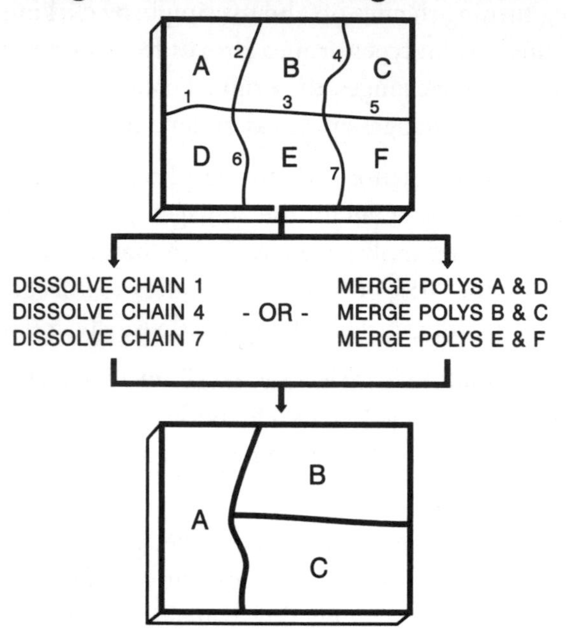

Image 2: Political Redistricting

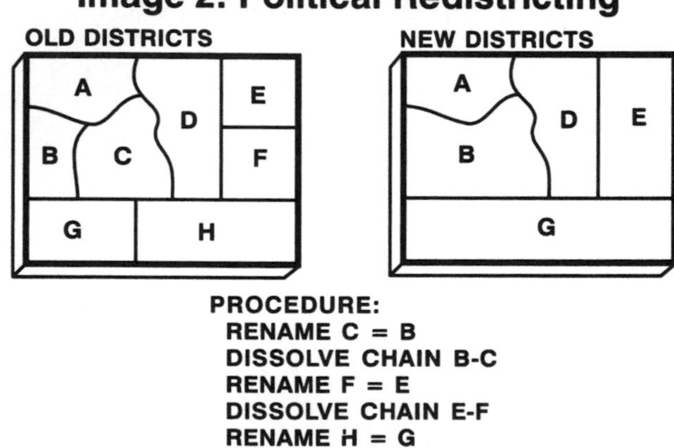

Fig. 7-11: The concepts of dissolve and merge.

Dissolve and Merge

Often in GIS there is a need to merge or combine features, such as small districts into larger provinces. Image 1 of figure 7-11 shows the process of reducing six districts into three new ones, using two similar methods. The first step is to dissolve specific borders. Districts A and D are combined by dissolving chain 1, B and C are joined by dissolving chain 4, and chain 7 is dissolved to combine E and F.

A more direct method is to instruct the GIS to integrate districts by merging selected polygons. This may involve no more than pointing to each district and giving the command to merge (which, in effect, dissolves the borders). A and D, B and C, and E and F are merged, as shown in the A-B-C map. As expected, a good GIS will reconstruct an appropriate database for each new district.

A political redistricting application is shown in image 2, in which existing districts are redesigned into a new set of districts. This is often needed because of shifting populations and other considerations. Follow the steps, which are similar to the procedures discussed for image 1.

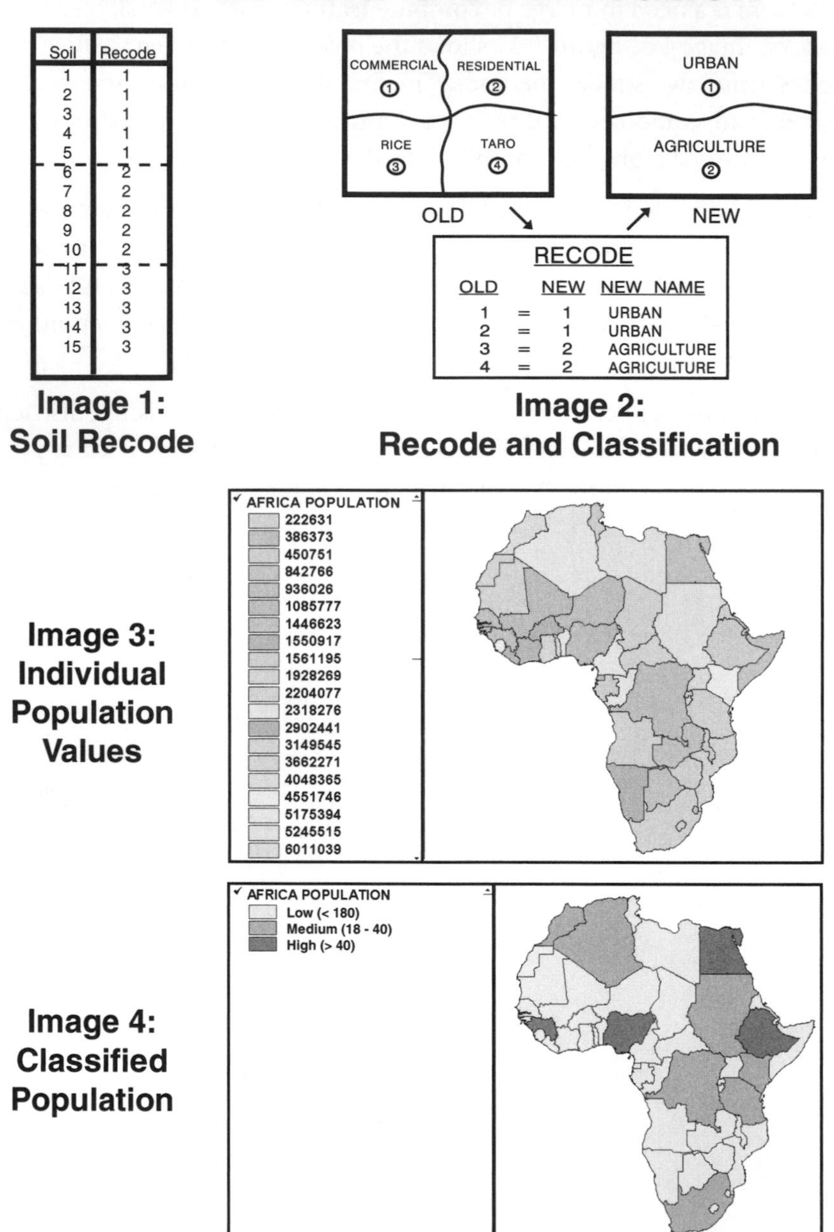

Fig. 7-12: Recoding and reclassifying African population.

Recoding and Reclassification

Field data typically takes the form of individual observations with specific identifications, such as plant names. Classification of data is important in most GIS applications; thematic mapping or other presentations of analysis are typically by class rather than individual record values. Classification is easily achieved by *recoding* data into groups or classes. This section discusses a few basics of classification, and Chapter 8 explores how classification is used in analysis.

In principle, recoding is a simple step: changing attribute numbers or names. Image 1 in figure 7-12 shows how 15 detailed soil types are organized into three general categories. Codes 1 through 5 are recoded to 1, codes 6 and 7 are changed to 2, and 11 through 15 are changed to 3. These could also be letters (e.g., Soils A to O reclassed to A-C). Now a three-class soil theme exists that may be less confusing than a 15-class theme, as well as more useful for some purposes. Whether working with a vector or raster system, changing numbers or names is rather easy to accomplish.

Image 2 shows an old theme of four land-use classes, which are generalized and classified into only two in the new theme. The recode operation is shown, which sometimes may be no more complicated at the keyboard than this demonstration. Both maps serve particular purposes. Remember that the old data still exist under the original names, and the new theme, created from the same data, has a new, unique name.

The two maps of Africa illustrate the value of reclassification. Image 3 has each nation's population displayed as a unique value and tone (color in the original, though there are too many for the eye to organize). The legend at left is not useful because it is merely a disorganized series of large numbers (a list too long to show in one view). Such a map does not serve a useful purpose outside of providing a record that is stored more efficiently in a database.

The image 4 map is more useful, showing three classifications of the populations. Patterns and relationships may be perceived and the numbers are reasonably understandable. Obviously, this is a much better inventory presentation than the image 3 map.

Recoding and classification are fundamental procedures in GIS. They are very useful preparation for many analytical operations. The next chapter surveys some of the basic analytical methods of GIS.

CHAPTER 8

BASIC ANALYSIS

Introduction

After inventory operations, the next step in using GIS data is to perform basic analytical operations. There is no sharp division between inventory and basic analysis because both yield a variety of data and information. Several fundamental GIS procedures that set up many analytical operations (notably recode, overlay, and buffers) are presented in this chapter. A sampling of other analytical operations is offered, but complete coverage of the many operations is beyond the scope of this book.

Database Recode
Data Generalization

Image 1: Generalization of Soils

Soils Database

ATTRIBUTE CODE	SOIL TYPE	NEW SOIL RECODE
1	A1Z	A
2	A2Z	A
3	A3Z	A
4	A3Y	A
5	B1Y	B
6	B2Y	B
7	C6H	C
8	C6J	C
9	C7J	C

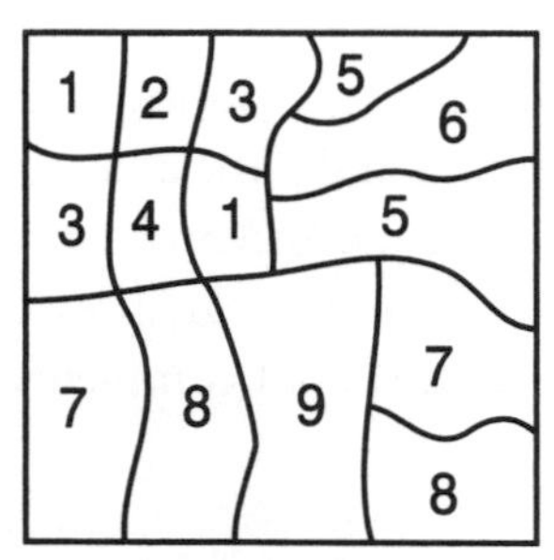

Map 1: Original Soil Map

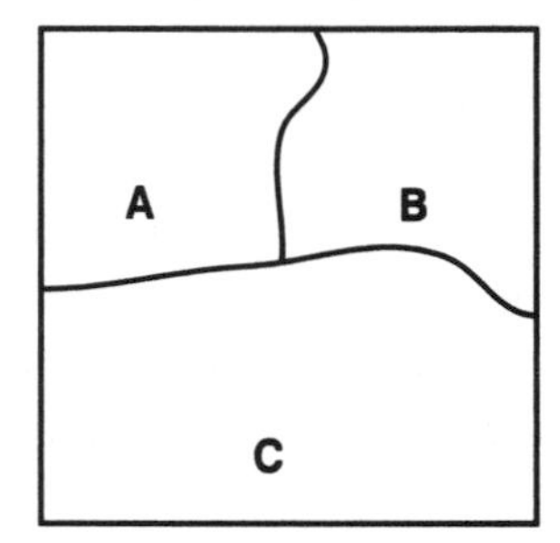

Map 2: Recoded Soils

Image 2: Recode for Agriculture

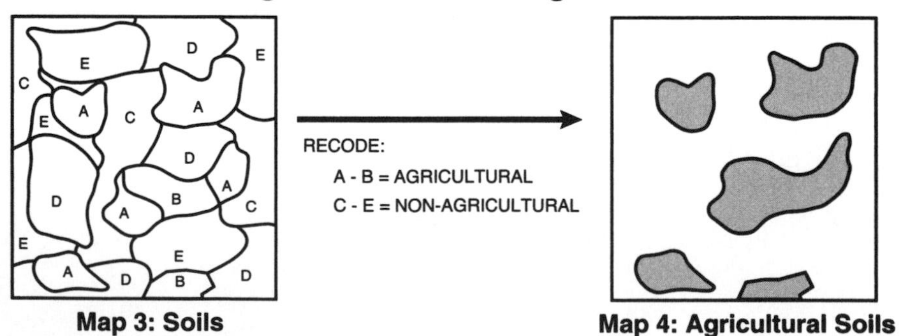

Map 3: Soils

Map 4: Agricultural Soils

Fig. 8-1: Recoding soils for spatial generalization.

Database Recode

Noted in the previous chapter is the usefulness of recoding and classifying data, both for general information and as a device for geographic analysis. This section deals with the recode operation as a basic analysis tool.

Recoding and Analysis

A theme with many features and classes of data may have a good database, but its order (or organization) may not be apparent. By reducing the large number of classes into a manageable list (and rearranging into a logical sequence), you can gain a better understanding of the data, perceive new relationships, and obtain new information. Generalization of data is also known as *data reduction.*

The soil theme in image 1 of figure 8-1 has nine classes scattered over an area (map 1). The distribution (or patterns, if they exist) are difficult to detect. In the database, however, the soil types are named with a major first letter (A, B, C), which indicates a particular soil group. By recoding the nine categories to three according to the first letter only, perhaps a better view of the major soils can be seen. Generalization effectively simplifies the theme database and map.

The Recoded Soils theme (map 2) shows an easily understood spatial distribution. The major soil types occur together as groups, and their spatial relationships are evident after generalizing the data.

Image 2 is a recode showing a similar reduction of soil data from original field classifications (map 3) to a generalized agricultural soil map (map 4). Classes A and B were interpreted as suitable for agriculture, whereas classes C and D are unsuitable. Reducing spatially complex themes to generalized patterns is a standard and very useful recode application.

Recoding is a common preparatory step for many GIS operations and one of the most used and useful GIS procedures. The steps involved in these tasks are rather simple, but the conclusions can be highly valuable.

Theme Display Recoding

Map 1:
Selected Themes
Nations
Cities
Lakes
Ecosystems

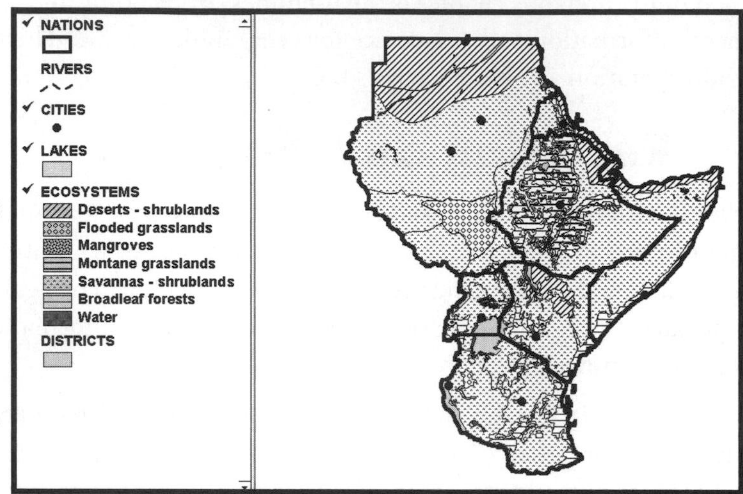

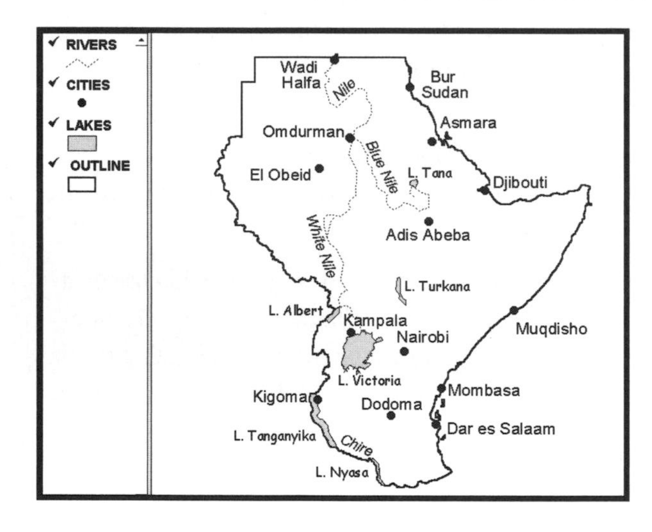

Map 2:
Selected Themes
Rivers
Cities
Lakes

Map 3:
Selected Features
Districts within 250 km
of Adis Abeba

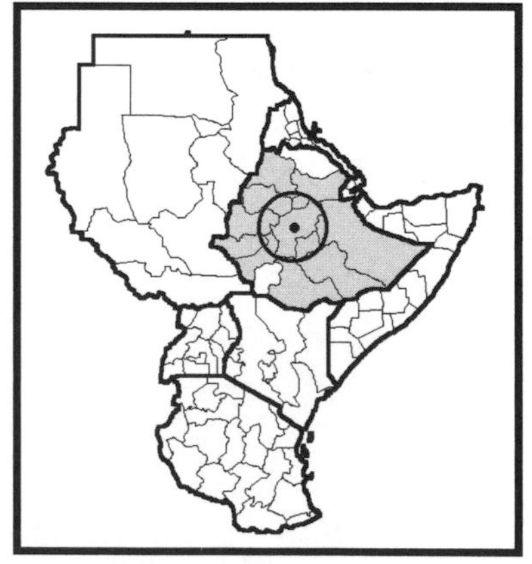

Fig. 8-2: Examples of theme selection and recoding.

Theme Display and Recode

Reducing a complex project display to a more simple one is an easy operation, using either display options or database operations. Basically, recoding the display is effective generalization, and is an initial step in spatial analysis. A common function of modern GIS display is the ability to turn themes and features on and off (ArcView uses a check box beside a theme's name in the legend frame). Because it is so easy to view or hide features, this is a quick method of exploring spatial distributions and relationships. Where are the cities in relation to the rivers and lakes? Simply turn on those features and do a bit of "eyeball analysis."

Customized views take very little effort, letting the user concentrate on the applications rather than operations. Figure 8-2 shows three versions of an East Africa project. There are too many themes in the legend frame to present in one display (although a large color map may work satisfactorily). Map 1 shows selected themes of nations, cities, lakes, and ecosystems, presenting initial relationships between the three (difficult to interpret at this small scale). Map 2 displays the rivers, cities, and lakes (note that only the themes used are displayed in the legend on this map). Each map was constructed in only a few minutes, which is basically instant map recoding (certainly impossible with manual cartography).

Sometimes it is desirable to reduce a theme to selected features, recoding the theme itself into sub-themes. For example, you might present only the highest population districts or those that exist in a certain area. Map 3 shows districts within 250 kilometers of Adis Abeba (Addis Ababa), Ethiopia (those that intersect a 250-kilometer circle around the city). They are recoded to a different value and darker display shade. These districts can be saved as a separate theme for subsequent analysis.

Overlay

Image 1: Visual Overlay

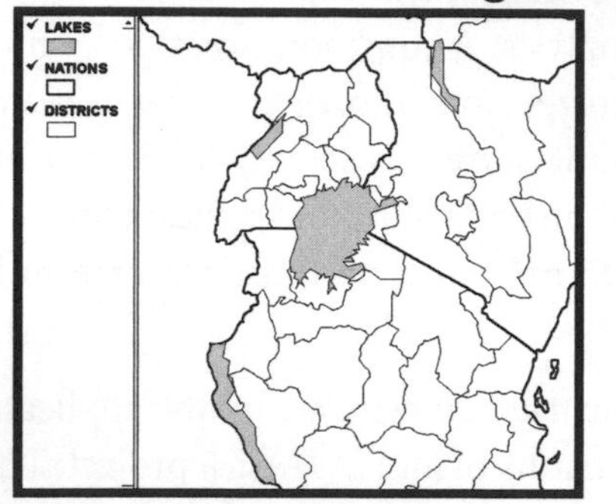

Lakes on Districts

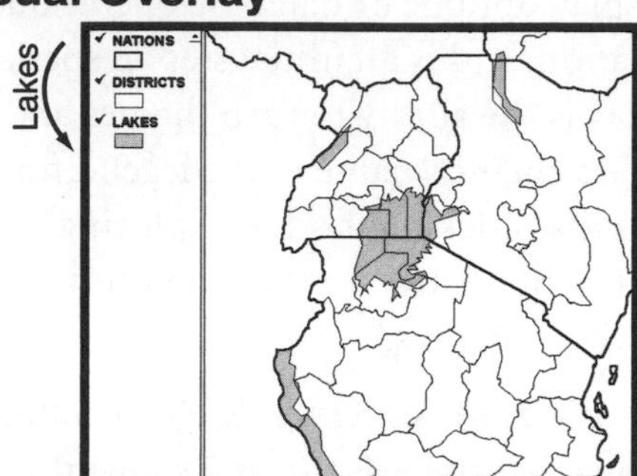

Political Divisions

Image 2: Data Overlay

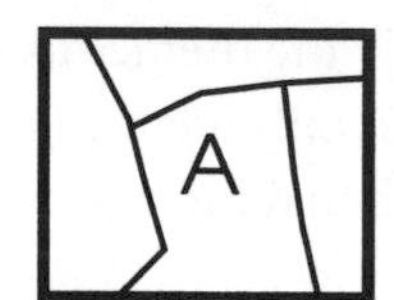

+ B =

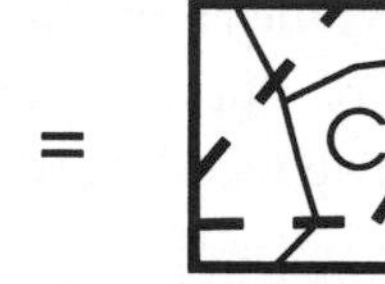

Image 3: Soil and Vegetation Overlay

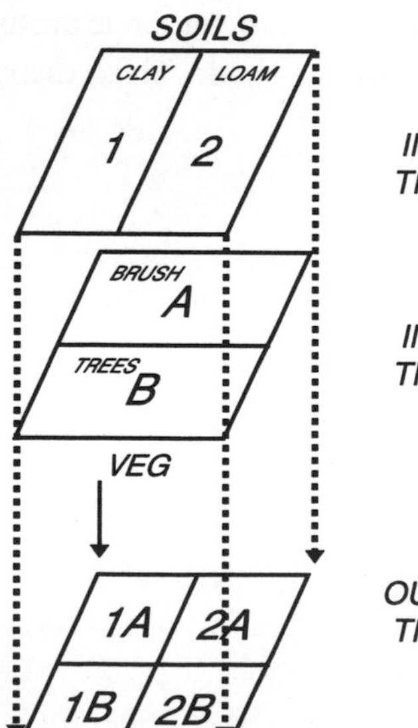

INPUT THEME 1

+

INPUT THEME 2

↓

OUTPUT THEME

SOILS

POLYGON	NAME	SIZE (Ha)
1	CLAY	2
2	LOAM	2

VEG

POLYGON	NAME	SIZE (Ha)
A	BRUSH	2
B	TREES	2

POLYGON	SOIL	VEG	SIZE (Ha)
1A	CLAY	BRUSH	1
1B	CLAY	TREES	1
2A	LOAM	BRUSH	1
2B	LOAM	TREES	1

Fig. 8-3: Examples of visual and data overlay.

Overlay

The sections that follow discuss overlay, a major GIS operation for combining features and themes. Examined are visual overlay, data merging overlay, overlay principles, intersect and union operations, overlay options, clip operations, the mask-and-replace technique, database merging, and appending themes.

Visual Overlay

The concept of GIS overlay is fairly simple: combining two or more theme files, usually in preparation for further analysis. There are two basic methods of overlay, both of which accomplish the same results. The first is visual, as demonstrated by turning themes on and off at the display to overlay them visually.

The position of the theme in the legend frame determines the order of display in ArcView. That is, the theme list at the top of the legend will be the one on top visually. Changing the order of themes in the legend list (by clicking and dragging the name) will change the visual placement. For example, image 1 of figure 8-3 shows two magnified East Africa views with different theme organizations. Map 1 shows Lake Victoria overlaying the district and state boundaries. Image 2 has the administrative boundaries over the lake in order to show complete political divisions, even on water. It was made simply by moving the lakes theme from the top to the bottom of the legend list.

Visual overlays are advantageous in many ways, but they are not permanent (unless made into a map). Analysis is largely by eye and therefore limited in capabilities. Making derived data depends on true merging of data.

Data Merging Overlay

There are many needs in GIS for combining several themes into a single one for permanent use. By merging the data rather than just visually overlaying, new themes with derived data are constructed. As image 2 shows, the basic process is to add theme A to theme B in order to make theme C. All line features are now in the same theme and database.

Image 3 shows the basic approach. An overlay of Soils (Input Theme 1) and Vegetation (Input Theme 2) displays the relationship between the two themes (Output Theme). Soil polygons 1 and 2 (Clay and Loam) are combined with vegetation polygons A and B (Brush and Trees). The input theme databases identify each polygon and give their corresponding attribute name and size, whereas the output theme database presents each combination of soils and vegetation. Polygon sizes have also been recalculated. Some GISs offer these products automatically in the overlay operation, but low-end ones require the operator to do some of the database work.

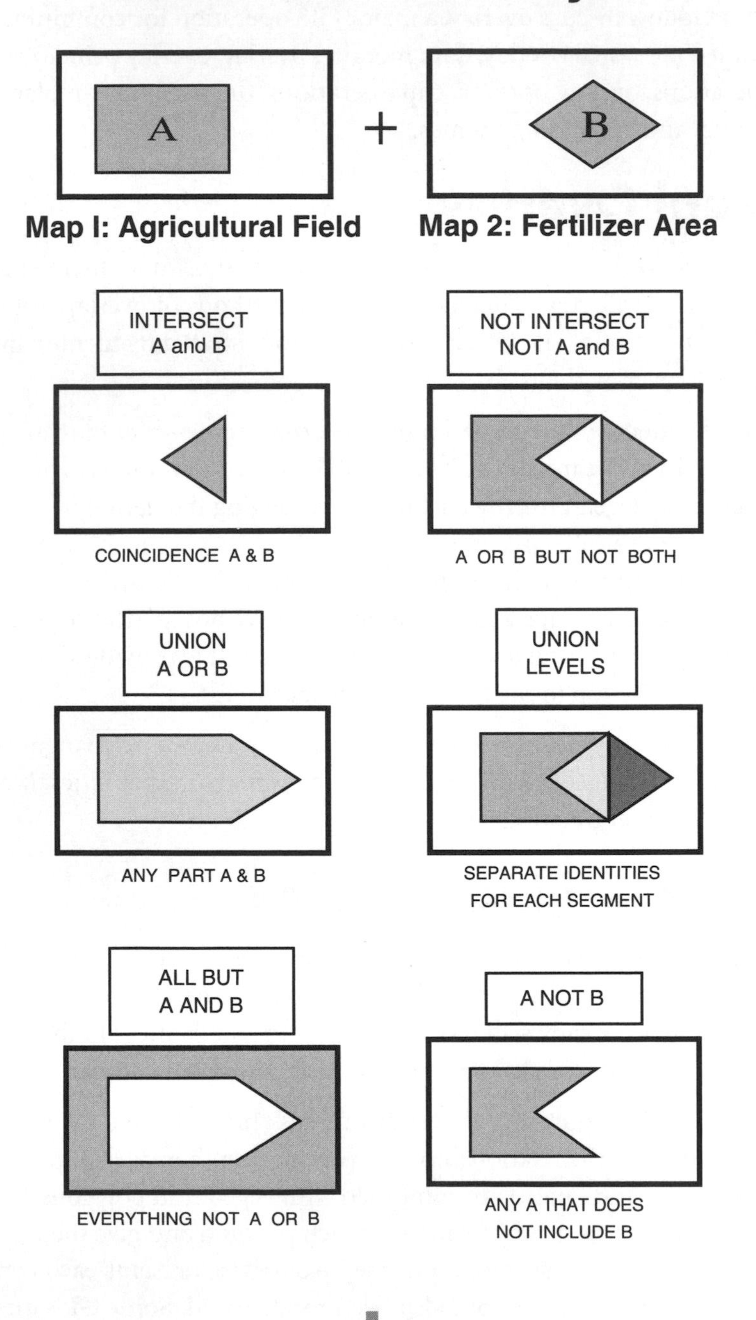

Fig. 8-4: Examples of Boolean overlays.

Overlay Principles

Overlay is a powerful GIS capability, yet today's sophisticated systems are making the term almost obsolete. Whereas a few years ago the GIS user had to actually place one theme or set of features on another to begin analysis of relationships, many of today's data-merging steps have been incorporated into single-command operations.

In modern GIS, we do not necessarily think "overlay" to integrate themes; rather, we consider more specific objectives, such as point-in-polygon association. Thus, although the overlay concept is still valid, it is used more in a general or generic sense, leaving the actual operations to be more descriptive and specific to the particular need of the moment.

There are various spatial relationship overlays that can be performed. Figure 8-4 shows a model of the basic overlay relationships, sometimes termed Boolean overlays because of the logic used. The application concerns partial spraying of fertilizer on an agricultural area from an airplane. At top, map 1 shows feature A (the agricultural field), and map 2 is the fertilizer area (feature B). The diamond-shaped fertilizer area covers only part of the agricultural field. The Boolean overlay options include the following:

- *Intersect, A and B:* The common area of both features, where the fertilizer contacts the crops. In some cases it is important to know where features in theme A overlap those in theme B. This identifies which areas of the crops received the fertilizer and which did not.
- *Not Intersect, A or B:* A or B, but not both. This includes where two features exist, but does not include where they overlap, the opposite of intersect. That is, all other agriculture and fertilizer, except where they occur together.
- *Union, A or B:* Combination of A and B; the full extent of both features.
- *Union Levels:* A, B, and their intersect. These are separate identities for each segment of the union. The application and effect of fertilizer on each part is different; therefore, each section should be identified.
- *All But A or B:* Everywhere except the union properties. The opposite of union.
- *A Not B:* Any part of A that does not include B. All crops that have not received fertilizer. It could show the reverse: any part of B that does not include A.

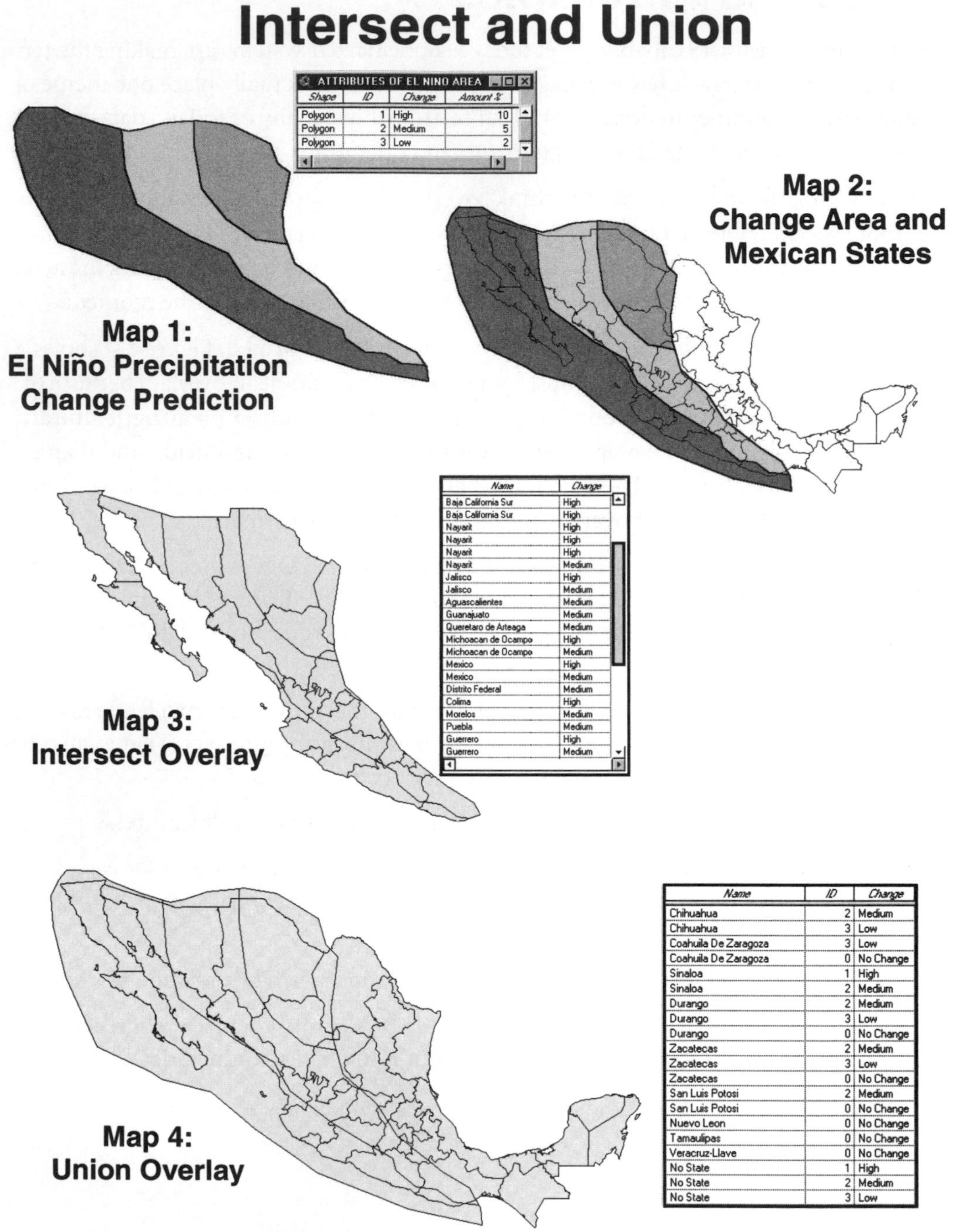

Intersect and Union

ATTRIBUTES OF EL NINO AREA

Shape	ID	Change	Amount %
Polygon	1	High	10
Polygon	2	Medium	5
Polygon	3	Low	2

Map 1:
El Niño Precipitation
Change Prediction

Map 2:
Change Area and
Mexican States

Map 3:
Intersect Overlay

Name	Change
Baja California Sur	High
Baja California Sur	High
Nayarit	High
Nayarit	High
Nayarit	High
Nayarit	Medium
Jalisco	High
Jalisco	Medium
Aguascalientes	Medium
Guanajuato	Medium
Queretaro de Arteaga	Medium
Michoacan de Ocampo	High
Michoacan de Ocampo	Medium
Mexico	High
Mexico	Medium
Distrito Federal	Medium
Colima	High
Morelos	Medium
Puebla	Medium
Guerrero	High
Guerrero	Medium

Map 4:
Union Overlay

Name	ID	Change
Chihuahua	2	Medium
Chihuahua	3	Low
Coahuila De Zaragoza	3	Low
Coahuila De Zaragoza	0	No Change
Sinaloa	1	High
Sinaloa	2	Medium
Durango	2	Medium
Durango	3	Low
Durango	0	No Change
Zacatecas	2	Medium
Zacatecas	3	Low
Zacatecas	0	No Change
San Luis Potosi	2	Medium
San Luis Potosi	0	No Change
Nuevo Leon	0	No Change
Tamaulipas	0	No Change
Veracruz-Llave	0	No Change
No State	1	High
No State	2	Medium
No State	3	Low

Fig. 8-5: The intersect and union overlay operations.

Intersect and Union

Two of the most useful overlays are the intersect and union operations. Intersect merges only the parts that share common space (where the two themes overlap), and union combines all of the features involved. Map 1 of figure 8-5 shows a hypothetical predicted El Niño precipitation change area in the region of Mexico. The expected changes are noted as High (10% decrease), Medium (5% decrease), and Low (2% decrease).

Map 2 shows how the prediction area corresponds to the Mexican states (thinner lines). Map 3 is the intersect overlay, with polygons constructed from the overlap, containing both Mexican and El Niño area attributes (as seen on the database). There are 22 states affected, but 43 polygons are constructed in the intersect. The database shows many states having several classes of change (high, medium, low). With this type of information, each state can foretell the potential precipitation changes affecting it.

Union provides the full extent of both input themes, not just the common areas. Map 4 presents the union overlay showing all of both the El Niño area and Mexico. Although it appears similar to map 2, union provides the comprehensive data merging rather than just visual display overlay. Because there are 69 polygons in the overlay, only the last 20 are displayed in the database here. Note that the ID numbers correspond to the change attribute, including 0 for no change. The three bottom records (No State) are the El Niño areas not occurring over Mexico. This type of overlay helps the national government to evaluate potential impacts of El Niño.

Overlay Options

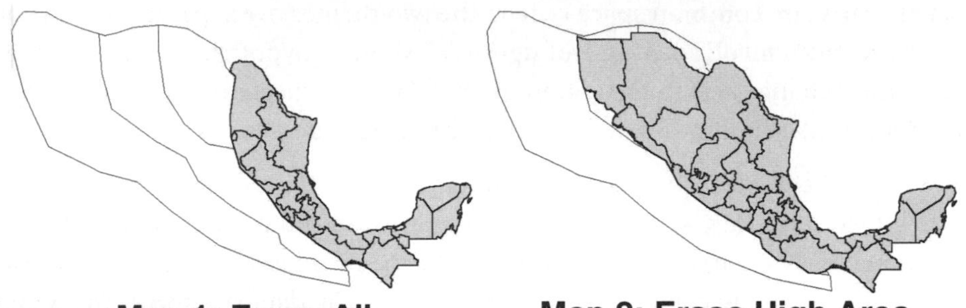

Map 1: Erase All

Map 2: Erase High Area

Map 3:Identity
All States Identified

Map 4: Select States Completely within Area

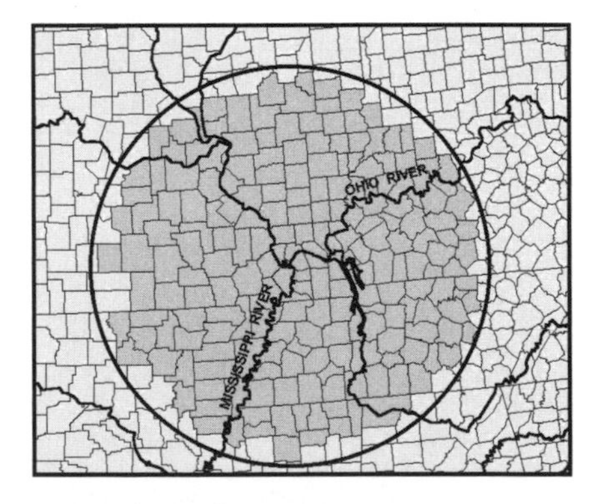

Map 5: Select Counties Completely within Circle

Fig. 8-6: Examples of types of overlay options.

Overlay Options

Other types of overlay and data merge are possible, with various accompanying spatial and attribute conditions. Erase (there are different names with different software) deletes the common area covered by two themes. That is, one theme erases intersecting parts of a second theme. Shown in map 1 of figure 8-6 is the El Niño area erasing all of the underlying Mexican states, an "Erase All" option. The remaining unerased states are shown in darker tones, and the El Niño area is outlined here for reference.

Selected parts of the erasing theme can perform the deletion. Map 2 shows only the High change area making the erase, leaving the Medium and Low area with the unaffected states. Perhaps the planners do not believe the Medium and Low changes are significant and want to know which parts of the country will be disturbed.

The Identity overlay combines both input themes, but the output area is identified by the first input theme. Map 1 displays the overlay, but with Mexico as the identifying and controlling theme over the El Niño area. This shows the entire country and where the El Niño effect will occur within it. Identity is similar to the intersect overlay (the additional unaffected states are included) and similar to the union overlay (but with the areas outside Mexico excluded). As in many GIS operations, these results could be achieved in other ways, but Identity presents options that make the operation easy and give the product specific qualities that usually take several more steps in other types of overlay and recoding.

Selecting complete polygons within an overlay area is another useful option. Maps 1 through 3 show the portions of polygons that spatially match input themes in various ways, but in another situation only those states that occur completely within the El Niño area may be needed. Map 4 presents only the complete states that are within the El Niño area.

Map 5 displays the U.S. counties that exist entirely within the 320-kilometer circle (buffer zone), centered at the junction of the Ohio and Mississippi rivers. Disaster management planning needs to know which counties are in the danger zone. The operation is an overlay of the counties within the circle feature, with the option of including only the counties completely inside the buffer.

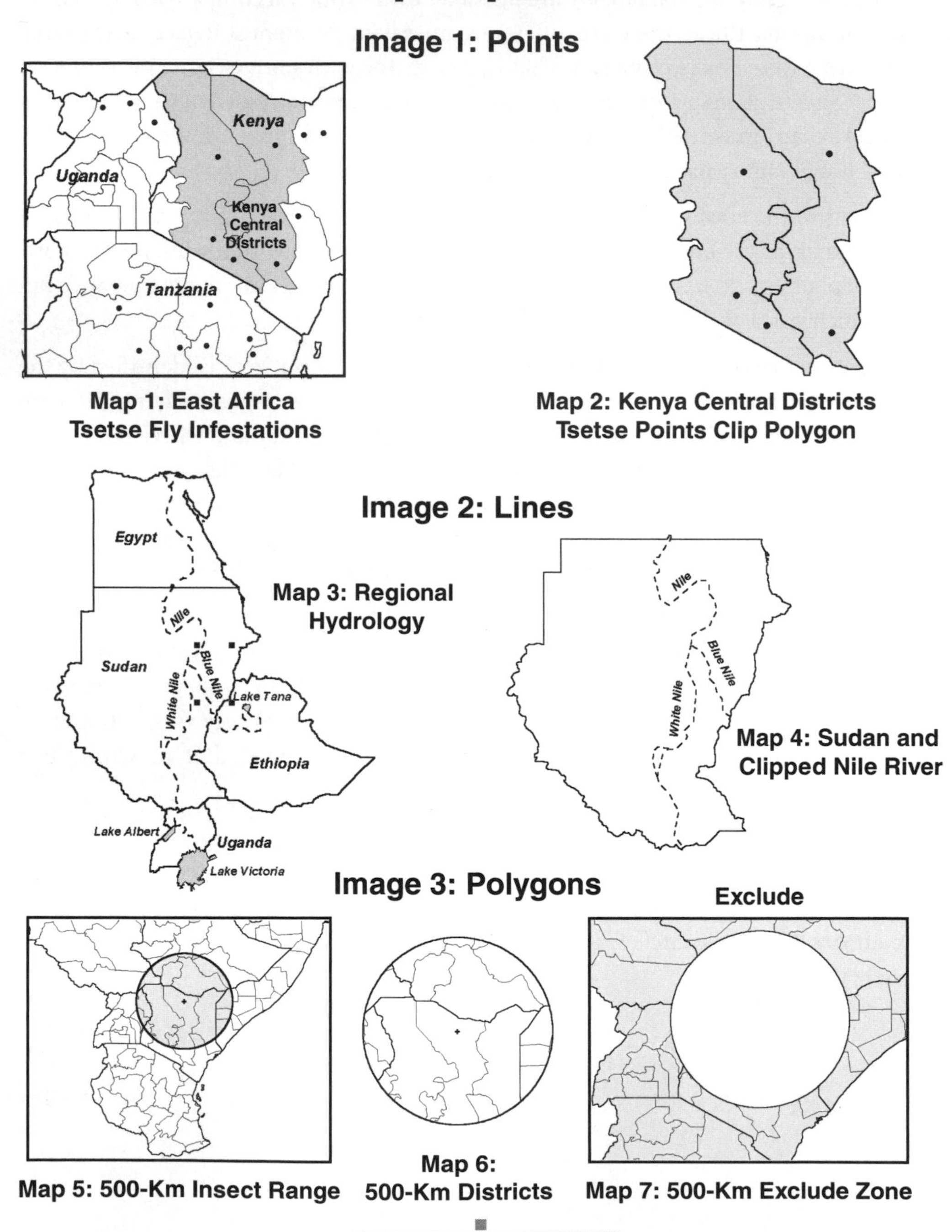

Fig. 8-7: Examples of clip operation options.

Clip Operations

Clip is an option that removes a selected part of one theme using another theme, selected features, or a graphic. In effect, it is an overlay operation that uses one part of a theme to select part of another by extraction (cutting and removal). Clip is particularly useful in that it selects features spatially, rather than by some attribute, because location of features is often the important aspect. Clip also saves the selections as a separate theme. The following examples present clip operations for points, lines, and polygons.

Point Features

Map 1 in figure 8-7 shows tsetse fly infestation points (dots) in East Africa, with Kenya's central districts selected (darker tone). Public health agencies need to know where these events occur in the central districts, and the selected districts are used to clip the tsetse fly infestation theme points, as shown in map 2. This can be an effective method of overlaying features based on location.

Line Features

Polygons can clip line features easily. The Nile River system stretches from Egypt to Uganda and Ethiopia (map 3). By using Sudan as a clipping polygon, only part of the river is selected (map 4). Some GISs permit this type of operation in a single operation, such as selection of Sudan and then clipping of the river.

Polygon Features

A graphic may be drawn for a particular purpose and then used as the clip feature. Shown in map 5 is a cross symbol in northern Kenya, indicating a particular dangerous insect location. The insect's range is 500 kilometers and the regional agencies need to know which districts are endangered. A 500-kilometer circle (buffer zone) is created and used as a clip feature on the districts. Map 6 displays the districts within the circle, a useful presentation for regional planners.

In some GISs, Clip offers Include or Exclude options. These allow the user to include an area and features of a clipped area, as in the illustrations, or to exclude them, leaving the remainder of the theme. Map 7 shows an Exclude clip of the 500-kilometer insect range, basically the opposite of the example previously discussed. Can you think of a good reason for this map in the tsetse situation?

Mask and Replace

Mask

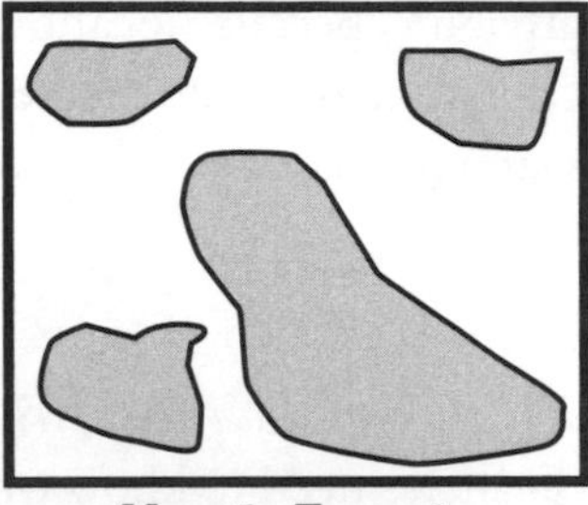

Map 1: Forests

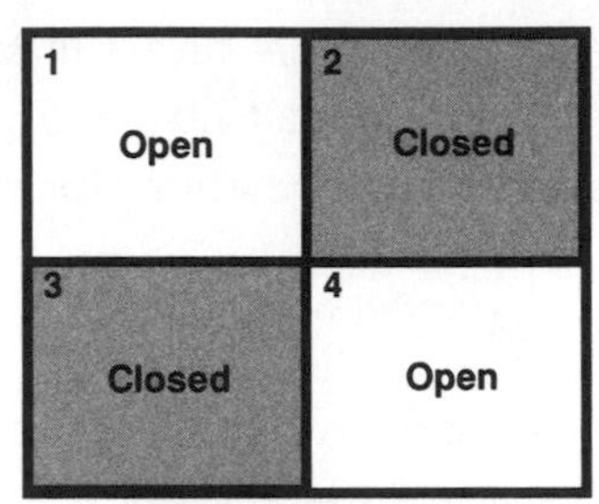

Map 2: Mask Available Areas

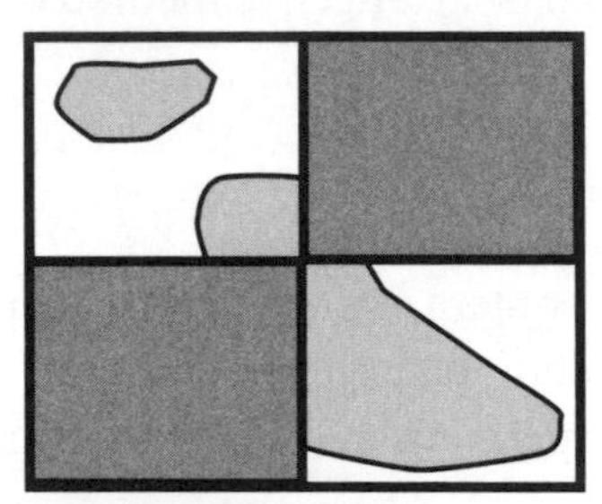

Map 3: Timber Permit Areas

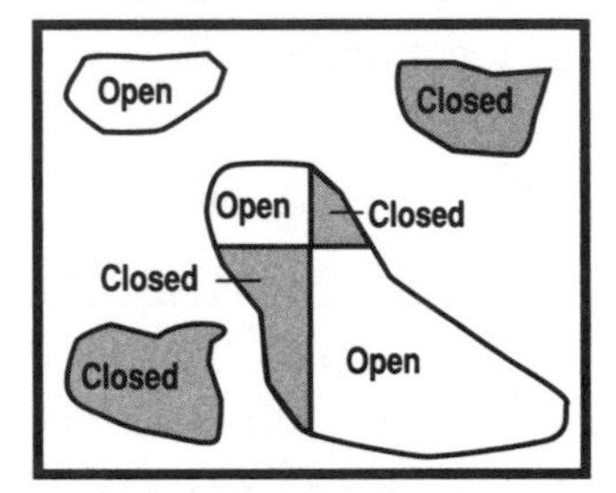

Map 4: Management Areas

Replace

Map 5: Vegetation: 1980 (Kudzu = dark tones)

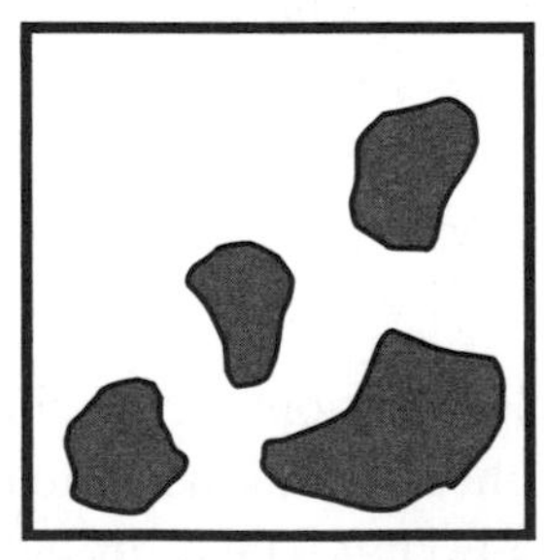

Map 6: Kudzu: 1990

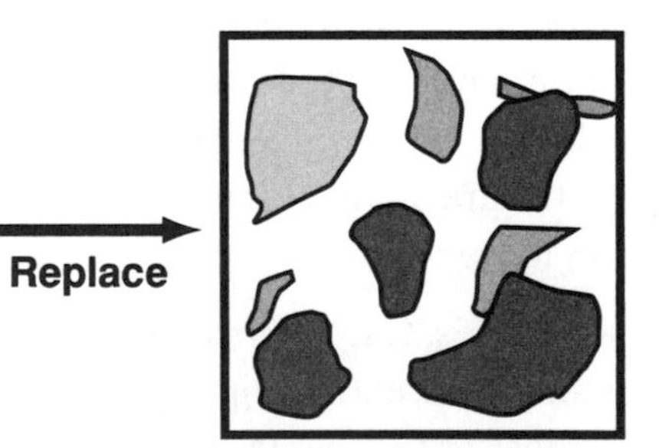

Map 7: Vegetation: 1990

Fig. 8-8: The concepts of mask and replace.

Mask and Replace

Mask is a type of clip operation in which a designated section or set of features from one theme is used as a "window" for selecting parts of a second theme. The window allows visual and data access to the features in it, while blocking out all other parts. For example, the forests in map 1 in figure 8-8 are under consideration for timber management, but sections 2 and 3 are under a protection policy, leaving only sections 1 and 4 available during this time, as seen in map 2. When the mask is applied to the forest area, only those parts in the available quadrants will be open for timber management permits (map 3).

Map 4 shows the timber management plan, with closed and open areas. *Replace* (also called *cover* in some GISs) is another type of clip in some ways, in that it transfers selected features from one theme to another, covering those in the second theme. Replace is ideal for updating features spatially without having to go through elaborate recoding and overlay operations.

The vegetation maps at the bottom show the expansion of kudzu, a particularly fast-growing vine in subtropical parts of the world. In map 5, kudzu is shown in the darkest tones and other, slower-growing vegetation is in light shades. The growth of kudzu was mapped in 1990 (map 6). Because no change in the other vegetation was observed in 1990, the kudzu polygons are replaced onto the 1980 vegetation map, updating the region to 1990 without having to rebuild the entire vegetation features (map 7).

Replace is essentially a convenient selected-feature overlay option. Of course, the replaced features cannot cover the entire map area because that would simply be a copy of itself.

Database Merging

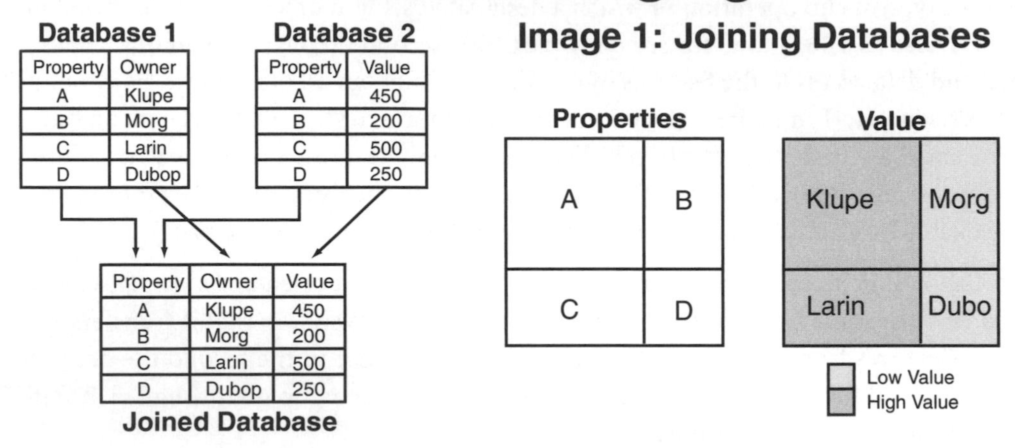

Image 2: East Africa Application

EAST AFRICA TABLE 1

Shape	COUNTRY	POP.	AREA KM2	CURRENCY	LANDLOCKED	POP. DENSITY
Polygon	Djibouti	450751	21637.641	Djibouti Franc	N	21
Polygon	Eritrea	3662271	121940.797	Birr	N	30
Polygon	Ethiopia	53142970	1132328.000	Birr	Y	47
Polygon	Kenya	25835250	584428.688	Schilling	N	44
Polygon	Somalia	9951515	639065.125	Shilling	N	16
Polygon	Sudan	27713420	2490409.000	Pound	N	11
Polygon	Tanzania	28386270	944976.875	Schilling	N	30
Polygon	Uganda	18144360	243049.906	Schilling	Y	75

EAST AFRICA TABLE 2

NATION	LIFE EXPECT.	FERTILITY	NAT. INCR.
Djibouti	48	5.8	2.3
Eritrea	55	6.1	3.0
Ethiopia	46	6.7	2.4
Kenya	49	4.7	2.1
Somalia	46	7.0	2.9
Sudan	51	4.6	2.2
Tanzania	53	5.6	2.9
Uganda	42	6.9	2.9

EAST AFRICA TABLE 3

Shape	COUNTRY	POP.	AREA KM2	CURRENCY	LANDLOCKED	POP. DENSITY	LIFE EXPECT.	FERTILITY	NAT. INCR.
Polygon	Djibouti	450751	21637.641	Djibouti Franc	N	21	48	5.8	2.3
Polygon	Eritrea	3662271	121940.797	Birr	N	30	55	6.1	3.0
Polygon	Ethiopia	53142970	1132328.000	Birr	Y	47	46	6.7	2.4
Polygon	Kenya	25835250	584428.688	Schilling	N	44	49	4.7	2.1
Polygon	Somalia	9951515	639065.125	Shilling	N	16	46	7.0	2.9
Polygon	Sudan	27713420	2490409.000	Pound	N	11	51	4.6	2.2
Polygon	Tanzania	28386270	944976.875	Schilling	N	30	53	5.6	2.9
Polygon	Uganda	18144360	243049.906	Schilling	Y	75	42	6.9	2.9

Fig. 8-9: Joining tables using common fields.

Database Merging

In figures 8-5 and 8-6, overlaying the El Niño area with Mexican states is largely a graphical approach involving two themes that have dissimilar spatial characteristics (different shapes and positions). The output product is a third theme with distinctive spatial features, made from the two input overlay themes. In many GISs, the databases also change according to the type of overlay. Graphics are important in this process because of the spatial diversity. However, when there are no spatial differences between themes, overlaying the graphics will not contribute to the data merging because there is nothing new spatially. Data merging can occur in the databases. This is the database approach to overlay.

The operation of merging databases typically uses a common field to make the join. For example, if both databases have the field "Name" referring to states, they can be joined easily, with the attributes from both databases linking to the record names. Image 1 of figure 8-9 shows two simple databases that have the same records under a common field (Property). However, the databases have different attributes (Owner and Value). Perhaps they came from different sources and now need to be combined.

The two tables are joined by using Property as the common field, making a database that contains the owner and the value of each property. This is a one-to-one merge, matching one record with another. A graphical overlay will produce nothing extra in terms of visual presentation (the spatial extent of properties is identical). However, two different maps can be made, one for property and one for value, and the overlay is evident on the value map because it shows both attributes. Value is displayed as a thematic map, with the lower values in the lighter shade.

Image 2 shows two East Africa databases, each with the names of the nations but with different attributes. Note that the field names for the nations are not the same (country versus nation), but they have the same format (size and data type) and record names, which is usually sufficient for joining them. A join will merge the data, with table 2 going into table 1 (hence, the table 1 country name remains). The lower database presents the merged tables. This is, in effect, an overlay of the two data files, but without the unnecessary graphics. Of course, several map displays can be constructed to present the new data, but they are not part of the overlay process.

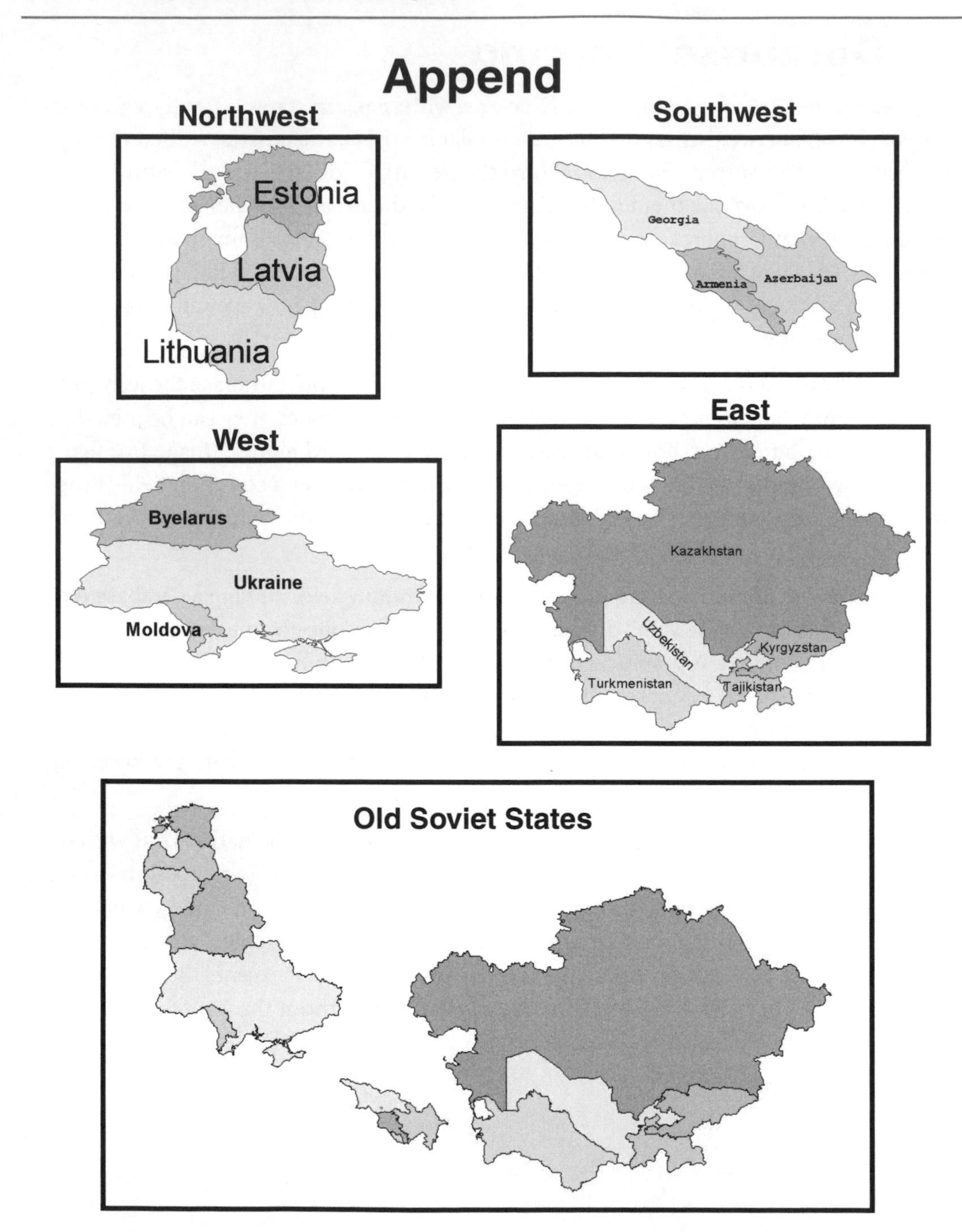

Fig. 8-10: Joining themes using the append operation.

Append Themes

Most of the overlay operations discussed previously emphasize themes that match spatially. That is, they cover the same area. GIS permits joining adjacent or partially overlapping themes. This is called *append*, or sometimes *merge* or *mosaic* (among other possible names from different GISs). In effect, the append operation adds one or more themes or maps to an existing one.

Themes can be perfectly adjacent to each other, separated, or overlapping in odd ways. The two requirements normally are that both be properly georeferenced and that they share at least one common item in the database in order to have a join item. Remember that georeferencing establishes real-world coordinates on a theme. Append is simply a matter of joining features according to their actual world locations. Differences in projections are normally not important.

When overlaying, GIS matches the coordinates, not tics, theme corners, or borders. Therefore, if coordinates are accurate in each theme, the actual shape of the theme is not important to the computer (though it might look strange to the eye). Actually, with proper georeferencing, one theme can be a small area of the other theme (fit inside or partially inside) and Append will join them.

Figure 8-10 shows four separate theme maps of the former Soviet Union non-Russian states. They are different scales and projections from various projects. Appending them into one "Old Soviet States" theme was easy and fast—a matter of a few simple commands. Of course, they fit together very well in a common projection because of their locations in the real world.

Map Algebra
Raster Cell Overlay

Image 1: Overlay Using Add

Input Theme A

1-1 3	1-2 3	1-3 4
2-1 0	2-2 1	2-3 0
3-1 2	3-2 4	3-3 6

+

Input Theme B

1-1 4	1-2 2	1-3 2
2-1 5	2-2 5	2-3 5
3-1 4	3-2 1	3-3 1

=

Output Theme C

7	5	6
5	6	5
6	5	7

Raster Cell 1-1: 3 + 4 = 7
Raster Cell 1-2: 3 + 2 = 5
Raster Cell 1-3: 4 + 2 = 6

Image 2: Map Algebra Application
Kenya Aridity

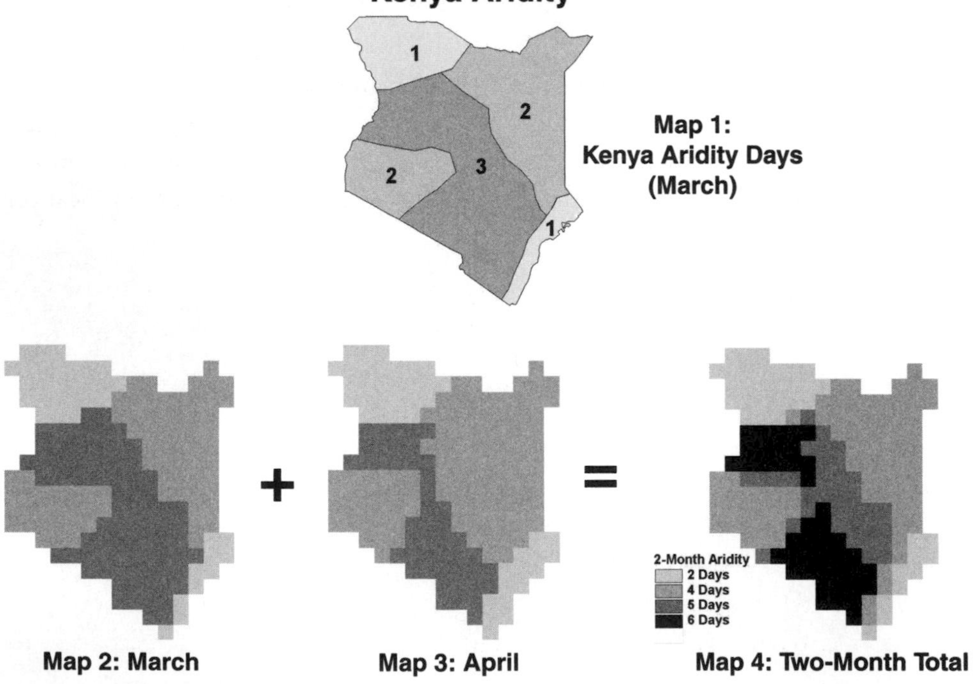

Fig. 8-11: An example of map algebra in an overlay using the add operation.

Map Algebra: Raster Cell Overlay

So far, overlay has been discussed as putting one theme on another or merging databases. However, systems that use coded attribute values, particularly raster structures, may need other types of operations to perform overlays. Recall that the raster system has particular advantages, such as easy modeling and compatibility with remote sensing imagery, but the cell-based GIS can require more management for some functions. Modeling, for example, is sometimes best performed using number codes for cells. Recode and overlays are important parts of the raster analysis process.

Instructions must be given on how to deal with the interaction of the cells in an operation. When cell 1 of theme A is merged with cell 1 of theme B, a mathematical operation is necessary to assign a code for cell 1 of the new output theme C. The use of mathematical operations is called *map algebra*, which usually includes Add, Subtract, Multiply, Divide, Exponentiate, and other operators.

For example, if the map algebra operation is to add, the value of cell 1 of the first theme is added to the value of cell 1 in the second theme and the sum is recorded in cell 1 of the output theme (A + B = C). Each cell is added to its counterparts in the other theme(s). The cells must be spatially registered between themes (have the same cell area). It is common for themes to have the same number of rows and columns of cells, but a smaller part of one theme may operate on part of another theme.

Image 1 of figure 8-11 shows a nine-cell raster view from two themes and their interaction when overlaying with an Add map algebra operation. The small number in the upper left of each cell is the cell number, referred to by row-column sequence, and the large number is the attribute value. These could be measures of precipitation for two months, with an objective of mapping the total rain over that period (Output Theme C). The sum for cell 1-1 area is 7 (presumably millimeters). Study the other cells and see how simple the process is. Raster overlays are very efficient and fast because the program has only the simple task of adding two numbers.

An applied example is presented in the image 2 maps of Kenya. The nation occasionally experiences drought, and good information helps to anticipate and manage conditions. Map 1 is a vector theme that shows "Aridity Days" polygons for March (number of arid days). The themes were gridded, with very large cells sizes (half-degree latitude-longitude) for demonstration purposes, but an actual project would use much better resolution when possible.

Although a two-month total could be easily performed in the vector format, more complicated numeric analysis and modeling are often best in raster. March (map 2) and April (map 3) arid days raster maps are added to make a two-month total (map 4). A full-year calculation would be equally easy in raster, but rather difficult in vector form.

Map Algebra
Multiply and Maximum

Image 1: Overlay Using Multiply

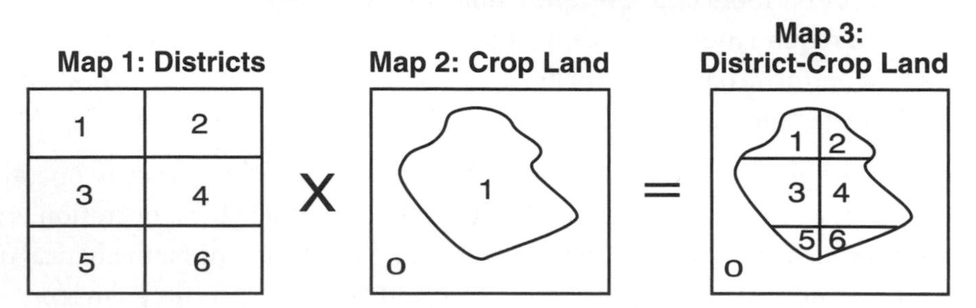

Image 2: Overlay Using Maximum

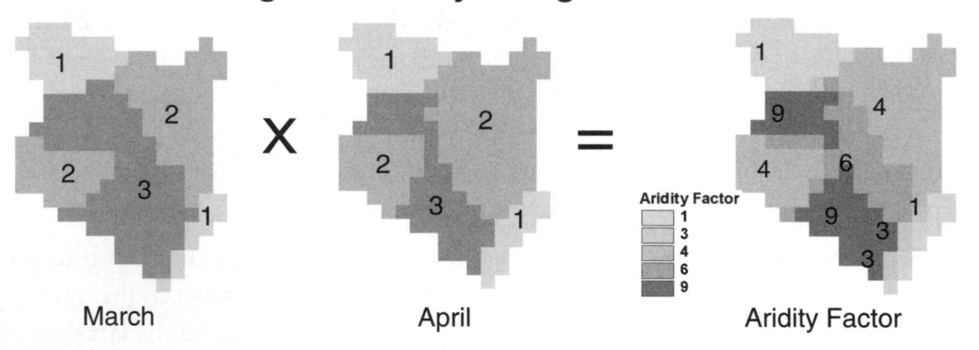

Image 3: Rainfall Cells

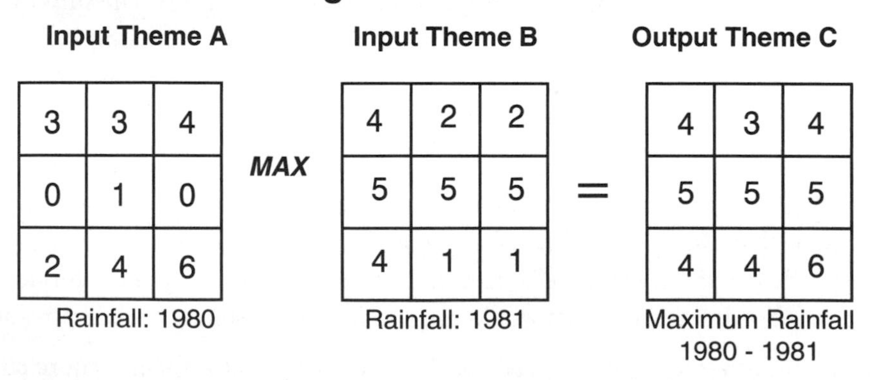

Fig. 8-12: Examples of map algebra using the multiply and maximum operations.

Map Algebra: Multiplication and Maximum

The multiplication and maximum functions are two other common types of overlay operations that make use of map algebra. Each has particular uses. Not discussed are other common map algebra operations, such as subtract and divide.

Multiplication

Thinking in terms of map algebra can make the overlay process efficient and convenient. Consider the application of cropland in political districts, shown in image 1 of figure 8-12. Not shown are the many raster cells that cover each polygon, each with a feature code. The project is designed to find which districts have cropland. The districts (map 1) are numbered 1 through 6 (perhaps codes for a set of names), and the cropland (map 2) is coded 1, with non-cropland = 0. There are several possible approaches to deriving the answer, but the simplest in this case is to make an overlay operation using Multiply (cell values are multiplied between themes).

Anywhere a district cell meets its corresponding non-cropland cell (value 0), the obvious multiplication product will be 0, but because cropland is value 1, each district will retain its original number (N x 1 = N). The results in the output theme are easily interpreted. Where there is no cropland (code 0), the districts are coded 0, and where there is cropland (code 1), the corresponding number of the district is produced in the output theme. Map 3, District Cropland, is easy to interpret because no further recoding is needed to indicate the districts.

Image 2 shows the Kenya aridity topic used in figure 8-11. In this demonstration, climatologists and resource managers found that adding days is not a good indicator of potential ecological impact, but multiplying the number of arid days between months is more telling. The Aridity Factor map shows the products of multiplying cells between March and April. Patterns can be seen even at this low resolution.

Other math options could be used in these tasks, such as Add or Subtract, but an additional recode of the results may be needed to get a comparable final output theme. To test this statement, work out an Add on the image 1 task. Map algebra permits overlay of more than two themes; Add and Subtract usually work satisfactorily in such cases, but Multiply and Divide can get difficult when using three or more.

Maximum

Determining the maximum number between two or more cells is very easy. The program compares the cell values of each theme, finding the highest number and placing it in the output theme value. It is that simple. Image 3 shows nine cells of rainfall that are interpreted for the highest value. The results are in the output theme C, showing the maximum recorded rainfall in each cell in the period 1980 to 1981.

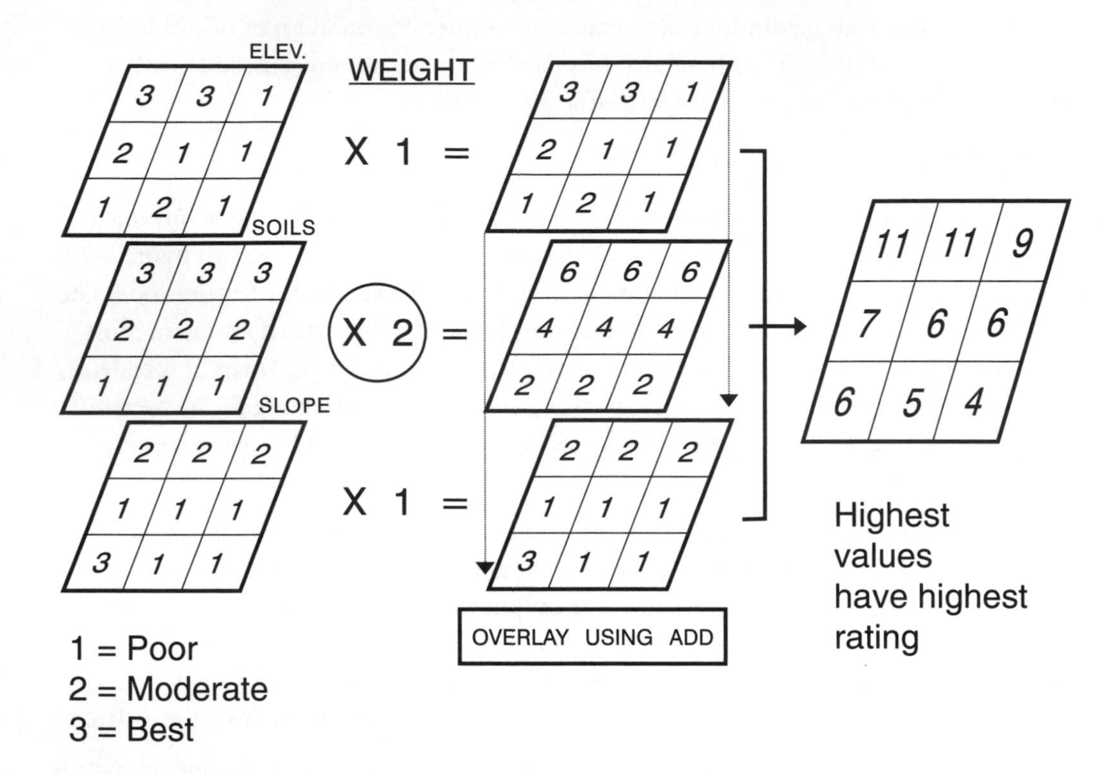

Fig. 8-13: An example of giving extra weight to themes.

Overlay Using Weights

Giving extra significance to some themes or features is often necessary, such as when one theme is more important than others in a particular application. The degree of importance is expressed as *weight*, or the factor by which the existing cell values should be increased (or perhaps decreased if the theme is less important than others). The user can take the step of recoding the important theme to an appropriate set of values in the data to get the proper attribute value weight, but this involves additional time-consuming, error-prone work. Many GISs allow a weighing factor in the overlay process; for example, asking if the included theme is to be increased in value.

In figure 8-13, an area is under investigation for possible agricultural development, and the best sites need to be found. All input themes have been recoded to Poor, Moderate, and Best categories (designated 1 through 3, respectively, on each theme). However, the Soils theme is more important than Elevation or Slope in this project, and its cell values are given twice (2X) the weight in the overlay. The themes are added in this case, with the highest ratings given to the highest values. The 2X weighing of Soils gave it extra influences in the analysis.

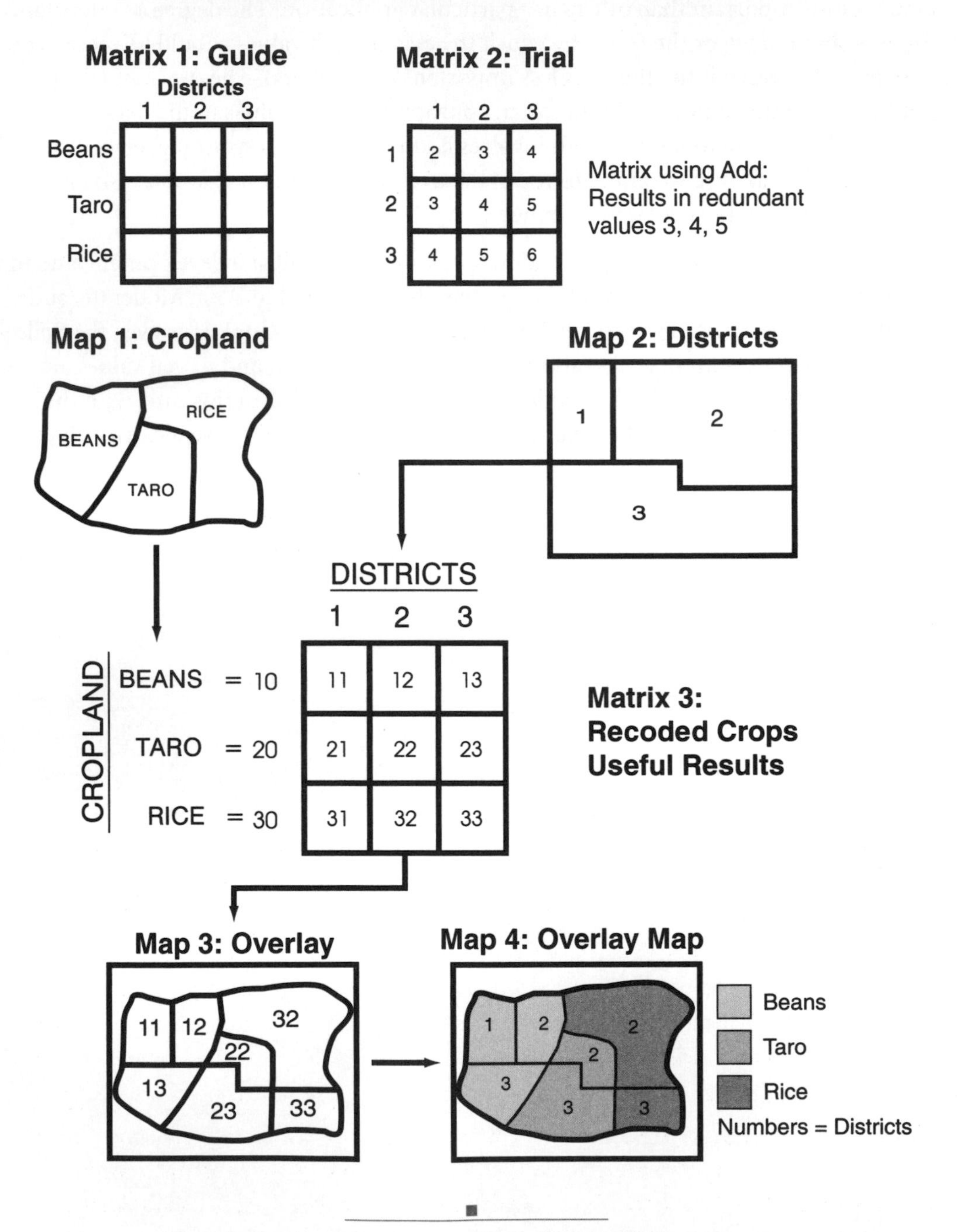

Fig. 8-14: Using matrix recode in an overlay operation.

Matrix Recode Overlay

As can be seen in the previous examples, the user must be careful with raster overlays because results can be confusing. It is easy to lose track of where output numbers come from. It is always a good idea to calculate the possible output values on paper, to know the results of each feature or value combination. When a given value is produced, the user should know how it was made. Ideally, each pair of input numbers should create a unique value (unless there are reasons to produce specific values from several input pairs). One of the best ways of ensuring control is to plot the overlay on a matrix table, with one theme as the columns and the other as rows.

The project shown in figure 8-14 is designed to overlay a Crop Land theme (map 1) with a Districts theme (map 2) to determine which crops occur in each district. A matrix is constructed with crops as rows and districts as columns (matrix 1 is a guide). Overlay with Add is used for simplicity in this example.

A regular sequence of numbers could be used for both themes (1through 3 for districts and 1 through 3 for crops), but notice in matrix 2 (trial matrix) that there are redundant resulting values; that is, 3, 4, and 5 are derived from two combinations of input values. If a 3 is presented in the output map, there is no way to know if it came from District 1 and Taro (crop 2) or District 2 and Beans. A more clever system is needed to make unique values. To ensure fast and easy interpretation of crop-district combinations in the final product, the crops will be recoded to 10, 20, and 30.

Note what happens in the matrix 3 with the new codes. Each tens number (11, 12, 13) came from Beans (the first crop), each twenties number (21, 22, 23) from Taro (crop 2), and each thirties number (31, 32, 33) from Rice (crop 3). The last digit in those numbers reflects the district number. It is useful coding: first number = crop, last number = district. Thus, 23 is the value for crop 2 and district 3. This is a fast and easy technique that assists in understanding the data.

This is a good system and the computer easily manages nine numbers. If there is no further analysis, the task is finished and no additional recodes are required. The overlay in map 3 shows the coincidences of each crop and district. Note that several combinations did not occur; for example, some crops are not in all districts. For map presentation, there can be a problem because the human eye has difficulty tracking that many colors or shades on a monitor or map. One tactic to employ in this case can be to combine members of a logical set, such as coding for all Bean values (11 through 13) to a given color, perhaps brown. Taro can be blue, and Rice will be yellow. This allows the eye to see all major colors as crops (map 4).

If each district value needs to be separated, particular shades or symbols can be used within these colors; for example, cross-hatch lines and/or various tones of each color. For simple presentation here, the overlay results are given in tones of gray for the crops (light = beans, medium = taro, and dark = rice), with numbers for the districts. A systematic display helps visual interpretation and aids in seeing patterns.

Overlay Codes
Planned Results

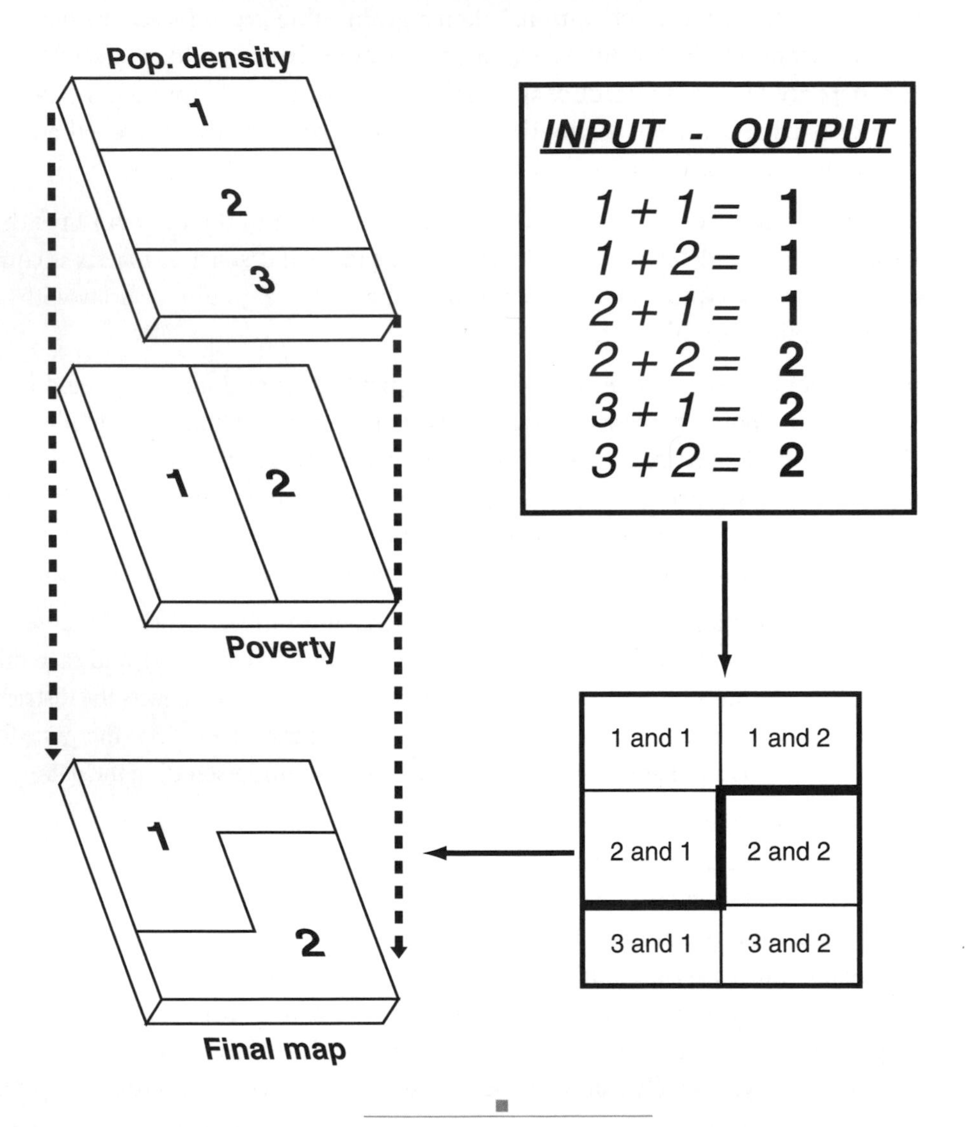

Fig. 8-15: An example of planned results for overlay codes.

Overlay Codes: Planned Results

Using a matrix to plan the raster output is convenient, but it does require some recoding preparation and mental gymnastics, such as deriving a clever coding system. It will be better if the desired final code can be entered and the program allowed to do most of the work. Some GISs permit assignment of overlay codes rather than relying on map algebra to perform calculations and determine results. That is, the GIS presents a set of choices such as the following:

```
Theme A code 1 + Theme B code 1 = ___
Theme A code 2 + Theme B code 1 = ___
Theme A code 3 + Theme B code 1 = ___
Theme A code 1 + Theme B code 2 = ___
Theme A code 2 + Theme B code 2 = ___
Theme A code 3 + Theme B code 2 = ___
```

The user enters the desired output code in the blank and the overlay is performed. The output codes do not have to be unique, and they can be used many times if needed. A matrix sketched on paper could be helpful in planning the coding, but this system is independent from recoding.

A simple application is shown in figure 8-15, with the top theme coded 1 through 3 (increasing population densities), and the Poverty theme coded 1and 2 (magnitudes of poverty). The user wants output codes of 1 and 2, taken from the combinations shown in the operations list box. Results will be as wanted rather than conforming to algebraic limitations. Several recoding steps are saved in the process. The box under the Input-Output list shows how the map will be coded, and the Final Map (lower left) presents the results.

Analysis Methods

Image 1: Analysis in the Database

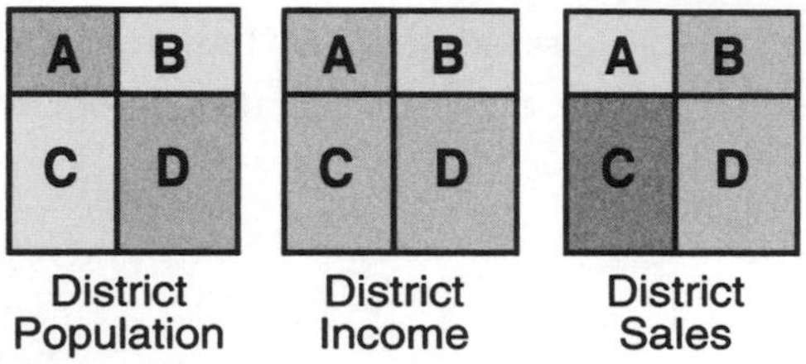

A B C D — District Population

A B C D — District Income

A B C D — District Sales

DISTRICT	POP	INCOME	SALES	MKT POTEN
A	2.2	11.2	22	7
B	1.0	9.8	51	9
C	1.7	10.1	41	5
D	2.1	11.1	50	5

Image 2: Analysis by Overlay

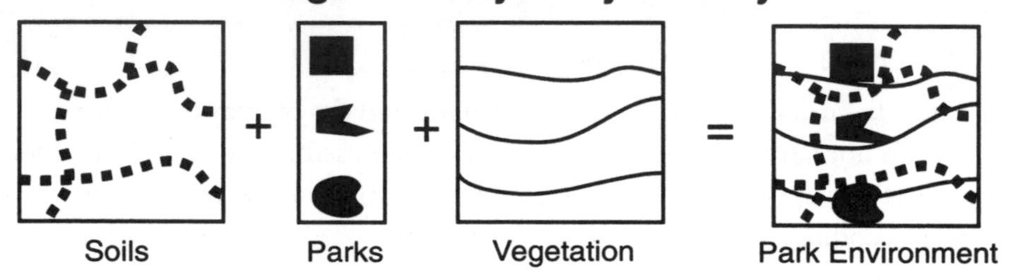

Image 3: African Example

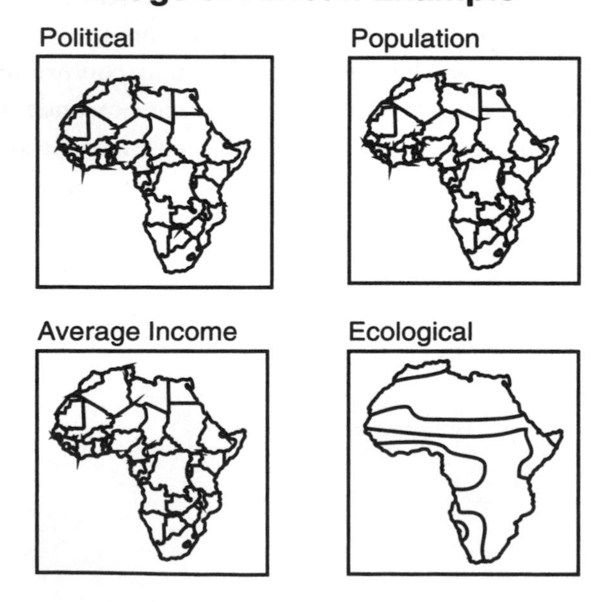

Fig. 8-16: Types of analysis methods.

Choosing the Analysis Method

Two fundamental approaches to basic overlay analysis have been discussed: database and graphic. Some low-end GISs have only raster structures and weak database capabilities that require manipulation of graphics to generate analytical output; for example, map algebra as the only method for overlays. However, most GISs have strong database and graphics integration, usually giving the option of taking either a graphics or a database approach to analysis. How does the user choose which option to take?

Each application has its special needs and circumstances, of course, but the first consideration should be to use the database when possible. This is consistent with the database approach in GIS, as discussed in Chapter 2. Using the database whenever possible is also logical, particularly when themes are spatially identical.

Three themes are shown in image 1 of figure 8-16. Each theme shows different attributes of the same four districts (A through D). The purpose here is to combine the attributes (population, income, and sales) to produce a market potential number for each district. Because each theme has identical spatial features (the districts), there is no need to use graphics overlays to make the combination; all of the work can be accomplished in the database. In the database, simply add a new attribute field for the derived market potential number ("Mkt Poten") and calculate the number from the existing attribute columns. This is fast, efficient, and accurate. Graphics can then be used to present the results in an effective manner.

If the themes and features are not spatially identical, the graphics approach may be needed, basically because the databases will not match. Image 2 shows soils, a smaller theme of parks, and vegetation, each spatially different. There is no common database structure, so a graphical overlay is used in order to combine them. The overlay can then be used for analysis. Powerful GISs may have alternatives for merging data, such as joining databases in various ways.

The illustration of Africa in image 3 demonstrates a typical set of themes for a project. Overlays and analysis involving the Political, Population, and Average Income components can be performed easily at the database because the features match spatially and are defined by discrete database records (by nation). However, to include the Ecological theme may require graphics overlay because of the uncommon polygons.

Buffers

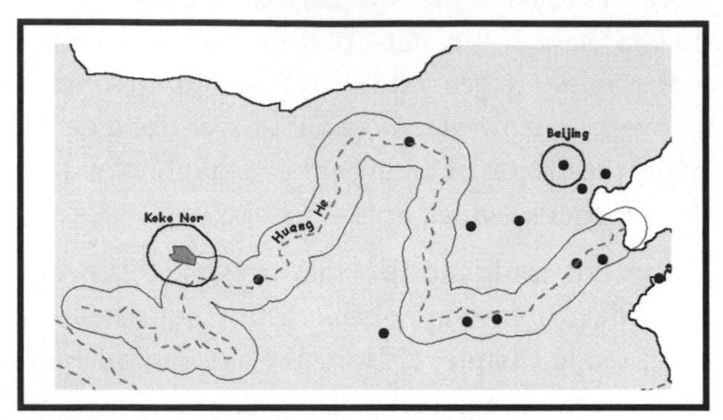

Map 1: 100-Km Buffers
Point, Line, Polygon

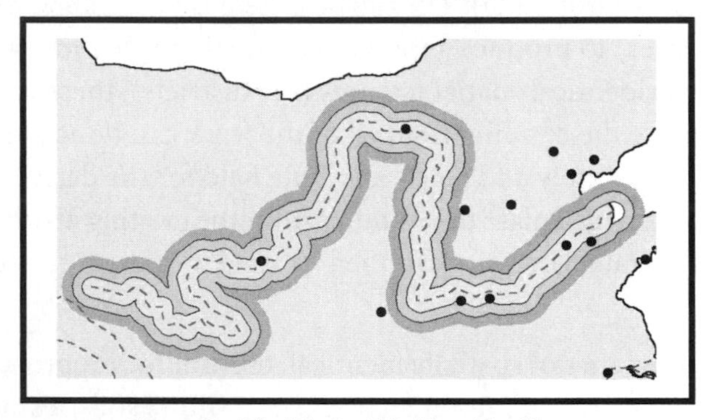

Map 2: Multiple 50-Km Buffers

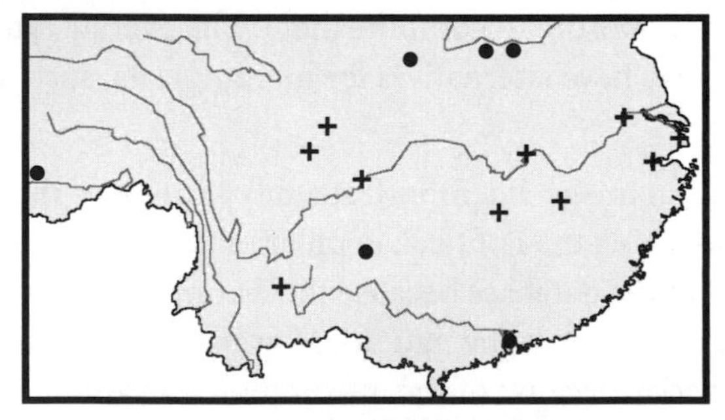

Map 3: Pseudo-Buffer Selection
Yangtze Cities, 150-Km

Fig. 8-17: Buffer examples.

Buffers

Building zones around features is a standard and very useful GIS capability (but difficult to do manually). *Buffer* is the common term for the zones, though other names are associated with this process, such as *spread*, *search*, and *corridor*. The operation can be easy: select the feature(s), give the desired distance, and the GIS builds the buffer outward from the selected feature or features.

Map 1 in figure 8-17 shows 100-kilometer buffers around each GIS feature type in China: point (city, Beijing), line (river, Huang He), and polygon (lake, Koko Nor). Note that GIS can provide buffers around only selected features; there are other cities and rivers not buffered. In addition, more than one buffer can be constructed. Map 2 is the Huang He with three 50-kilometer buffer zones, each colored differently. They can be used as individual or collective features for additional analysis; for example, as an overlay to show which cities are within 50 to 100 kilometers of the river. Many GISs support variable distance buffers, and each zone can have different distance increments.

Some GISs have "pseudo-buffer" operations that permit selection of features within a given distance of other features without actually constructing a buffer. This is a very useful shortcut that bypasses recoding, buffering, and overlay. The operation consists of selecting the focus feature or features (river), selecting the theme that has the target features (cities), and then activating the "select within distance of" option that asks for the real-world distance. Map 3 shows cities within 150 kilometers of the Yangtze River selected (cross symbols). All of the target features that exist within the noted distance are selected, ready for additional analysis or for saving as a separate theme.

In some systems, the selected feature to be buffered must be in a separate theme before a buffer can be constructed; for other systems, the central feature is simply selected and buffering is completed. In simple raster structures, the number of cells must be given to build the buffer, but most GISs accept real-world distances as standard input. Buffering is a routine operation, along with Recode and Overlay, that often serves as a preparation operation for more advanced analysis.

Buffer Applications

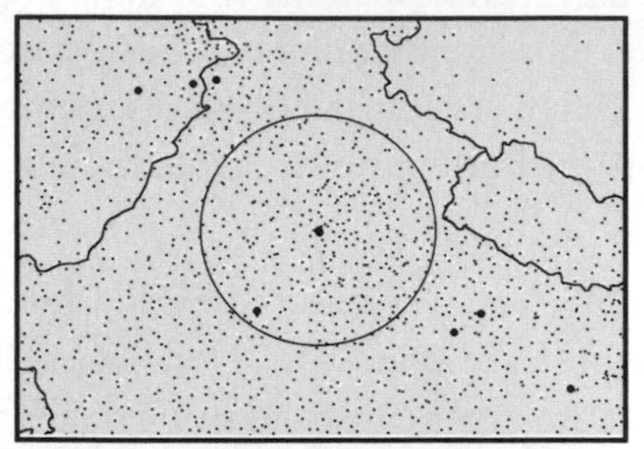

Map 1: New Delhi Region
300-Km Buffer

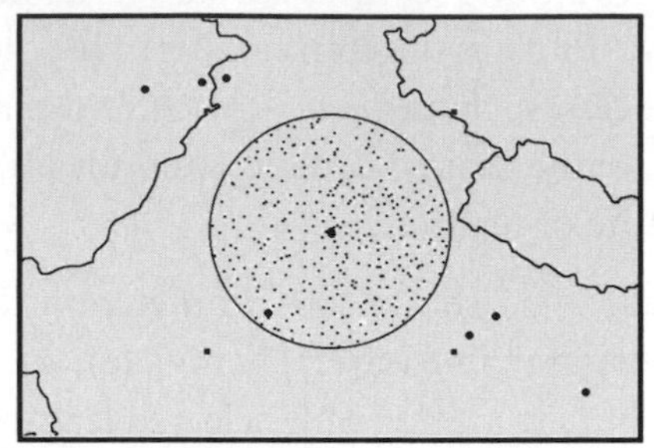

Map 2: 300-Km Urban Selection

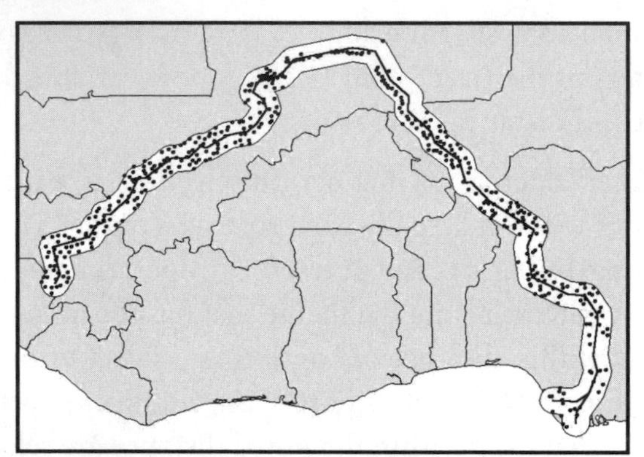

Map 3: Niger River Flood Zone
Endangered Towns and Villages

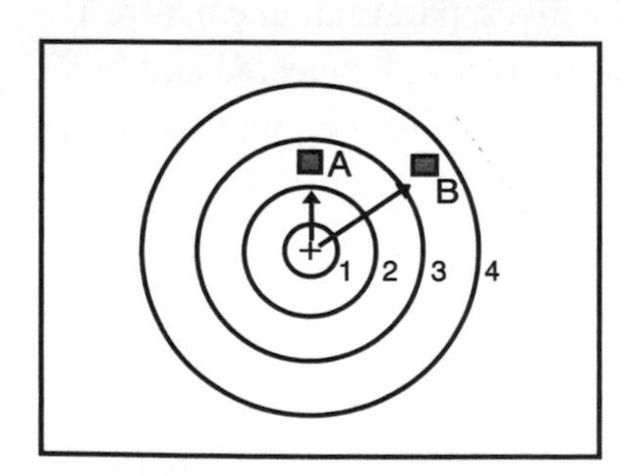

Map 4: Postal Zones
Cost of Mailing

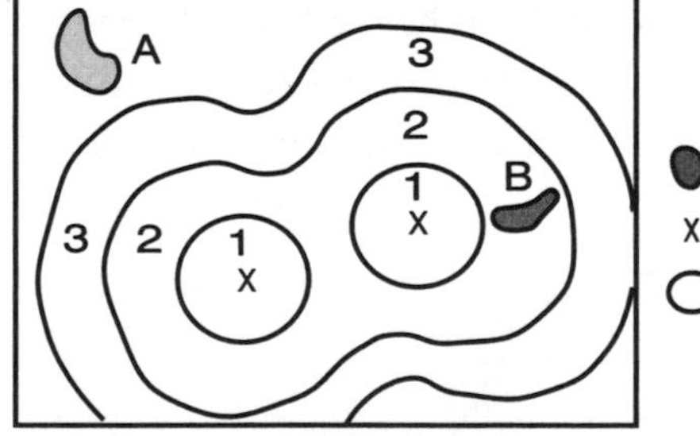

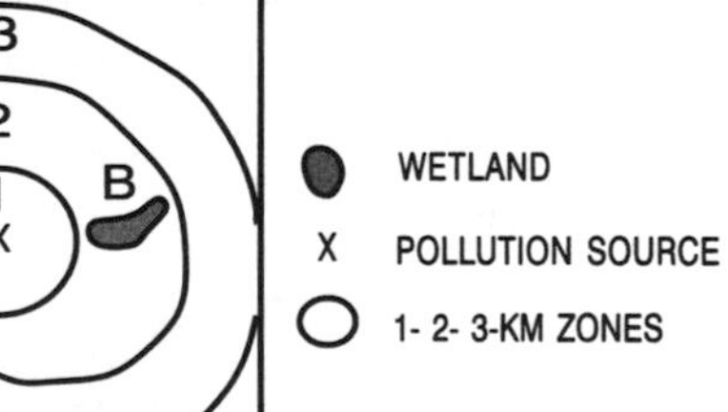

Map 5: Pollution Zones
Decreasing Impact

Fig. 8-18: Examples of buffer applications.

Buffer Applications

Buffers have many uses, mostly dealing with distance from selected features. Three simple applications are shown here, followed by more sophisticated use in the following material.

Buffers are excellent for selecting features within a given distance of another feature(s). Map 1 of figure 8-18 shows a 300-kilometer buffer around New Delhi, India, used to select towns and cities for a business development project. The operation required selection of New Delhi as the source feature for buffering, creation of the buffer (usually a few easy clicks to choose buffer type and distance), and then an overlay step to produce map 2. Graphics were not needed until the presentation; most of the work can occur without visual reference. From the database, additional attributes of the urban sites can be queried for further analysis if needed.

A second application deals with flood hazard management (map 3). Although most of West Africa is arid, floods can have devastating effects, displacing thousands of people. It is very important that agencies involved with disaster relief (or better, disaster prevention) should know which villages are prone to flooding under certain conditions. The towns within 50 kilometers of the Niger River are located using a buffer. As previously, there are only a few easy steps in the operations.

GIS can assist in two major ways in this type of application. First, before the flooding, the user can run a series of "what-if?" operations, putting in various flood condition buffers to detect which villages are prone to damage. With that set of data and maps, preventative measures can begin in order to help avoid major problems.

A second way is to have the landscape themes ready for analysis when a disaster occurs. GIS can perform complex analysis for a selected area, large or small, very quickly (within minutes), permitting rapid and efficient response. Lives and property can be saved—a wonderful application of GIS technology.

Because buffers are created outward from the central feature or features, it is easy to build concentric zones. In fact, this is how some GISs (particularly raster systems) first constructed buffers. When a buffer operation begins, the GIS radiates one-cell increments outward to the edge of the theme or to a selected distance. Recoding the cells to selected buffers is fairly easy; for example, at 10-kilometer zones.

Map 4 is a distance calculation for postal rates. The cost of mailing increases outward in zonal increments from the central post office. Mailing to destination A (zone 2) will be less expensive than mailing to destination B (zone 3) because of zone distance differences.

Map 5 shows distributed pollution zones, with progressively less hazardous conditions away from the two sources (X). Wetland B is clearly in a more hazardous situation than wetland A. This is an example of gravity models and "distance decay." The center has the highest intensity (or magnitude, such as pollution), and the effects decrease outward, measured by buffer zones.

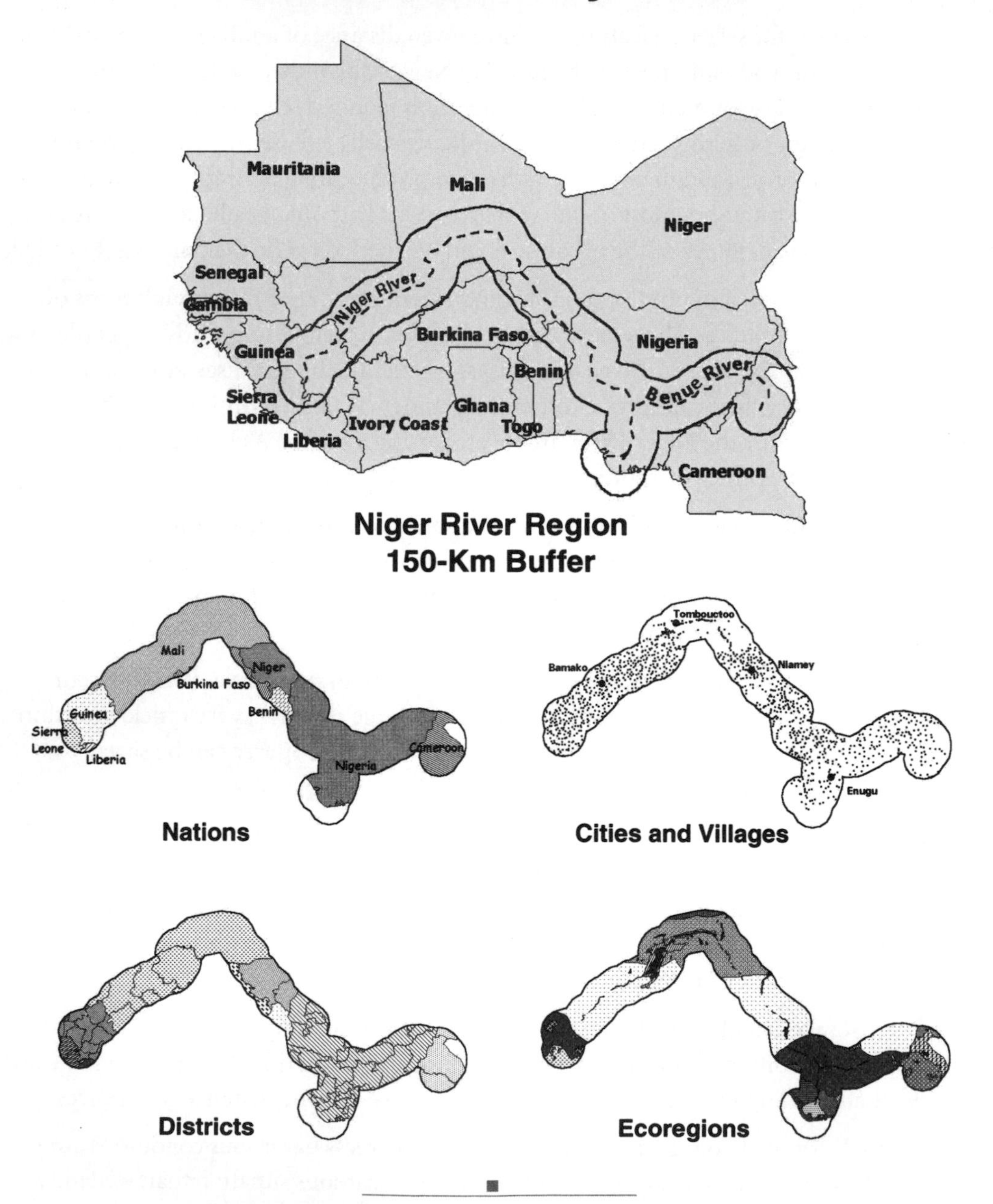

Fig. 8-19: A buffer applied to the Niger River system.

Buffer Application: West Africa River Region

Ecologists and regional planners understand how rivers may be natural regions that largely ignore or defy administrative boundaries. Watersheds are natural planning areas that can include only parts of the overlying political polygons. The Niger River in West Africa is a fragile system in a climatically and culturally dynamic part of Africa. Much of it passes through the Sahel, a delicate continent-wide swath of grassland that suffers frequent terrible drought, displacing and killing many people. Planners must have good information if rational responses are to be made. GIS is an ideal technology for gathering, analyzing, and presenting such information.

Figure 8-19 shows maps of a Niger River information system (which includes the Benue River in the southeast), built from a 150-kilometer buffer. The buffer is an odd-shaped one that is difficult to construct manually, but to the GIS, shape and size are irrelevant. These maps, including the initial buffer, took less than one hour to build, with most of the time spent on cartography because the buffer operation is very fast.

The maps deal with basic regional data, and a comprehensive information system can have dozens of theme maps. If the data sets are available, the procedure can be almost a "production-line" process of calling up the needed theme, making selections and recodes if necessary, constructing the buffer, and finishing the presentation. The monochrome presentations here are greatly improved by color, of course.

Spatial Analysis

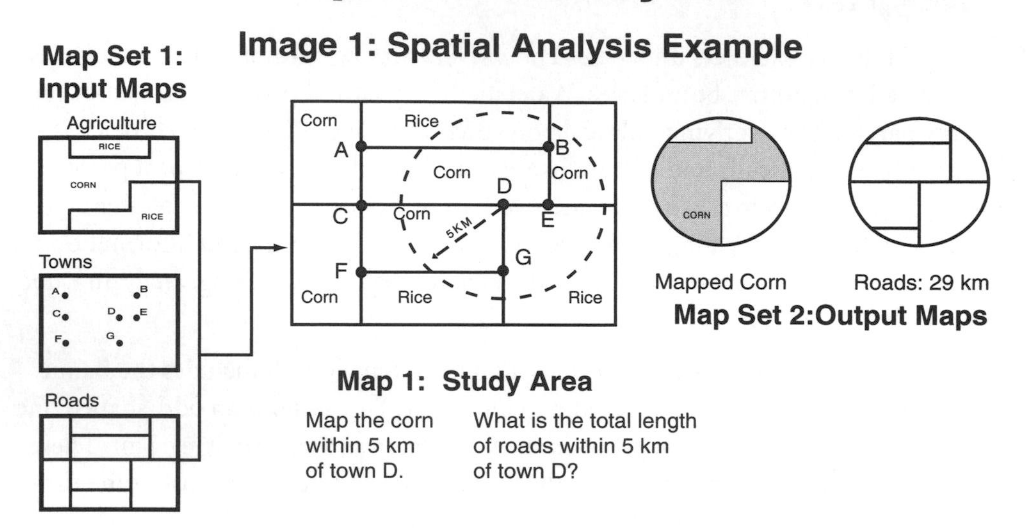

Image 2: Australia Project

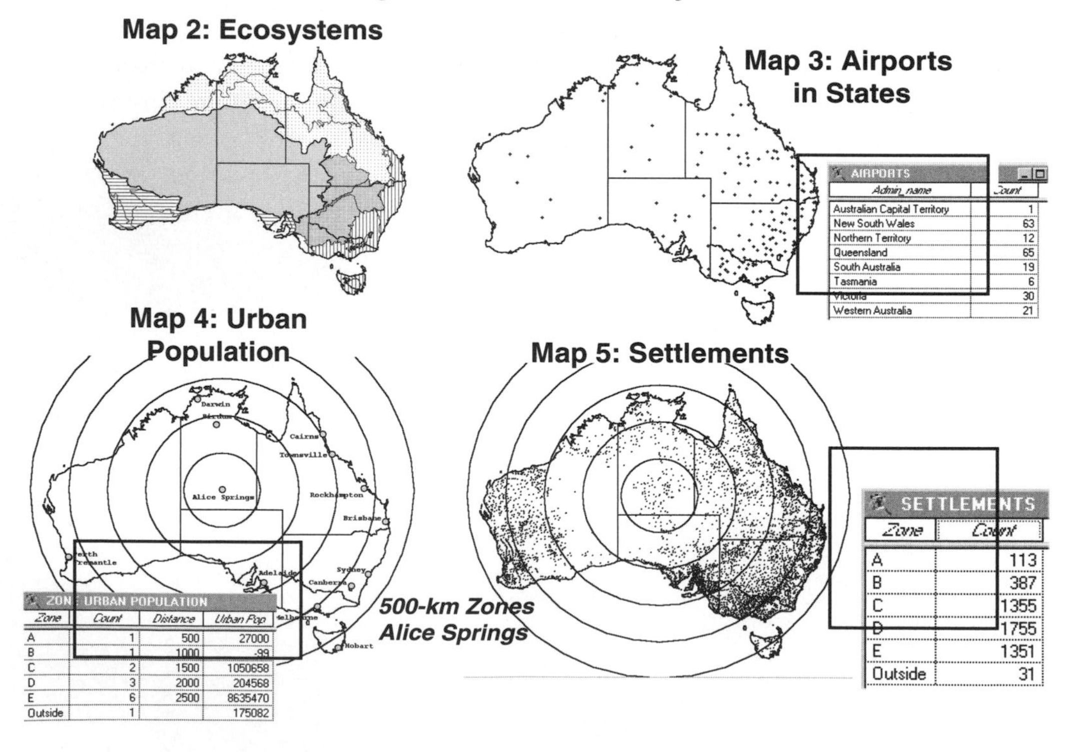

AIRPORTS

Admin_name	Count
Australian Capital Territory	1
New South Wales	63
Northern Territory	12
Queensland	65
South Australia	19
Tasmania	6
Victoria	30
Western Australia	21

ZONE URBAN POPULATION

Zone	Count	Distance	Urban Pop
A	1	500	27000
B	1	1000	-99
C	2	1500	1050658
D	3	2000	204568
E	6	2500	8635470
Outside	1		175082

SETTLEMENTS

Zone	Count
A	113
B	387
C	1355
D	1755
E	1351
Outside	31

Fig. 8-20: The concept of spatial analysis.

Spatial Analysis

There are numerous procedures and techniques using the operations discussed in this chapter, but the basic focus is on providing information for and starting spatial analysis. Image 1 of figure 8-20 combines some of these steps in a simple multiple spatial analysis demonstration. First, an overlay of Agriculture, Towns, and Roads (map set 1) creates map 1 of the study area. These input maps can be recodes from other data sets. Then a 5-kilometer buffer is applied around town D to address proximity characteristics, such as the spatial distribution of corn and the length of roads, tasks expressed in text beneath map 1. The mapped corn is shown in map set 2 (clipped from the buffer), with the fields dissolved into a single polygon. The roads within the buffer are mapped, also, and a total is given.

A more elaborate Australia project application is given in image 2. Four thematic maps (maps 2 through 5) dealing with various spatial analysis tasks were constructed from a number of overlays and then recoded for presentation.

Map 2 is an ecosystems theme with the states, but the desert is given prominence by dissolving the subdivisions into a single feature and giving it a uniform gray tone for clear visualization. The relationship of the desert expanse to the other ecosystems and to the states becomes apparent. Map 3 displays airports. The location and number of airports in each state is important in Australia. The map is made with recodes, an overlay, and analysis of the number of points (airports) in each polygon (states). A summary table presents the results.

Maps 4 and 5 deal with 500-kilometer zones (A through E) from Alice Springs, in the center of Australia. Map 4 shows the number of cities and their collective population in each zone (-99 indicates no data for Birdum). This type of analysis can be important for transportation planning. Map 5 determines the number of settlements in each zone. The summary table presents the results.

These sample Australian maps are a few of the analytical possibilities with the available data. As will be seen in the next chapter, there are many more spatial analysis operations in GIS, for a wide range of applications.

Reports and Graphs

Image 1: Report and Graph

Agricultural Report

District	Ave Field Size	Min Yield	Max Yield	Mkt Distance
A	128	55	123	54
B	220	65	111	34
C	145	52	98	45
D	200	50	113	66

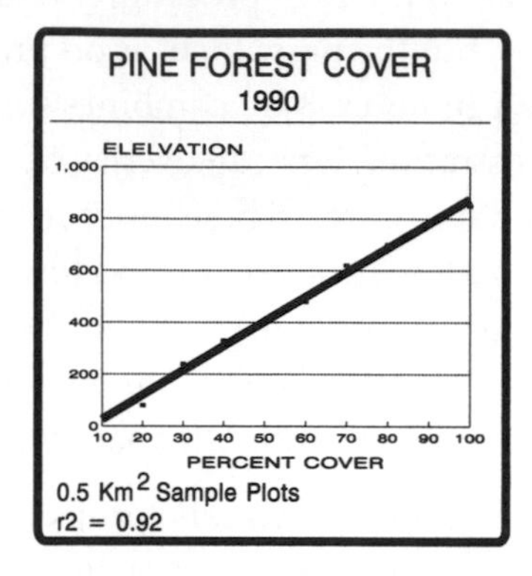

Image 2: Arizona Demographics

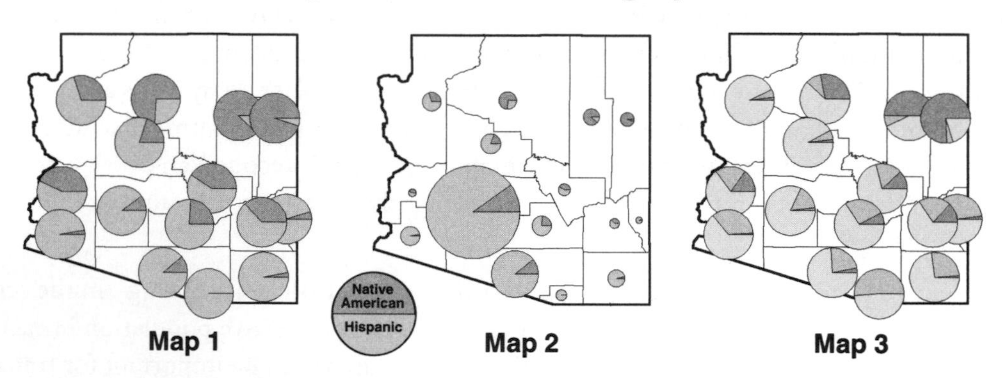

Image 3: Australia Population

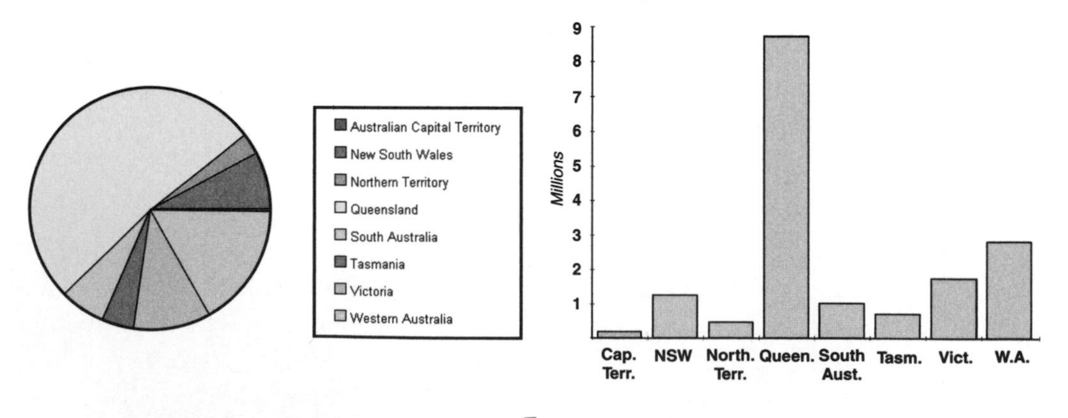

Fig. 8-21: Types of graphs and reports generated for statistical reporting.

Statistical Reporting and Graphing

Graphs and statistical reports are useful, and even recommended, with many analytical operations. They provide summary information, new information, and possible insights to aspects that graphics do not provide. Data visualization is a rapidly growing technology that offers a wealth of tools for analyzing and presenting information. There are many types of reports and graphs that can accompany GIS projects, and the examples here are only a very small sample.

A simple list of the average size of crop fields, with their minimum and maximum yields, along with the range of distances to market (all generated by GIS analysis), can be valuable in an agricultural survey project (image 1 in figure 8-21). The Pine Forest Cover graph plots the percentage of pine forest cover against elevation, taken from a number of sample points. It is possible that spatial and ecological relationships may emerge with this type of visual tool. The points fall very close to the regression line (line of best fit), and the correlation is calculated at 0.92, a very good indicator that density of vegetation increases predictably with elevation.

Such a plot is fairly easy to make and the results can be significant. The plot permits prediction and helps to explain what is occurring on the landscape; for example, the possible controlling reason for pine forest cover may be elevation (a cause-and-effect relationship if it is true).

Image 2 shows the use of pie charts for exploring ethnicity in Arizona, USA, counties. Arizona is in the southwest, on the border with Mexico and is known for its Native American heritages. Map 1 shows the balance between Native Americans and Hispanics in each county. The use of these charts can lead to insights about population distribution. For example, it appears that Hispanics outnumber Native Americans in most counties except a few in the northeast.

Map 2 has the same ratios as map 1, but the charts are sized according to county population, which may lead to other insights and questions. Map 3 includes Caucasians (lighter areas), giving the demographics another, more complicated, view of the state's population. These charts are a simple start in the project and many types of graphs can be used to explore the numerous demographic factors of a state.

Image 3 compares two standard charts for Australian population information. The pie chart presents the proportion of national population by each state, with an attached legend (normally in color for better interpretation). The bar chart presents the same information in a different format. One advantage is that there is a scale for determining approximate population numbers. Users must decide which is best for a given purpose or project.

GIS reports can be simple text and graphics, or they can include documents with maps, many types of graphs and plots, various tables, pictures, and other visualization tools.

However, because GIS is taking advantage of the Internet and other technologies, the old-style documents (largely paper-based) are giving way to more modern reporting, such as web pages with links to relevant and informative web sites, animation, and various other new ways of presenting information. Hardcopy documents may never disappear, but GIS is adopting new forms of information delivery. Chapter 12 addresses the future of GIS to emphasize the rapid changes underway.

CHAPTER 9

ADVANCED ANALYSIS

Introduction

GIS OFFERS MANY OPTIONS FOR ANALYZING DATA, from simple display for visual investigation to highly sophisticated, algorithm-rich techniques on the cutting edge of the geographic sciences. There is no sharp line between the basic analysis operations of the previous chapter and those presented here as "advanced" procedures.

This chapter demonstrates only a few of many analytical possibilities available in GIS. Even an incomplete review could take a hundred pages, but the essential idea is to show the types of analysis and applications GIS offers. Four basic categories are examined here: proximity analysis, graphics operations, terrain analysis, and network analysis. Many others could be explored.

Proximity Analysis

How far is A from B?

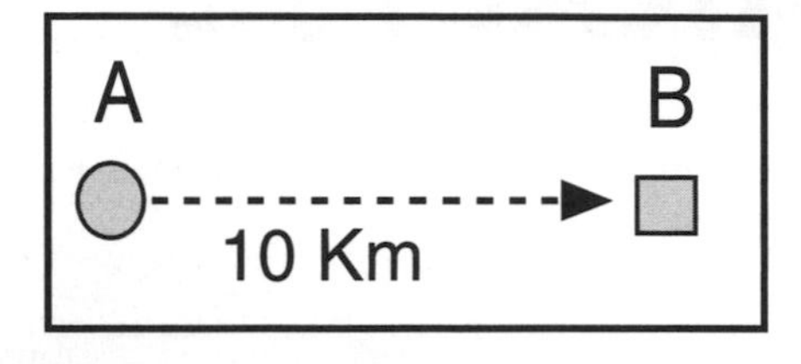

1) *Distance:* A to B = 10 km
2) *Time:* at 10 km/hr.,
   A to B = 1 hour
3) *Cost:* at $10.00/hr.,
   A to B = $10.00

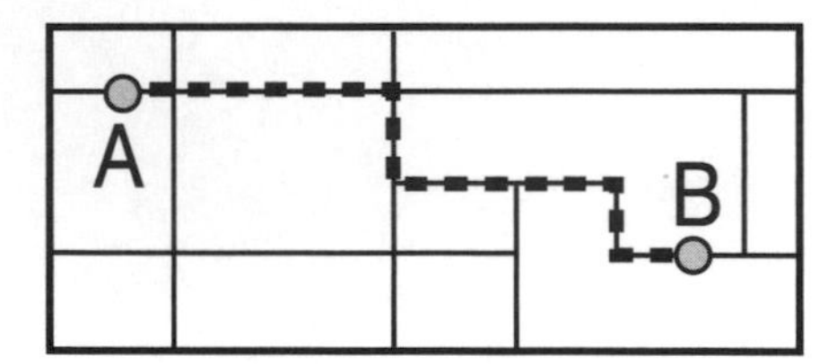

Shortest Route A to B

How many X features are within 5 km of feature Y?

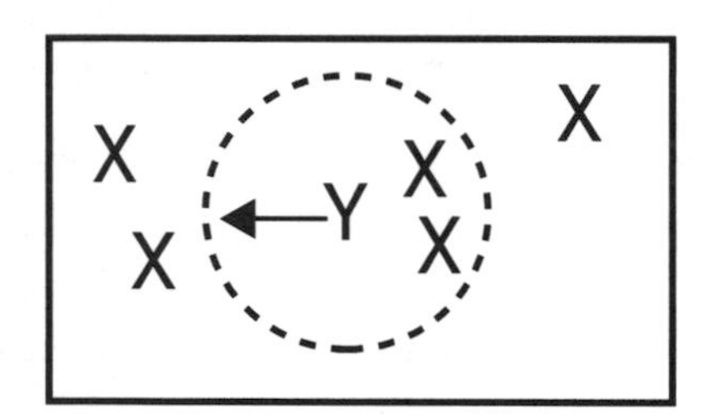

What is the nearest K feature to feature L?

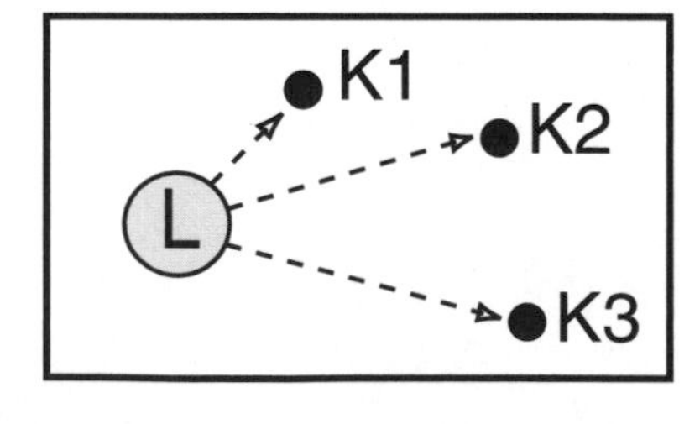

Fig. 9-1: Examples of proximity analysis.

Proximity Analysis

The term *proximity analysis* can mean any procedure that performs "neighborhood," vicinity, or distance analysis. Such procedures range from simple distance measurement and buffer methods (previously discussed) to more advanced operations and multiple-step methods using several analytical procedures. GIS offers numerous proximity analysis techniques. The basic framework for proximity analysis includes several types of spatial relationship questions, a few of which are represented in figure 9-1.

- *How far is A from B?* Acquiring the distance between features is a common operation. Some GISs allow simple point-and-click selection of features for determining distance (that is, pointing to the starting feature or location and then to the target or destination feature). Distance can be expressed in terms of space, time, or cost. All three of these factors are related. Conversion of spatial quantities to cost values is easily performed (dollars per kilometer x distance = total cost of trip). Time is calculated as the rate of distance per unit of time multiplied by distance. The user enters the cost per unit (money or time), designates the points to be connected (start, midpoints, and end), and the program runs the route using cost as the quantitative unit. Finding a shortest or least-cost route is a standard option.
- *How many X features are within 5 km of point Y?* Selecting a given class or type of feature within a given distance of another feature can be useful, such as the number of communities within a noise zone. Distance from points, lines, or polygons may be used. This can be a simple buffer selection or part of a more advanced analytical operation.
- *What is the nearest K feature to L?* Finding the closest feature of a certain type is very useful in many applications. This combines distance and identification in a search for the nearest feature of a given class. In this case, the program first locates all K features and then calculates their distance from L, reporting the one with the shortest distance.

Distance is a principal spatial quantity that relates to many GIS functions. Almost all of the concepts and operations discussed in this chapter use space as a primary factor, and various forms of proximity analysis directly and indirectly.

Nearest Features

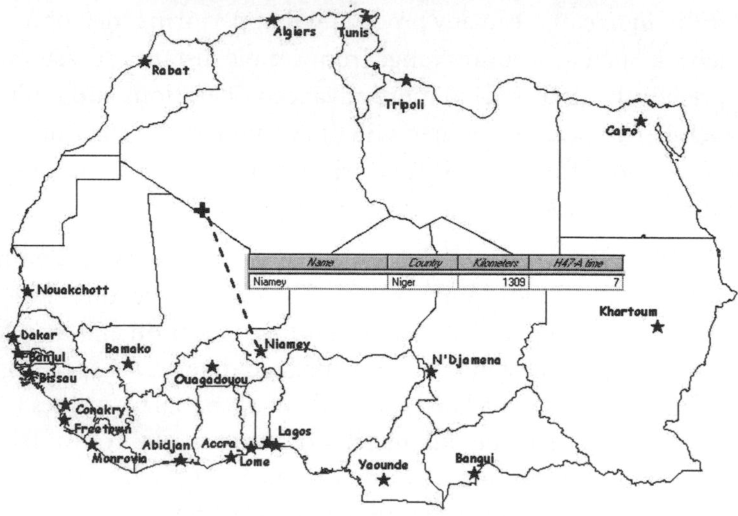

Name	Country	Kilometers	H47-A time
Niamey	Niger	1309	7

Map 1: Rescue Point Nearest Facility

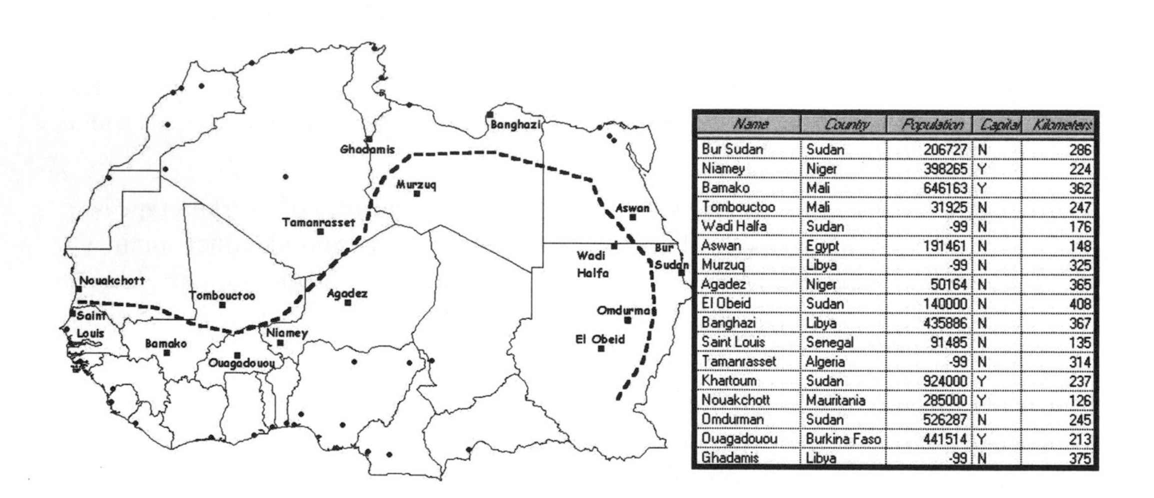

Name	Country	Population	Capital	Kilometers
Bur Sudan	Sudan	206727	N	286
Niamey	Niger	398265	Y	224
Bamako	Mali	646163	Y	362
Tombouctoo	Mali	31925	N	247
Wadi Halfa	Sudan	-99	N	176
Aswan	Egypt	191461	N	148
Murzuq	Libya	-99	N	325
Agadez	Niger	50164	N	365
El Obeid	Sudan	140000	N	408
Banghazi	Libya	435886	N	367
Saint Louis	Senegal	91485	N	135
Tamanrasset	Algeria	-99	N	314
Khartoum	Sudan	924000	Y	237
Nouakchott	Mauritania	285000	Y	126
Omdurman	Sudan	526287	N	245
Ouagadouou	Burkina Faso	441514	Y	213
Ghadamis	Libya	-99	N	375

Map 2: Expedition Service Points Nearest Cities

Fig. 9-2: Applications of the "nearest feature" query.

Nearest Features

Calculating distance between features or finding the nearest features are very useful GIS operations, especially when database attributes can be incorporated. *Near* is a generic GIS proximity analysis procedure that determines the shortest distance from one feature or features to a set of others; for example, from a central city to the nearest surrounding city. Near can operate with points, lines, or polygons.

Map 1 in figure 9-2 shows a simple but valuable application. A distress signal was sent from a remote part of the Sahara Desert, North Africa (the + symbol on the map), and rescue aircraft are based only in capital cities (in this hypothetical scenario). Because time is critical, the closest city should dispatch the rescue. Several capitals seem to be about the same distance, but Near finds the closest one (Niamey, Niger) and determines the distance between it and the rescue point. Distance can be expressed as flying time, and the database reports seven hours for an H47-A helicopter that cruises at 200 kph (plus time for takeoff and landing).

With a complete database of rescue facilities, Naimey may be found to lack the resources needed for this particular mission. GIS can quickly move to the next candidate city, Ouagadougou, Burkina Faso (misspelled on the map), at 1,337 kilometers and about 7.5 hours flying time.

GIS can find the nearest cities along a track, also. An adventure drive across the Sahara needs periodic refueling and supplies. Map 2 shows the planned trek as a dashed line. The program determines the closest cities that can operate as supply bases. Selected cities have a square symbol, and other cities are represented as small dots. The database presents useful information, such as the name and country of each city, whether or not it is the capital, and distance from the route.

There are many applications of Near, particularly in the emergency response field. Some GISs have other versions of this capability, as discussed in material to follow.

Spider Diagrams

Point-Point Distances

Travel Distances
0 - 1000
1001 - 2000
2001 - 3000
3001 - 4000

Map 1: Agadez Conference Travel

Capital	Kilometers	Fly Time
Niamey	761	1.2
N'Djamena	981	1.5
Ouagadougou	1190	1.8
Lagos	1304	2.0
Lome	1320	2.0
Porto Novo	1421	2.2
Yaounde	1535	2.4
Accra	1583	2.4
Bamako	1843	2.8
Bangui	1850	2.8
Abidjan	1878	2.9
Tripoli	2029	3.1
Monrovia	2399	3.7
Conarky	2462	3.8
Tunis	2507	3.9
Freetown	2521	3.9
Algiers	2526	3.9
Nouakchott	2645	4.1
Rabal	2663	4.1
Bissau	2691	4.1
Khartoum	2735	4.2
Banjul	2751	4.2
Dakar	2774	4.3
Cairo	3040	4.7

Distance and Flying Time to Agadez

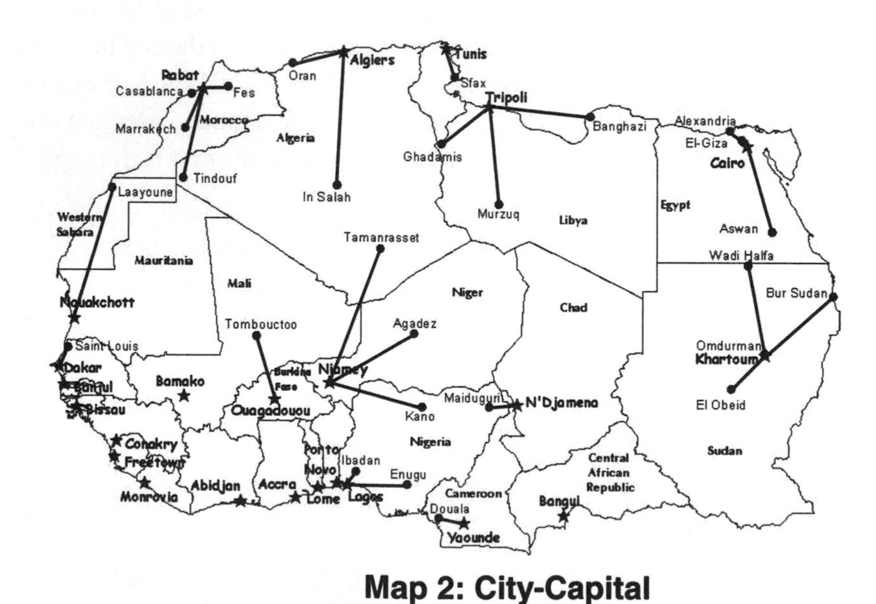

Map 2: City-Capital Distance Relationships

Fig. 9-3: Distance relationships using spider diagrams.

Spider Diagrams

Spider diagrams are special operations that compute (and show) point-to-point multiple distances. They draw lines from each feature or location to its nearest "source" (feature with certain attributes). Basically, one or more types of point features are connected to another type of feature, as defined by properties from the database.

A North Africa conference travel analysis is shown in map 1 of figure 9-3. The conference is being held in Agadez, Niger, because of its central location. Flying distance and directions from the other regional capitals are determined by the spider line procedure and are displayed on the map. The diagram initially makes lines between the features, but to give added information they can be depicted as classes according to some attribute.

Here the lines are shown in different symbols for each 1,000-kilometer increment, to indicate different distance categories (see legend). Database information includes the city, distance, and flying time, based on an average cruising speed of 650 kph. Records are arranged from the shortest to the longest distances to Agadez. GIS can provide a complete set of supporting information with only a few operations.

Map 2 presents a simple analysis of the shortest distance from non-capital cities to capitals in North Africa. Because distance is often a major factor or influence in service areas, the regional planner wants to know if each nation's capital serves its cities efficiently from a geographical standpoint. Phrasing the query another way: Which capital is nearest each city? The spider diagram operation links each non-capital city to the closest capital, regardless of country.

Because of the size and shape of North African countries and the distribution of their cities (as well as the huge Sahara Desert barrier), most have cities near their own capitals. However, several nations have at least one city that is closer to another capital, such as Algeria, Mali, Chad, and Nigeria. (Western Sahara, a disputed territory, does not have a capital on this map.) International commerce, geopolitics, and even social movements may be affected by such distances and the potential for border crossing. GIS can play a role in the analysis of regional urban geography.

Distance Selection

Bali Tourism

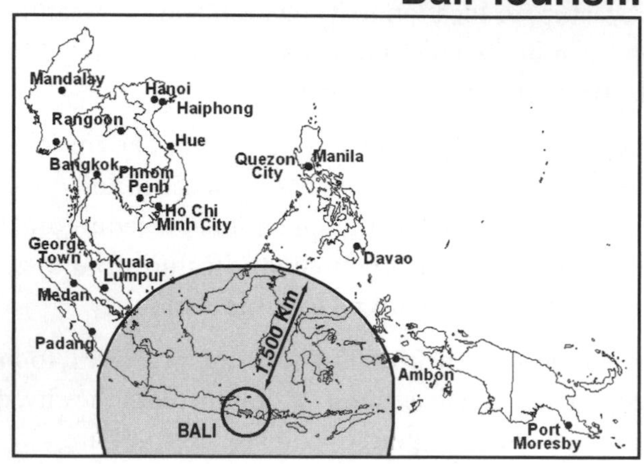

Map 1:
Bali Cities: > 1,500 Km
Buffer Exclude

Map 2:
Bali Cities: 1,500 - 3,000 Km
Distance Select
(No Buffer)

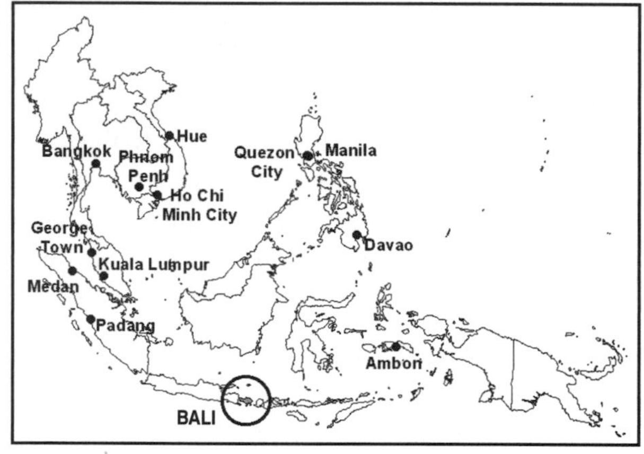

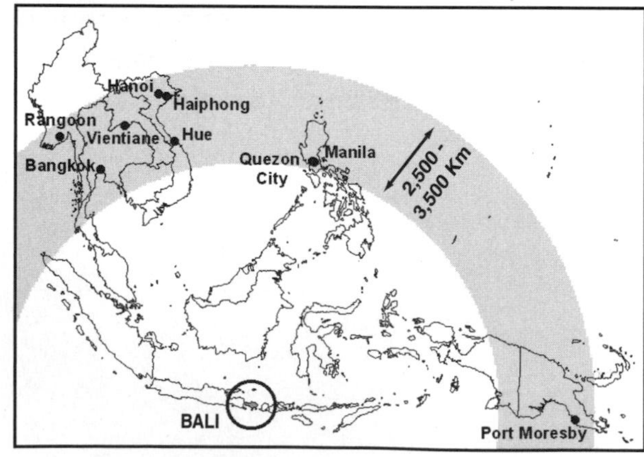

Map 3:
Bali Cities: 2,500 - 3,500 Km
Buffer Overlay

Fig. 9-4: Buffers used in distance selection.

Distance Selection

GIS usually offers several ways of accomplishing most tasks. Selection of features based on proximity characteristics can be achieved with vector-based selection options or raster data buffers (many GISs offer both). For simple initial selection, perhaps the vector operations may be best. If the project involves substantial modeling and analysis, raster data may serve well. A few techniques are described here, though they should be somewhat familiar from previously discussed GIS operations.

Figure 9-4 shows three maps in a Bali, Indonesia, regional tourism assignment. As a major world tourist destination, Bali does well in the global market. However, as Pacific Rim nations gain economic strength, Indonesia (in this scenario) wishes to reach out to its own region in order to be more inclusive of neighbors and to promote a new tourism infrastructure. In trying to decide which types of aircraft to purchase for the program, the middle-range cities (from Bali) are to be considered. Map 1 shows the cities over 1,500 kilometers from Bali; those closer are in the immediate "neighborhood" and must be served by an existing short-range airplane.

A buffer is applied from Bali that excludes the surrounding cities within 1,500 kilometers. Then the cities further out are selected and analyzed for additional relevant tourism factors. Note the slightly oblong shape of the buffer, which reflects Bali's shape. Interestingly, by selecting the cities over 1,500 kilometers from Bali, the database also easily identifies cities within that distance.

One type of aircraft being considered has maximum efficiency between 1,500 and 3,000 kilometers. The planners need to know which cities are within that range. In map 2, city selections are based on distance from Bali rather than using buffers (although the ultimate results would be the same, of course).

Map 3 shows a raster-based buffer making a similar selection of cities between 2,500 and 3,500 kilometers (to evaluate the market cities for a similar aircraft). There are several ways of making this type of buffer. Recall that many raster systems build radiating concentric zones from the target out to the edge of the theme or display, coding each raster cell with its distance from the origin. By recoding that theme, a middle zone can be made. That is, values 1 to 2,499 are recoded as 0, values 2,500 to 3,500 are recoded as 1, and those above 3,500 are also recoded as 0, leaving just the needed buffer to overlay the cities. Some GISs offer the option of creating the buffer by entering the desired distances and letting the program do the recoding.

There are other ways of choosing and analyzing features based on distance, and good GISs offer a rich set of operations. With the database, cities can be organized and classified according to other attributes, such as population or income. Marketing can be targeted to specific cities and groups of potential customers so that transportation can be planned efficiently.

Aggregation - Clustering

Grouping Close Features

Image 1: Aggregation of Tree Theme

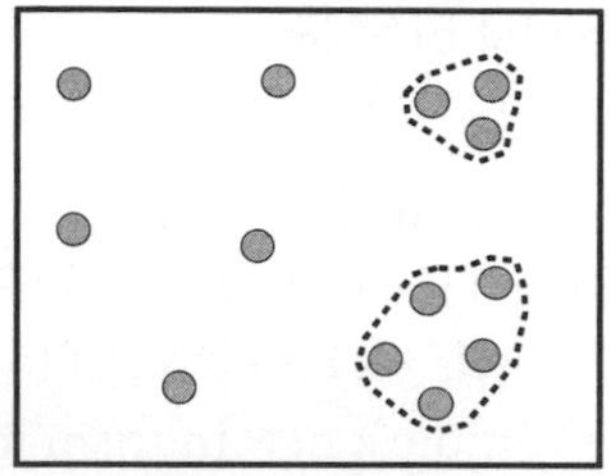

Map 1: Trees
Individual Features

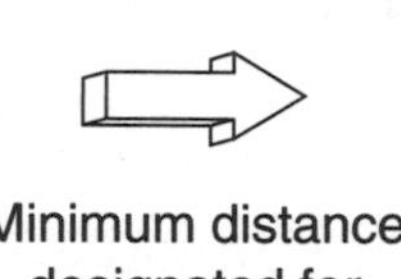

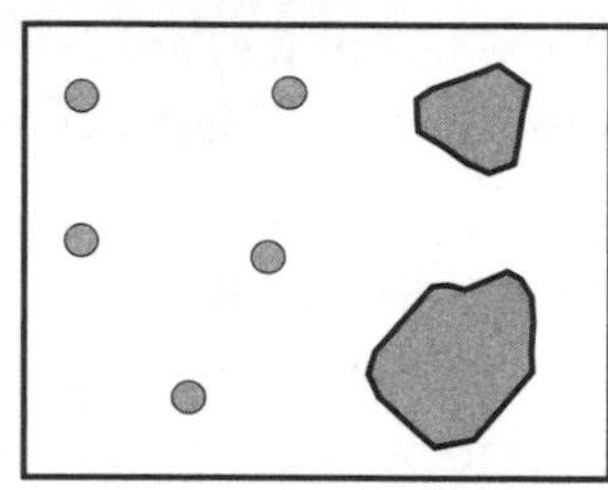

Map 2: Vegetation
Clustered Features

Image 2: Generalization of Agricultural Landscape

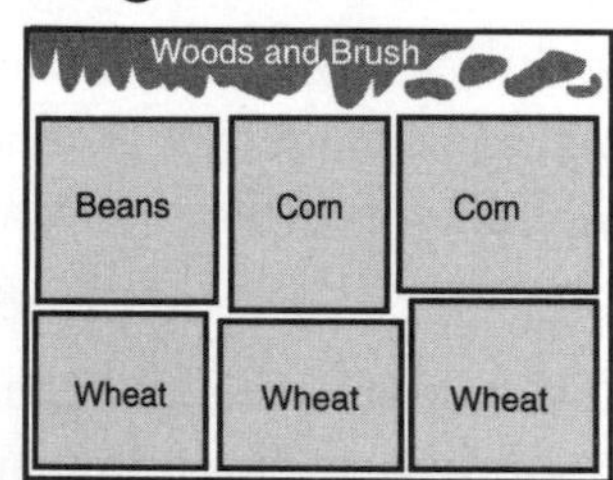

Map 3: Detailed
Agricultural Landscape

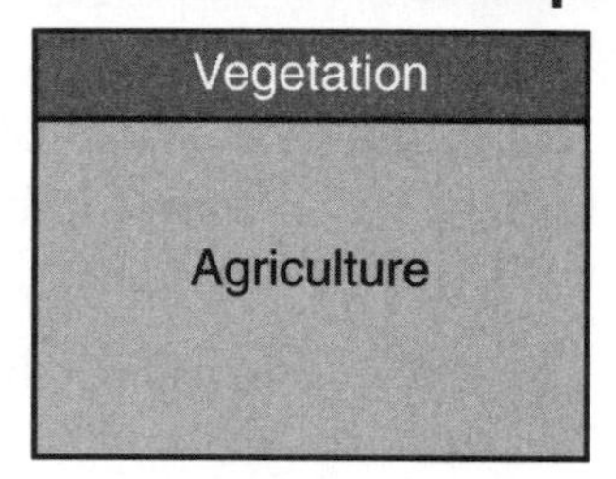

Map 4: Generalized
Landscape Themes

Image 3: Crater Clustering

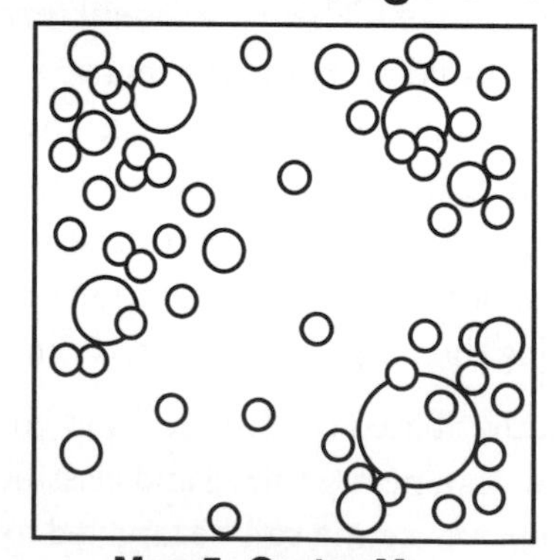

Map 5: Crater Map

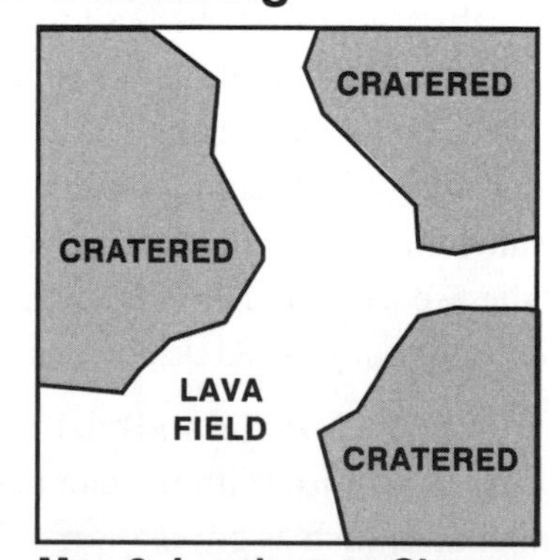

Map 6: Landscape Classes

Fig. 9-5: Examples of aggregation and clustering.

Aggregation

Aggregation groups features that are very close to each other. If features are within a specified distance, the program will combine them into a single unit. This is a generalization process that helps to reduce visual confusion. Map 1 in image 1 of figure 9-5 shows point features (trees) scattered around an area. Those on the right side of the map are so close together that they should be aggregated (clustered) into a single polygon. The user gives a maximum separation distance and features that are within the distance are grouped into polygons. The results are shown in map 2.

Aggregation helps reduce the small space between important features, as an aid to interpretation and mapping. For example, small access paths or narrow roads, all part of the agricultural landscape, often separate agricultural fields. Map 3 of image 2 shows six fields, separated by small spaces. Aggregation merges them into a single unit, which is recoded to a more general label of Agriculture (map 4). Vegetation has been grouped and generalized as well.

Some landscapes have numerous small, discontinuous features that give the appearance of a speckled (spotted) terrain, such as craters on the moon or karst sinkhole regions (limestone solutions) on Earth. The craters are not important individually, but when clustered they characterize the unit of land. These land units are also known as *photomorphic regions* because they are defined by the principal morphology (structure and appearance).

Image 3 is a portion of the moon, showing cratered terrain and smoother, lava field ground, though neither land unit is purely homogenous (map 5). Aggregation of craters (within a certain distance of each other) helps to define the landscape units (map 6). For additional analysis, the new polygons are easy to include in overlay operations with other mapping features.

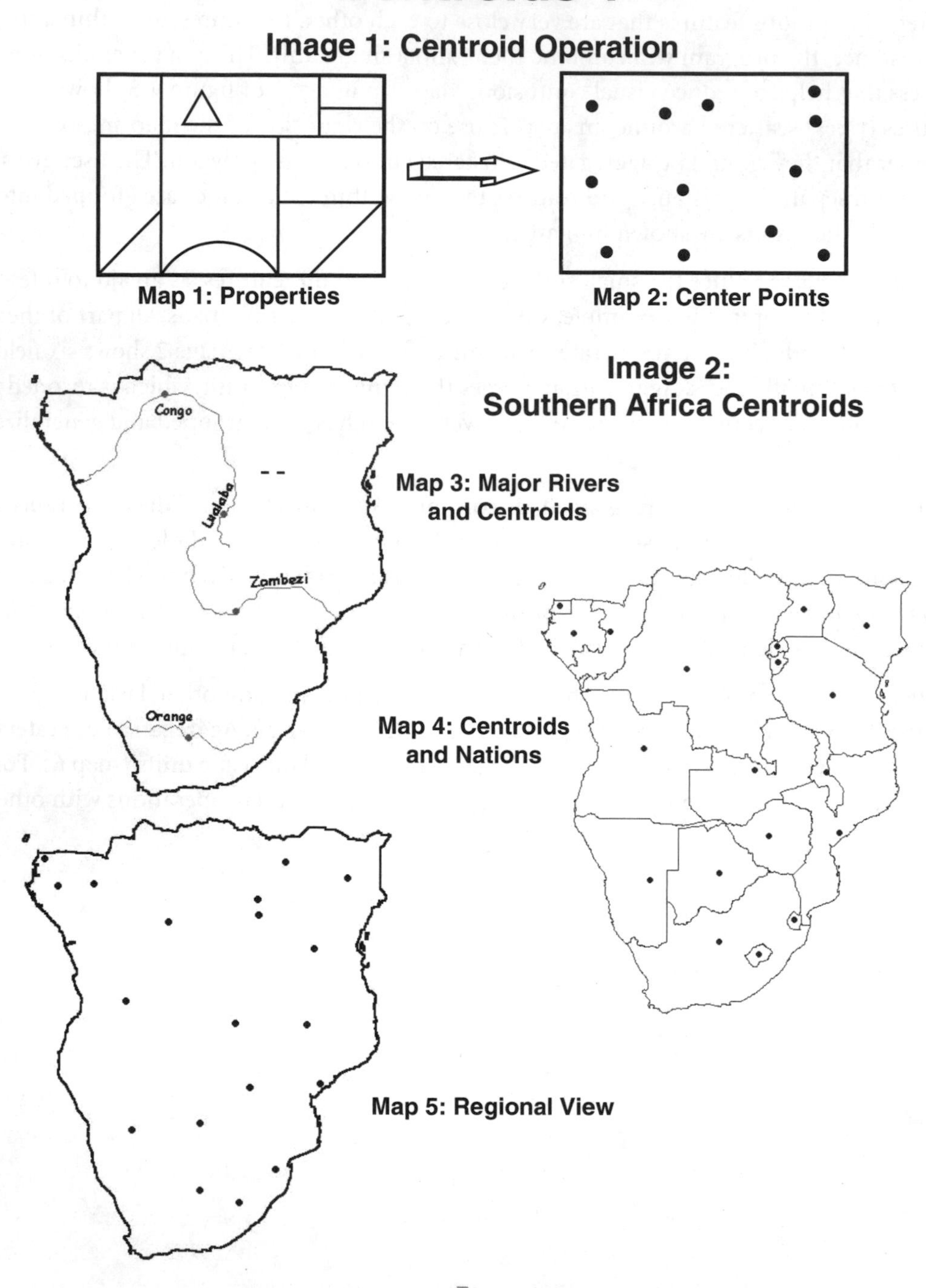

Fig. 9-6: Applications of centroids.

Spatial Operations

Spatial operations use spatial characteristics for analysis and mapping. Two of the most important are *centroids* and *Thiessen polygons*. Centroids associate the size and shape of lines or polygon features, whereas the Thiessen operation expands points into polygons.

Centroids

Lines and polygons can be represented by their center points, called *centroids*. The spatial centers are computed and a single point replaces the feature. All attributes (e.g., population or elevation) are transferred to the centroid's database. Center points usually are the mean X and Y coordinates. For example, a feature stretching from 20.0° and 20.4° east longitude (X) and 10.5° to 10.7° north latitude (Y) will be centered at 20.2° N, 10.6° E. Map 1 in image 1 of figure 9-6 shows a set of properties (polygons) that are transformed into representative centroids, shown in map 2. Note that the eye can appreciate the distribution better with points than with the confusing polygons.

Centroids can have several applications. First is the utility of knowing the geometric center of a polygon; for example, for the best location to represent an area. Placing a new capital of a district or locating headquarters for a sales area can begin with centroids. In addition, centroids can simplify an area of mixed-size and mixed-shape polygon to a more efficient visual and database structure, such as the maps in this illustration.

Spatial analysis is often easier with point features than with polygons. Using points rather than polygons for some analytical operations may be more convenient, such as cluster analyses that calculate the density of features.

Southern Africa is used as a demonstration region in image 2. Map 3 displays center points on line features. A centroid operation is applied to the major rivers of the region, producing point features (the rivers are superimposed for visualization, but they are not necessary; the Lualaba is a main branch of the Congo). The point locations can be important to hydrologists, water resource managers, and tourism activities.

Map 4 shows centroids applied to the Southern African nations. Note that most nations have center points in their approximate visual center, indicating a relatively geometric shape (equal east-west and north-south extents), but others have irregular spatial configurations. Spatial analysis can be applied to the distribution of points and to the shape of nations to gain insights on geographic processes, such as population migration and economic development factors.

Map 5 is the regional display of centroids (without confusing national borders). There are numerous spatial analysis operations that can be applied, such as density measurements, distributions, and overlay with other features. The next section shows experiments with centroids and capital city linkages.

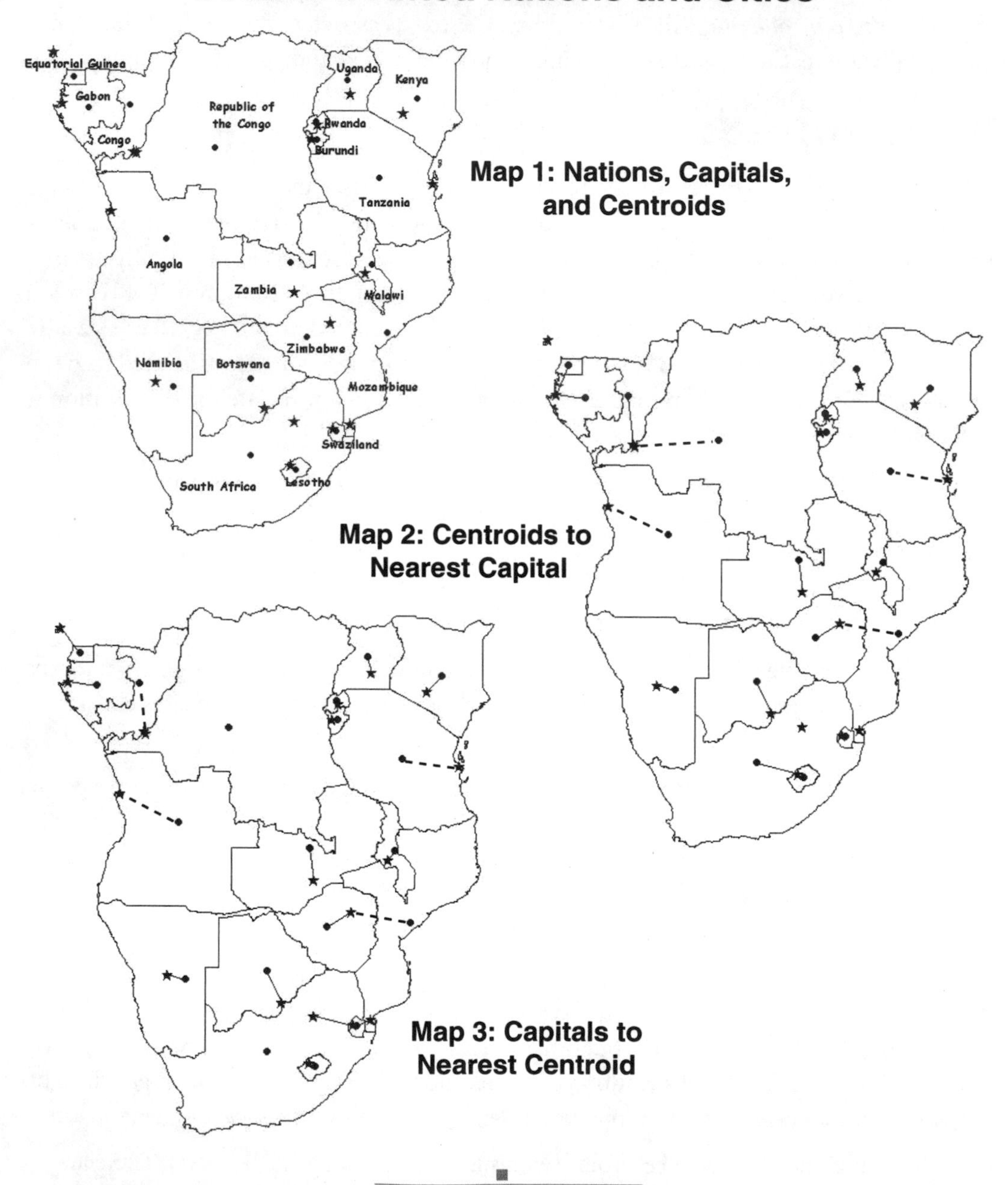

Fig. 9-7: Centroids applied to nations and cities.

Although it is too simplistic to say that capitals should be in the center of nations or even in the center of populations, there are concerns that unequal power and distribution of wealth may be influenced by the capital city location. This may or may not be valid, but an examination of the spatial relationships of centroids and capital cities can be illuminating.

Map 1 of figure 9-7 is a reference map with country names, capital cities (star symbols), and centroids (dots). Map 2 shows centroid-to-capital-city relationships; that is, the nearest capital city to each centroid. Spatial statistics calculate the average distance in kilometers, with plus or minus kilometers in the first standard deviation. The observed distances are reasonably close in some nations (such as Namibia, Uganda, and Gabon), but at least four are beyond the standard deviation of distance, shown with heavy dashed lines (Republic of the Congo, Angola, Mozambique, and Tanzania). Mozambique, South Africa, and Equatorial Guinea have centers closer to other nations' capitals.

Map 3 compares capitals to centroids; that is, the nearest center to each capital city. This is not necessarily the reverse of the first map; compare the maps to see how they differ. Although the same four nations are outside the norm in terms of distance, the connections are different for the Republic of the Congo (its capital is closer to Congo's center, although overlap with Congo's line makes interpretation difficult), and South Africa's capital now connects to a different nation (Swaziland rather than Lesotho, as in the first map). These maps can lead to questions and insights not ordinarily recognized.

Centroids are very easy to generate (usually just a command), and are fairly simple in concept and interpretation, yet they offer a wealth of spatial data considerations. It is probably true that centroids are underutilized GIS elements.

Thiessen Polygons

Image 1: Thiessen Polygon Operation

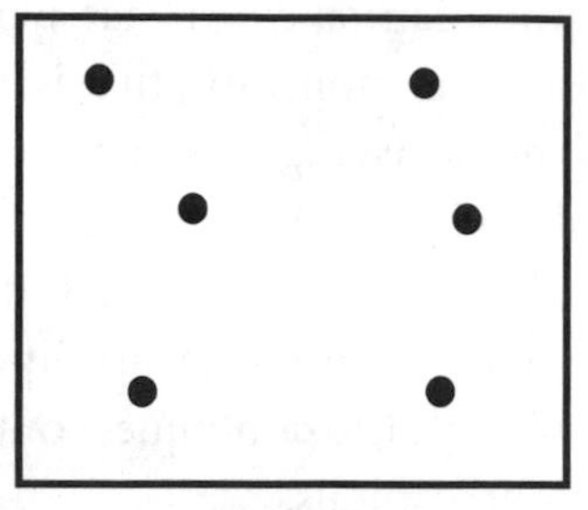

Map 1: Points

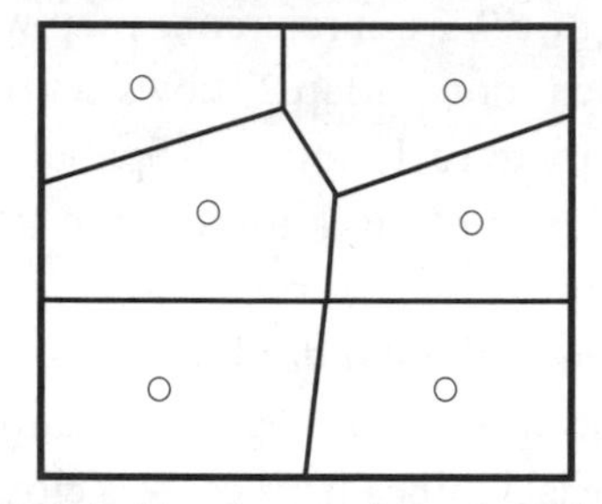

Map 2: Territory of Each Point

Image 2: City-Town Thiessen Mapping

Map 3: Cities and Towns

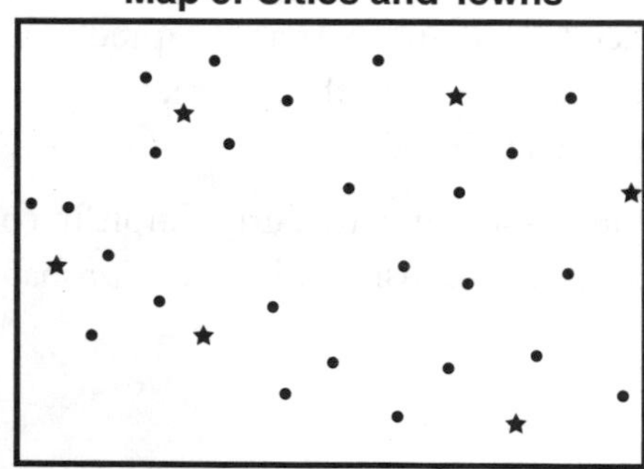

Map 4: City Points

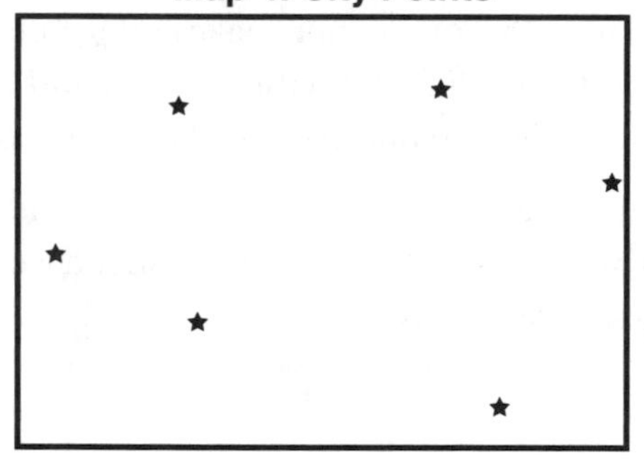

Map 5: Thiessen Polygons

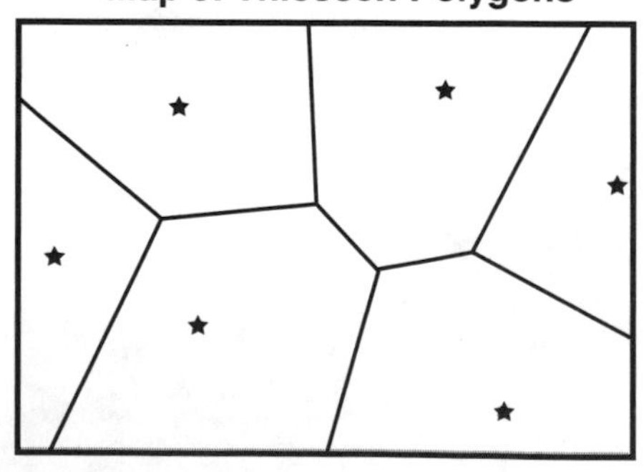

Map 6: Towns per City Hinterland

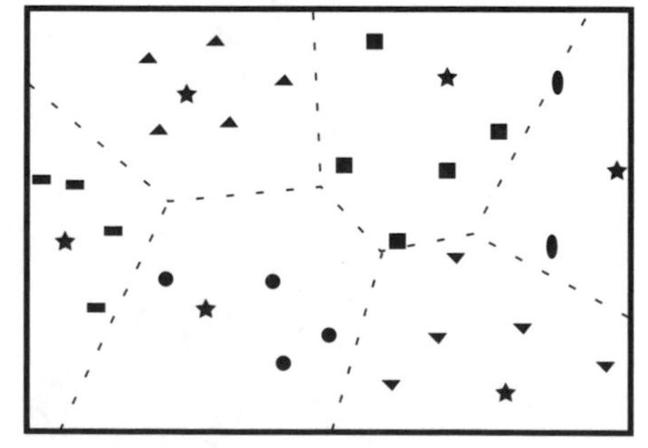

Fig. 9-8: Examples of Thiessen polygon application.

Thiessen Polygons

The opposite of centroids are equal areas around points, termed *Thiessen polygons* (also called *proximal polygons*). They are the "territories" of points. The program extends each point's area until it meets the next one coming from a neighbor point or until it runs into a theme edge. The boundaries are an equal distance (or halfway) between two points. Image 1 of figure 9-8 shows points in map 1,and Thiessen polygons created from those points in map 2. Like the centroid operation, building Thiessen polygons typically involves only a single command.

Image 2 presents four maps of a theme (settlements) to be converted to Thiessen polygons in order to define their "hinterlands"—areas of influence or service (or market area for businesses). Map 3 shows the cities (star symbols) and towns (dots). The city points are separated (map 4) and then used to construct the Thiessen polygons (map 5). The polygons could be individually shaded to indicate some attribute; for example, each city's population class.

Map 6 combines the cities and towns with the new polygons to show complete information (derived data). The surrounding towns are coded according to a city's polygon, using different symbols. This gives a larger view of hinterlands and market areas, from which further analysis can be performed. Actually, other factors will interplay in the final hinterland delineation, such as road infrastructure, other commercial establishments, other land use, and nonspatial factors (advertising, reputation, and pricing strategies), but the Thiessen polygons provide a good starting foundation.

Stores, branch banks, and other commercial establishments can use Thiessen polygons for analysis of service areas. Rain gauges can be placed on the landscape for recording precipitation, and Thiessen polygons, with topography, are used to help determine the area served by each device.

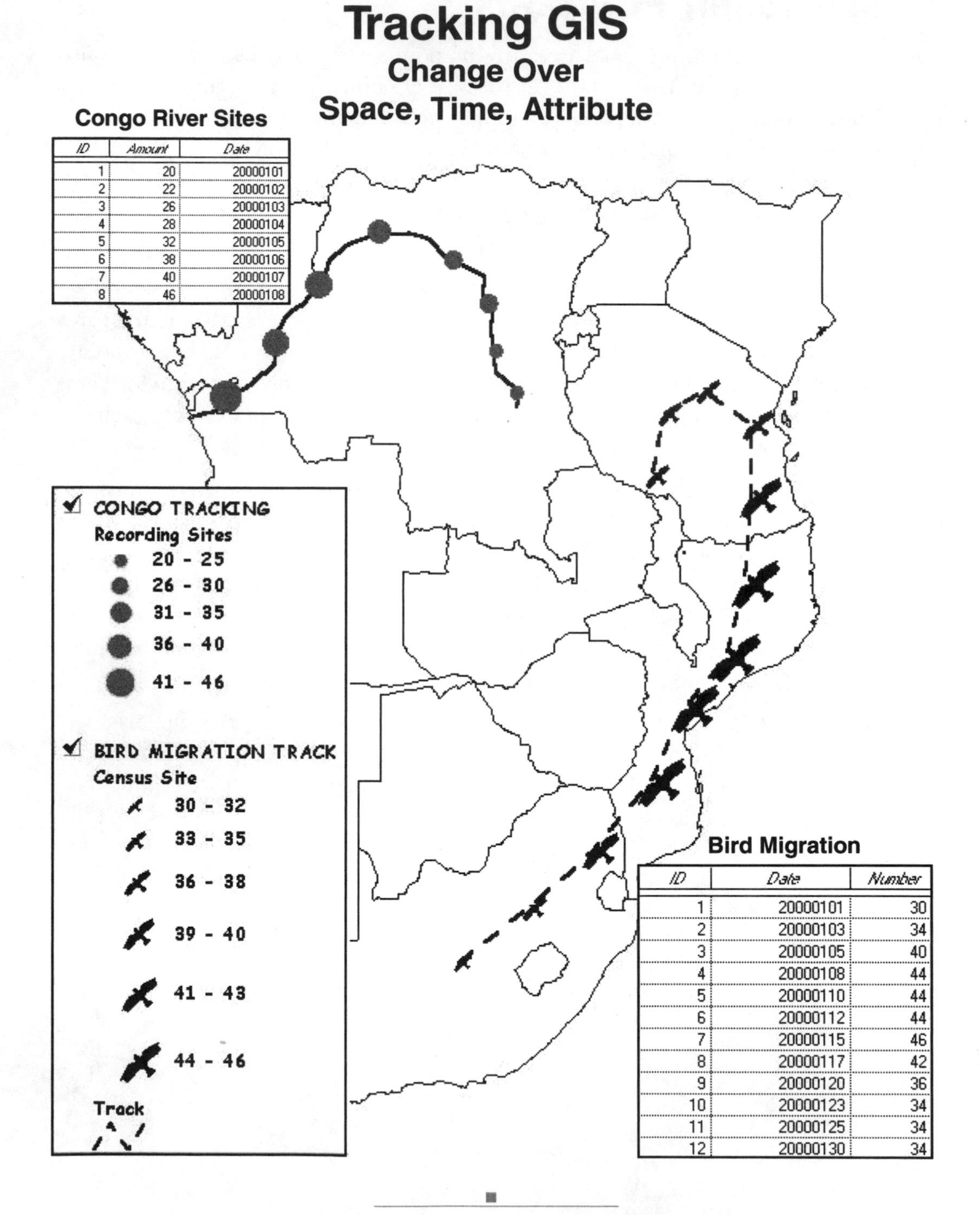

Congo River Sites

ID	Amount	Date
1	20	20000101
2	22	20000102
3	26	20000103
4	28	20000104
5	32	20000105
6	38	20000106
7	40	20000107
8	46	20000108

Bird Migration

ID	Date	Number
1	20000101	30
2	20000103	34
3	20000105	40
4	20000108	44
5	20000110	44
6	20000112	44
7	20000115	46
8	20000117	42
9	20000120	36
10	20000123	34
11	20000125	34
12	20000130	34

Fig. 9-9: Example of GIS tracking.

Tracking GIS

Tracking GIS is a relatively new technique that combines several steps into one procedure. This technique, depicted in figure 9-9, tracks feature locations and their status as they change through space and time, and then builds connections to incorporate all of the information into one theme. Previously, GIS was able to incorporate point data across a map (noting differences in location, time, and other attributes), but the process required the user to build the connections and to construct themes with special symbols and linked information.

Additional analysis was dependent on ad-hoc programs or operations. Tracking GIS provides an efficient and effective technique for integrating temporal and spatial data, even in real time (from GPS or other input sources).

Two hypothetical examples in Southern Africa demonstrate the basic functions. The first deals with a feature or process traveling down the Congo River (upper left) and increasing in "amount" over measured time. This could be a pollutant (increasing in toxicity) or a passenger boat (taking on passengers). Read the corresponding table and note the changing Amount over eight days (the date format is YYYYMMDD—year at 2000, month at 01, and day numbers 01 to 08). The symbols are sized relative to the Amount measure (in five classes). This type of information, especially in real time, can be very useful to the tracking and analysis of hazardous events.

The second example shows an endangered bird migration from South Africa, along the southeastern region, to Kenya. It tracks the 30-day migration period and the population dynamics of the flock. The population begins with 30, peaks at 46, and eventually ends with 34, a sad attrition for an endangered species. The bird map symbols seem a bit silly, but for some audiences, they may be appropriate. They are scaled according to the flock number.

Tracking GIS is a powerful tool that can do much more than is explored here. It is a new type of GIS and shows great potential as to its development and future application.

Terrain Analysis

Image 1: Volcano Island

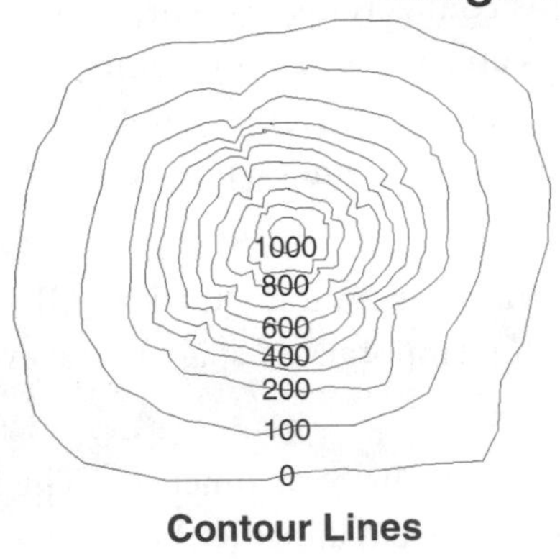

Contour Lines

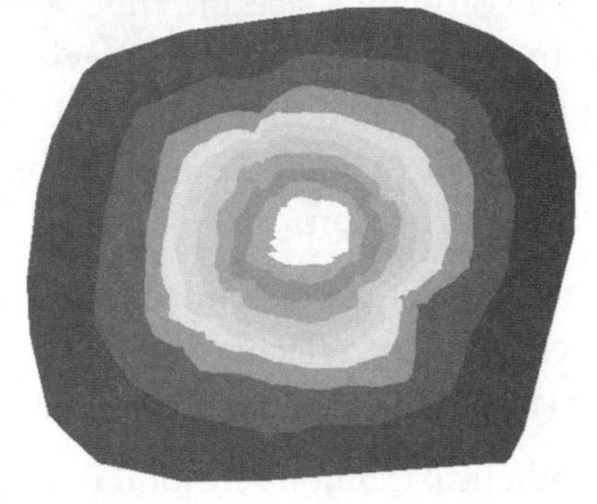

Shaded Elevation

Image 2: Central Kentucky, USA

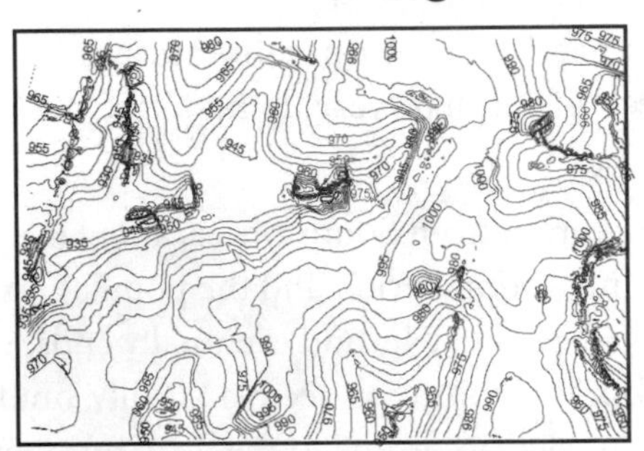

Contours: Central Kentucky

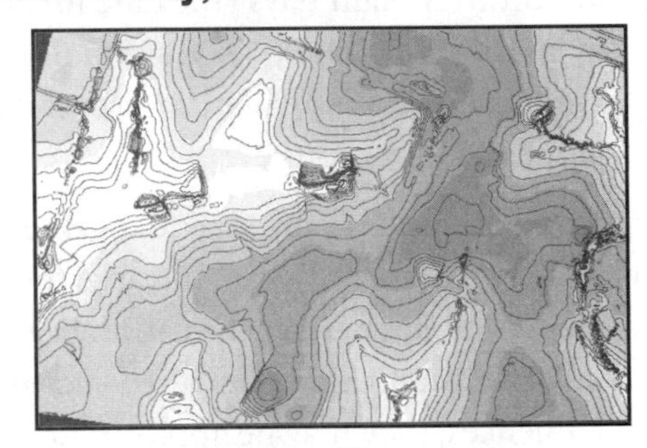

Shaded Elevation

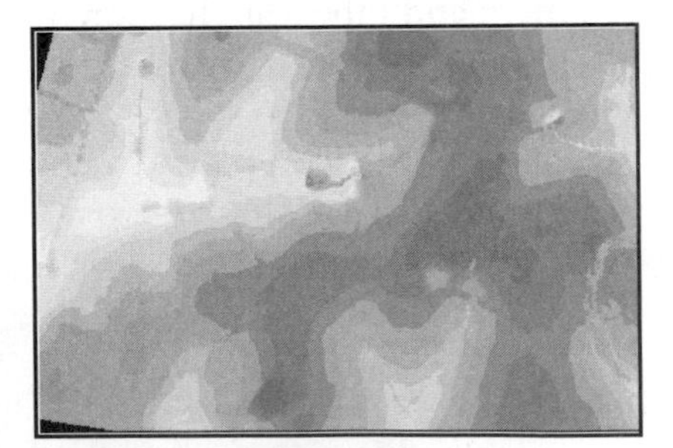

DEM: Digital Elevation Model

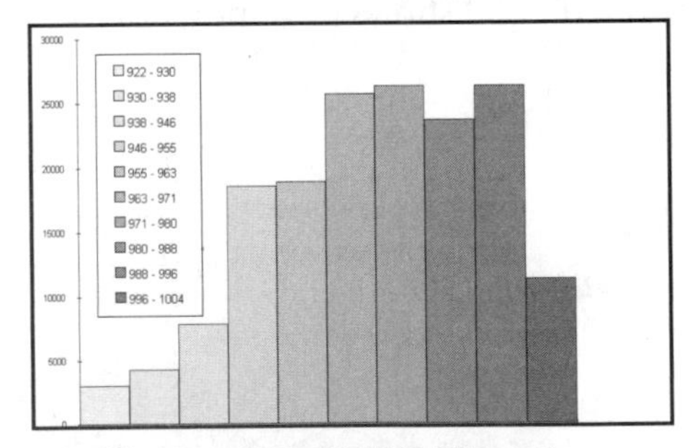

Histogram of DEM Elevations

Fig. 9-10: Techniques used in GIS terrain analysis.

Terrain Analysis

Most maps and GIS data deal with two-dimensional (X-Y) perspectives (e.g., flat maps depicting a region). However, GIS is also very good at incorporating the Z values, as well, such as elevation or any raised value of an attribute (e.g., population). X, Y, and Z are the principal spatial data coordinates in GIS. The sections that follow illustrate a few of the various terrain display and analysis options, most of which are largely based on X-Y-Z data.

Elevation Analysis

Land elevation is the most common Z-data, typically measured and expressed as height above mean sea level. Contour lines show elevations; that is, each line represents the same height above mean sea level everywhere along its length. GIS reads elevation points or contour lines and then displays the terrain more realistically in three-dimensional (3D) perspective. To demonstrate the various 3D terrain options, two examples are used in this section: a simple hypothetical conical volcanic island (easily visualized) and a more complex hill region of Central Kentucky, USA.

Image 1 of figure 9-10 shows the contours of the volcano (named Volcano Island), presenting the classical circular lines that decrease in size as they progress up the cone. GIS created the elevation map on the right, with 100-meter shaded polygons. Both are useful representations of the land, and the contour line data will be used in creating 3D versions of the island.

Image 2 uses four maps of the Central Kentucky region, a rolling mid-latitude landscape with hills and valleys. Map A shows the complicated contour line configuration, which is rather difficult to read except by experienced interpreters. The shaded-elevation map B is a better representation, showing higher elevations in darker tones. With a little practice, the landscape can be visually interpreted.

Map C is a DEM (digital elevation model) that is a raster version of the contour data. Each cell (small enough to be almost invisible except along some polygon edges) is a single elevation value. There are similarities with the standard shaded elevation map, though the two have differences. The raster format offers some display and analytical advantages, as will be seen. Map D is a bar graph, presenting the number of cells in each derived increment (a histogram).

All of these presentations are useful for most land-based projects. DEMs are particularly popular data sets because they are reasonably accurate (with a large number of cells) and can be employed in a variety of GIS operations.

Profiles

Image 1: Volcano Island Profile

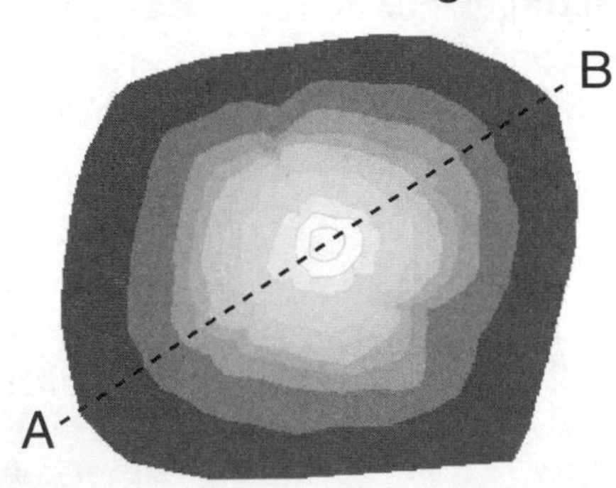

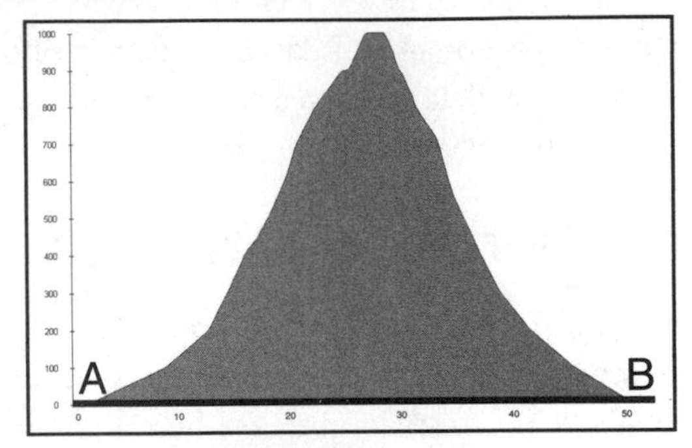

Image 2: Central Kentucky Profile

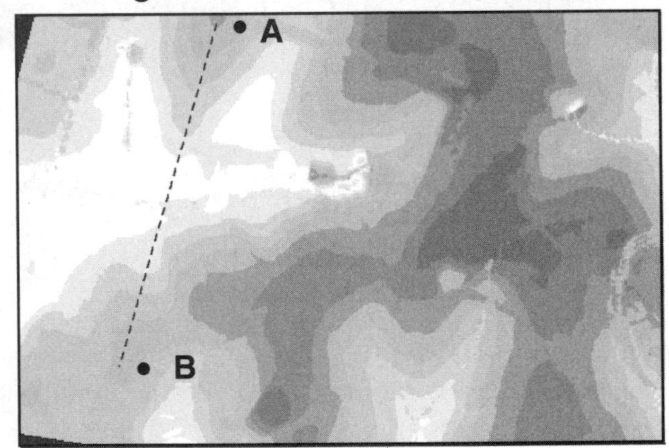

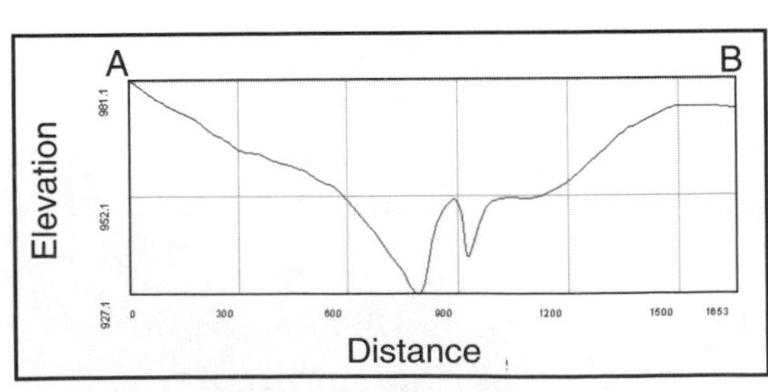

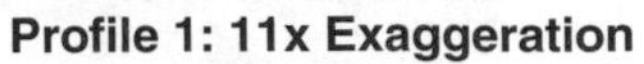

Profile 1: 11x Exaggeration

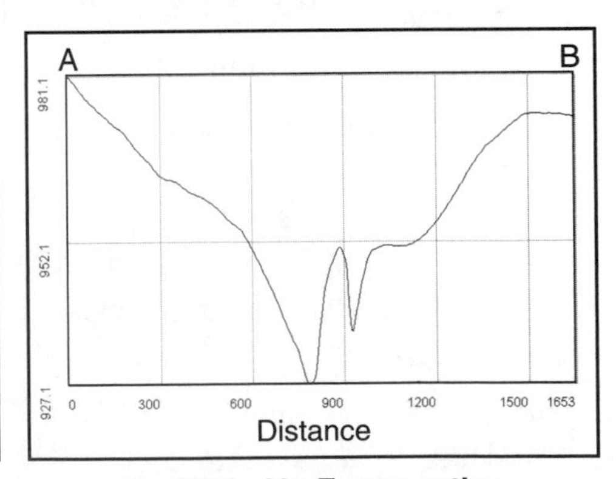

Profile 2: 19x Exaggeration

Fig. 9-11: Examples of GIS profiling.

Terrain Profiles

Profiles are easy and useful GIS operations for terrain analysis. A profile is created by drawing a line across an elevation theme display, reading elevations along the line, and then plotting the "slice" from a horizontal perspective to show the shape of the land. Profiles are very effective mechanisms for understanding landscape form and process. Geologists, for example, can interpret past terrain evolution through a few profiles of a region. Fault lines can be traced, river morphology (shape and form) described and explained, and potential land cover estimated.

Volcano Island (image 1 of figure 9-11) is a simple cone, and almost any line bisecting the island will result in a standard triangular profile. Line A to B in image 1 is the profile line, and the graph at right shows the shape from a side view. The vertical scale is usually exaggerated in order to show subtle changes. A 1:1 graph (no exaggeration) would be almost flat, and difficult to interpret. Therefore, vertical scales are exaggerated to reveal more information.

Image 2 is the Central Kentucky profile, shown on the map from A to B. It begins on a high elevation at point A, dips down into a valley, and goes back upslope to a plateau at point B. Two versions of the profile graph are shown, each with a vertical magnitude. Profile 1 uses 11x vertical exaggeration and profile 2 uses 19x. The graphs present identical data, but with different perspectives. The 19x version can reveal subtle changes and small elevational variations, whereas the 11x is more realistic. Both can be used in a project presentation.

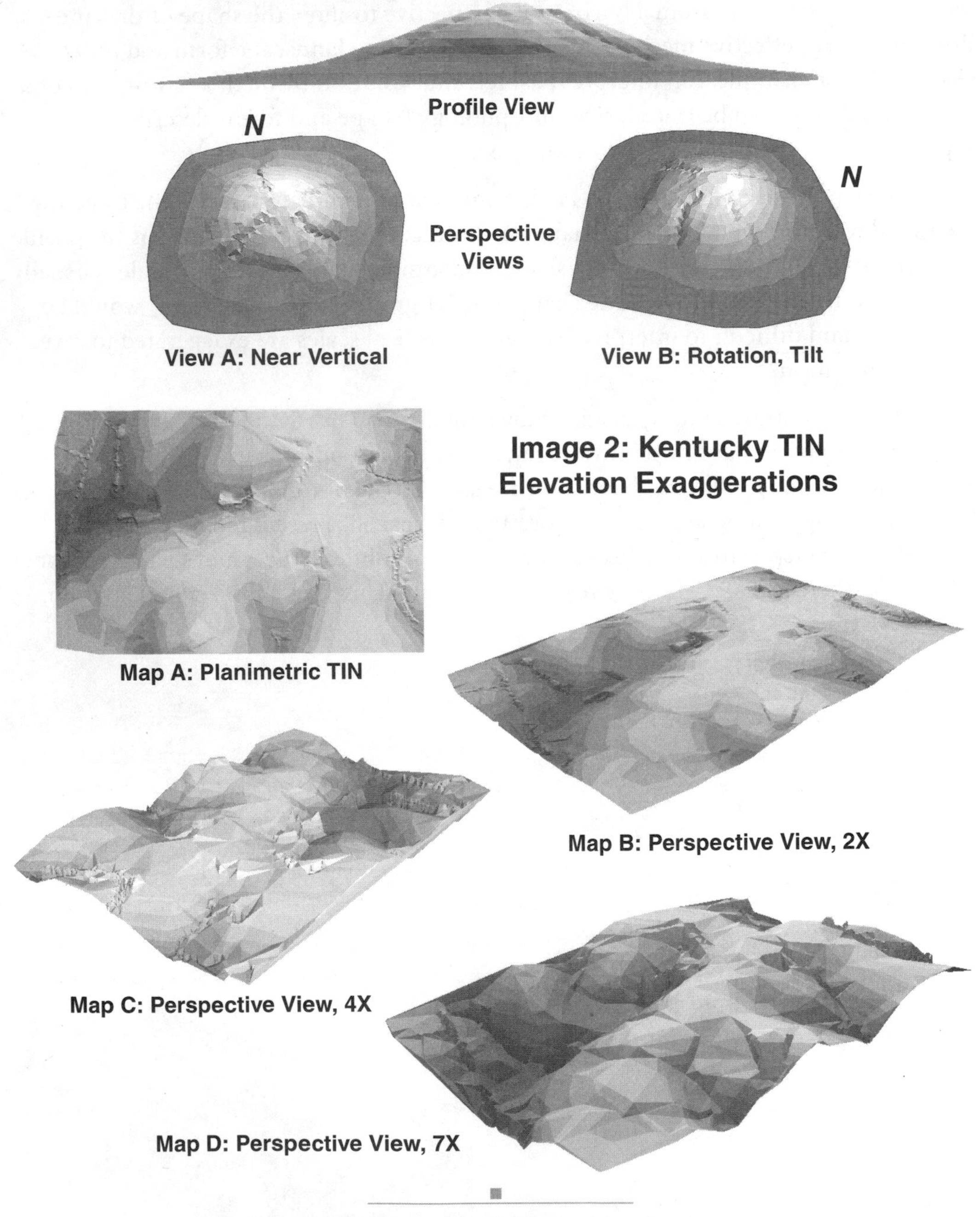

Fig. 9-12: Examples of 3D view applications.

3D Views

Most standard and GIS maps are in flat (two-dimensional, 2D) format, but presentation of landscapes as three-dimensional (3D) views is very useful, both as communications to specialist and general audiences and as an initial exploratory device. The conversion from 2D to 3D involves geometric and trigonometric calculations of elevation points to make surface-fitting triangles called TINs (triangular irregular networks).

The triangles cover the region and represent the terrain so that a 3D perspective can be provided. The mathematics of the process are not important here, but TINs are special data representations that do not integrate with standard raster or vector data formats very well. However, they are excellent visual descriptions of the landscape.

GIS permits great flexibility in observation angles and perspectives. Image 1 of figure 9-12 shows several 3D views of Volcano Island. The profile view at top differs from the previous profile graph because the map is actually made into a 3D view and turned sideways. Landscape features are visible and this scene appears almost realistic.

View A of image 1 is an overhead perspective from near vertical, showing most of the island and its features, but with a significant 3D look that gives the viewer a better understanding of the island than would be possible from a flat (planimetric) map. Any view less than 90 degrees overhead is termed oblique, meaning from the side. View B shows the island rotated about 90 degrees clockwise, and from a lower viewing angle. Perhaps it is a better perspective for certain features, such as the deep valley on the east side and the peak area.

Image 2 is a set of Kentucky TIN scenes (maps A through D), beginning with a planimetric (flat map) view (map A). Some of the triangular TIN structure can be seen. Different perspectives are created by rotating the TIN and raising the Z values by selected exaggerations. Maps B through D demonstrate how different perspectives reveal (or suppress) various landscape features.

Valleys and elevational variations may be more visible with higher exaggerations, but the TIN triangles are more apparent and distracting. Study the maps B through D to see the advantages and disadvantages of each example. Can you tell which of the three views is oriented (rotated) differently from the other two? (Answer: Map C, the 4x view, is turned 180 degrees from the others.)

Slope and Aspect

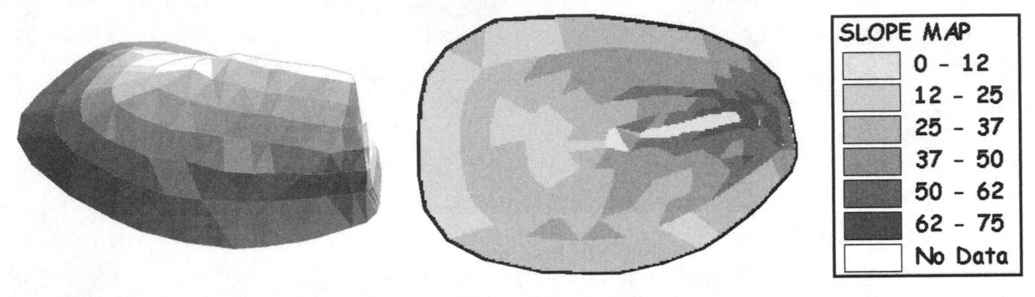

Image 1: Slope Map

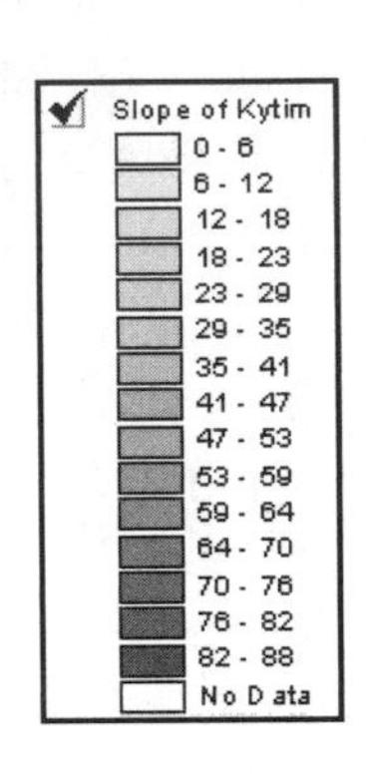

Image 2: Slope Map Kentucky

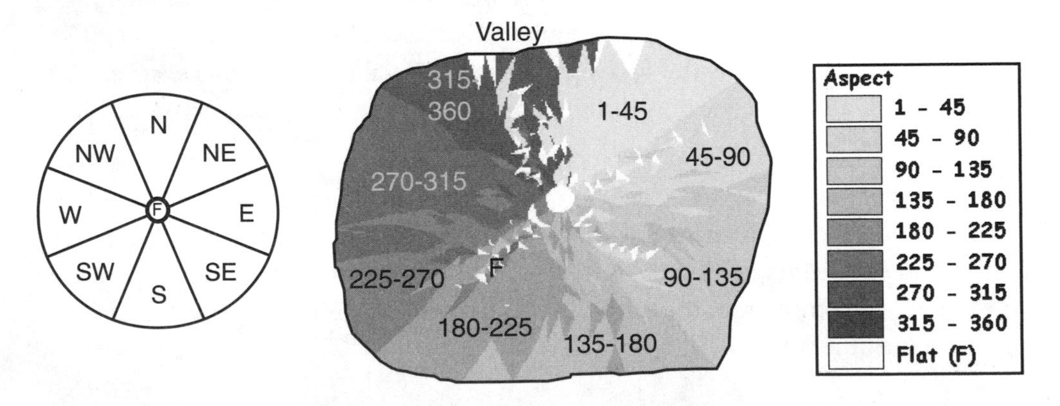

Image 3: Aspect Map Volcano Island

Fig. 9-13: Analyzing slope and aspect.

Slope and Aspect

Because TINs conform to the surface, they can be used to calculate the degree of slope and aspect (direction characteristics) of the land. Slope is very important in many applications because it determines steepness, indicates environmental conditions, and influences the land cover and land use. Slope is one of the most critical information needs for resource managers and urban planners.

Image 1 of figure 9-13 shows a demonstration mountain feature, shaped somewhat like a shoe, with a steep cliff on the right and gradual incline on the left, but with a sharp ridge at the top (the TIN triangles are obvious in places). It is easy to see where the steep and more gradual slopes exist. The slope map expresses slope magnitude in degrees of steepness (progressively darker tones) and the map should be easily interpreted.

The very steep cliff on the right side is represented by dark slope tones, but the gentle slopes on the left have lighter tones. The ridge at top is flat. (The triangles are even more evident on the slope map, even where they generalize too much. A higher density of elevation points would make much smaller and more effective triangles.)

The demonstration mountain is used because Volcano Island will show a common slope on most of the land, thereby displaying a single color or tone, which may be informative but not very visual. The Kentucky slope map is presented in image 2, but the complex landscape is difficult to interpret. Light tones dominate, so the land must be relatively shallow and gently sloping, though it appears to be very steep in several places. Perhaps this map is not informative for a general audience, but experienced users may find it useful.

Slope maps assist urban planners in determining which locations are safe for residential development and which should be left undeveloped. Resource managers use them for erosion assessment, and recreation businesses can search for ski areas.

Aspect is the direction a slope faces, expressed either in cardinal directions (north, east, and so on) or in degrees of azimuth (from 0 to 360 degrees). Aspect is an important ecological characteristic; for example, as a major influence on the type and nature of vegetation. Northern slopes in the northern hemisphere do not receive as much insolation (solar energy) as do the southern-facing slopes. Therefore, the temperature is lower, the humidity is different, and the magnitude of biologic decomposition is slower. Aspect may be important to residential developers, as well. Houses with significant sunshine may be more desirable than those facing in other directions.

Image 3 shows an ideal aspect model of a volcanic cone, with the obvious slope-facing directions indicated. Compare it to the aspect map of Volcano Island. The tones indicate 45-degree increments of aspect. Much of the island is relatively consistent in aspect except along the rugged valley to the north, where a jumble of aspects exists. There are several flat areas, such as the summit of the cone.

Slope and Aspect Options

Image 1: Shaded Relief

180° Illumination

325° Illumination

Image 2: Flow Direction

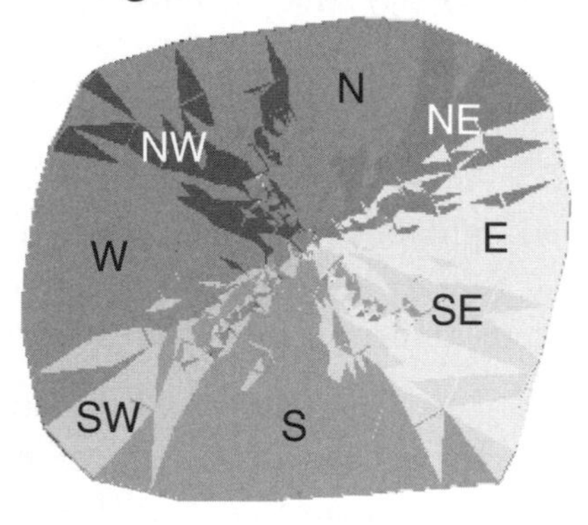

Image 3: Runoff Slope

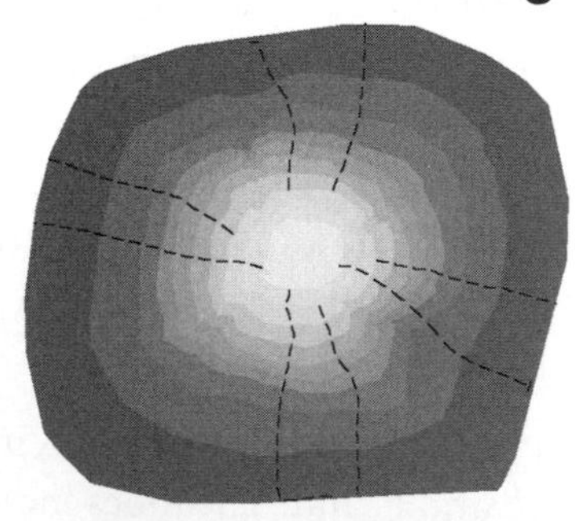

Steepest Downslope

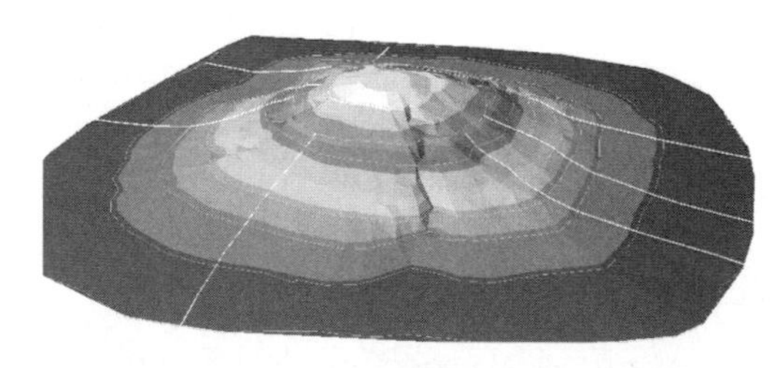

3D View

Fig. 9-14: Slope and aspect analysis options.

An aspect map of the Kentucky area is not presented because it is very complicated and not easily interpreted. This is consistent with rough and uneven topography, where a mix of aspect polygons tends to confuse. Zooming into a particular area often helps to reduce the visual clutter.

Slope and Aspect Options

Slope and aspect data offer several other applications in GIS. Image 1 of figure 9-14 shows how different illumination angles present different shaded relief landscapes. Sun illumination direction determines which features are highlighted or suppressed, such as deep valleys or small rises. The island has several appearances, depending on which side of the volcano is illuminated. Not shown are changes in sun angle (from low to high), which also affect landscape feature visibility.

A low-tilt view with a low sun angle from the south will show selected features much differently than an overhead view with a high sun angle from the east. GIS permits easy visual changes for testing various conditions.

Slope and aspect also determine the direction of downhill movement, such as precipitation, lava flows, or even landslides. Similar to aspect, flow direction mapping, as demonstrated in image 2, can be very useful to managers and planners. If the volcano is active, flow direction of potential lava eruptions can be predicted, leading to better hazard event management decisions. A landslide at a particular location on the slope may have predictable paths downslope, certainly a place to avoid development.

GIS can determine the downhill path of an object or event. Image 3 is a runoff slope map showing the path of water or moving objects as determined by gravity and topography. The points near the top of the volcano were entered interactively by the user (mouse point and click). Then the GIS calculated their downhill paths, presenting results with a dashed line. In effect, this shows the steepest downslope vector from the origin (point) to the lowest topographic point that can be reached. It is possible that a line will extend only a short distance, being stopped by intervening upslope topography. The 3D view (a slightly different version) helps the user better visualize the process, which can lead to information useful to flood or lava flow response management.

Shaded Relief Views

Map A: Ganges Plain and Himalayas

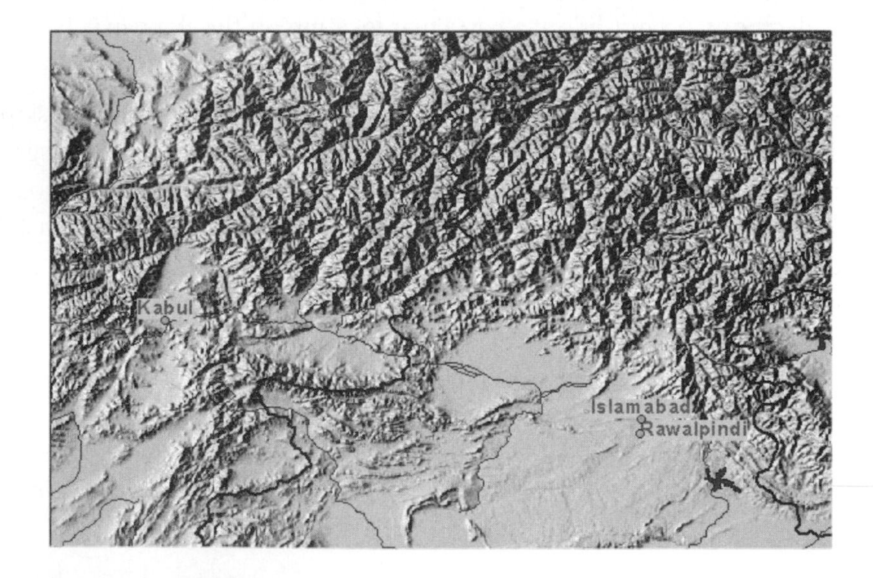

Map B: Pakistan and Hindu Kush

Fig. 9-15: Two shaded relief views.

Shaded Relief Views

Illumination applied to 3D data creates shaded relief views, as noted. Shaded relief actually means that one part of the topography is illuminated and the other is in shade, with intermediate parts in gray tones. To the eye, this gives an appearance of reality, making landscape interpretation easy and promoting further use of the data. Differences in topography can be very important spatial factors in many applications, and shaded relief views are ideal data visualization presentations.

DEMs are a common data format for most parts of the world. They are convenient and easy to use. Shown are two DEM data sets of South Asia, presented in shaded relief views. Map A of figure 9-15 demonstrates the dramatic topographic contrasts between the flat Ganges River plain of northern India and the adjacent Himalaya Mountains to the north. Although standard maps with contours or other 2D devices are clear to specialists, the region is easily appreciated by everyone viewing this type of presentation.

Map B is similar, showing a more rugged region in northern Pakistan, Afghanistan, and central Asia. Several physical provinces can be mapped, and regional geology is interpreted without difficulty. Note the long linear valleys in the mountainous area, probably fault lines created by the forces building the Himalayas. A slightly different terrain is evident in the northwest corner of the image, perhaps the edge of interior Asia, on the "backside" of the Himalaya Range. These views provide a wealth of data and are very useful for numerous applications.

View Analysis

Image 1: Line of Sight

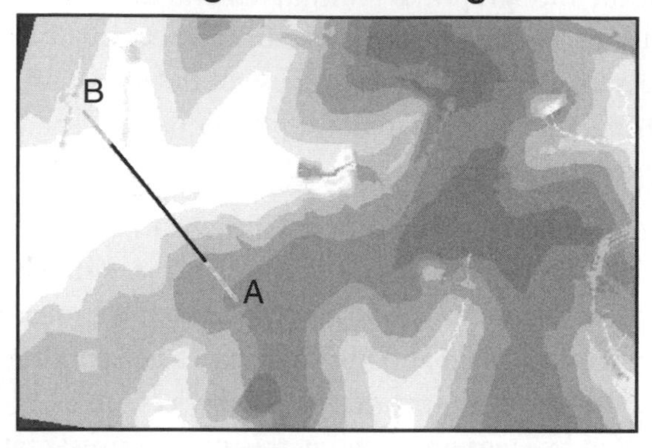

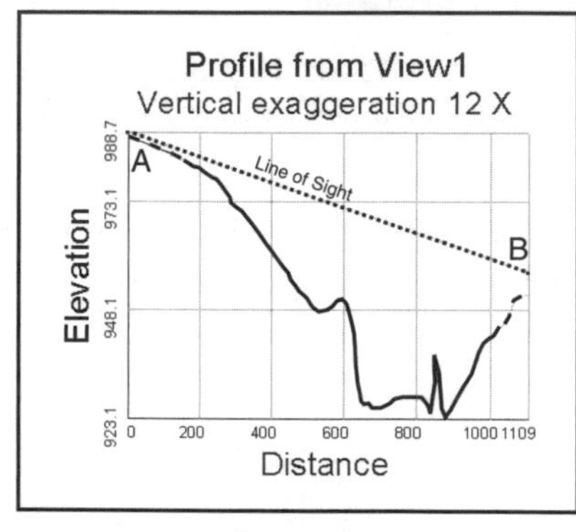

Visibility Profile 1

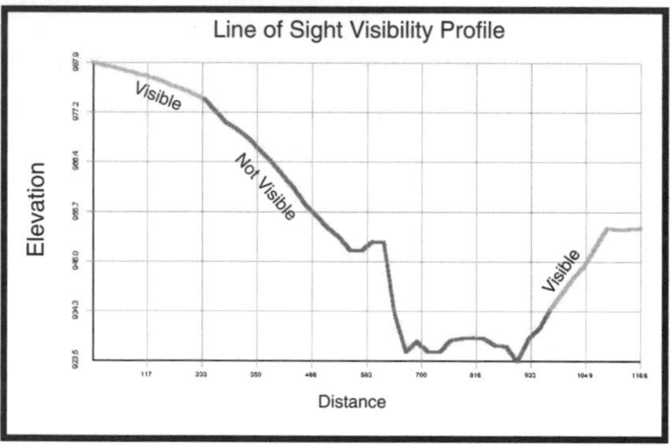

Visibility Profile 2

Image 2: Viewshed

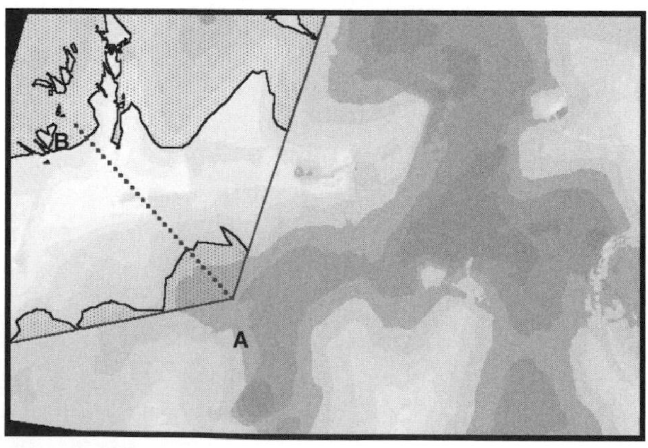

Fig. 9-16: Examples of view analysis.

View Analysis

Another useful GIS terrain analysis operation is the calculation of views from a given point under varying conditions. This is called *view analysis*, which has an objective of determining what can and cannot be seen from a specific location. There are numerous applications, such as resource management, hazard response, and urban development.

One aspect of view analysis is calculation of the view along a specified line to a certain point (from A to B), often termed *line of sight.* Image 1 of figure 9-16 presents a line-of-sight map and two view profiles. The map shows point A origin looking toward point B. An experienced map reader can visualize the changing topography between the points, but the computer provides detailed data in the form of topographic profiles under the A-B line of sight. The program also permits input conditions, such as the height of the observer (as in a fire tower or on the ground) and height of a target object to view. In this case, the question is: Can a person at 2 meters elevation at point A see an object 10 meters in elevation at point B?

GIS reads the topography and specified heights to calculate what can and what cannot be seen. Visibility profile 1 displays the line of sight, underlying topography, and viewing conditions. The topography line (normally in color) has light dashes to indicate visibility, and a dark solid to display what is out of sight. Note this is consistent with the A-B line on the map, which also indicates what can and cannot be seen from A. It appears that the observer can see across the valley to B, but she is too far from the edge to see down into the valley. Perhaps this is acceptable, or perhaps the observer must move closer to the valley in order to improve the visibility.

Visibility profile 2 is a slightly different graph of the same line of sight. It presents the visible (lighter line) and invisible topography. This is a more readable graph, but unlike profile A does not include the actual line of sight. Also note the differences in X and Y lengths, indicating two different vertical exaggeration presentations.

Image 2 is a viewshed, the entire area that can be seen from a given point under the same viewing conditions as image 1 (points A to B and height of observer and target). In addition, the field of view around the observer (to her left and right) can be selected (azimuth angles). Point A still looks toward B (dotted line) and the field of view is shown by the solid line, 120 degrees wide from the observer at A. The pattern on the land close to the observer and across the valley indicate the areas that can be seen, whereas the lighter, slightly transparent pattern in between shows what is not viewed.

Whereas a line-of-sight operation addresses a specific vector for testing the visibility of a particular feature, viewsheds are more comprehensive explorations of what can be viewed from given spot. Both techniques are useful, of course, and each has particular applications.

Surface Features

Image 1: Rivers and Lava Flow Features

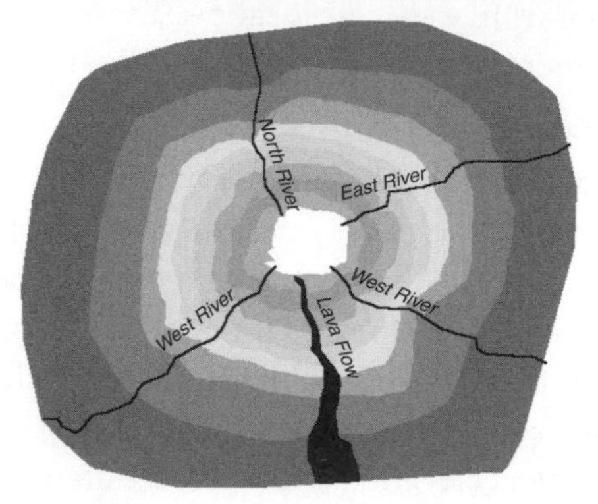

Image 2: 3D Views with Extruded Features

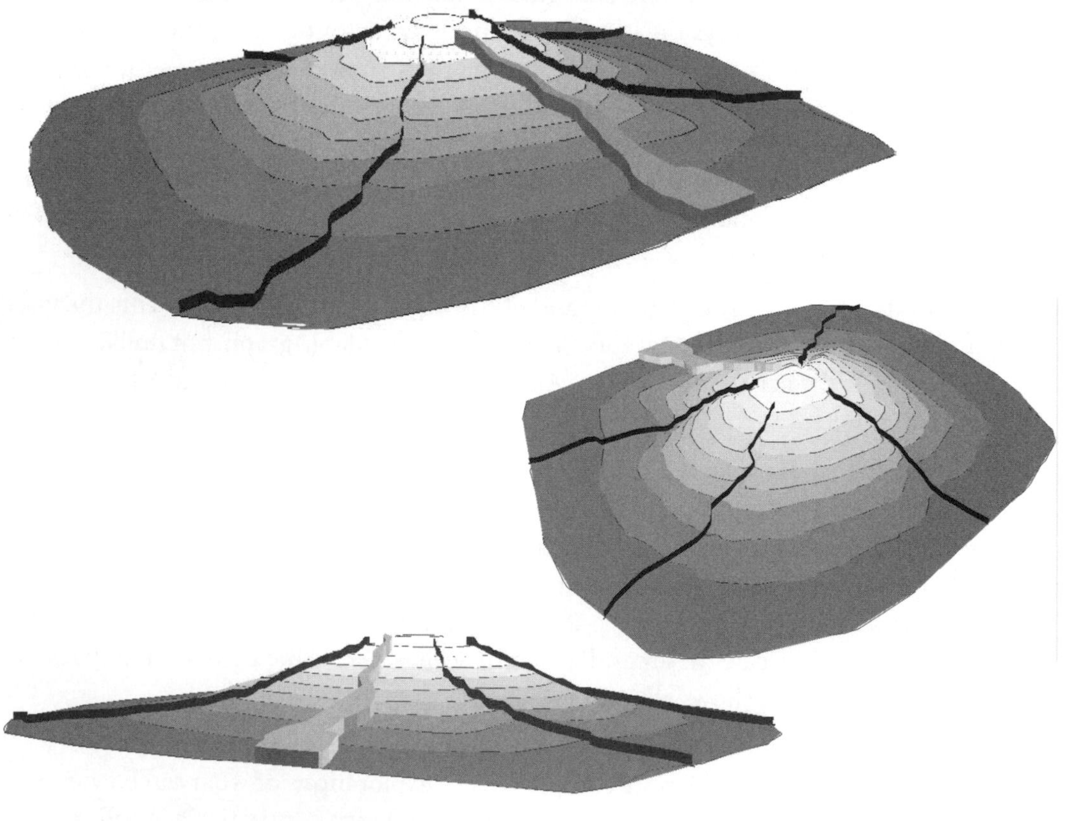

Fig. 9-17: Surface feature views and analysis.

Overlays and Additional Features

TIN data is not directly compatible with standard raster or vector formats, but fairly simple conversion of feature data into 3D files can create overlays. Image 1 of figure 9-17 shows the addition of four rivers and a lava flow on Volcano Island. The features here are regular raster data (derived from vector digitizing), but they can be converted to 3D files to make TINs (sounds complicated, but the process is not difficult). However, even a 3D perspective of the elevation data will still show the rivers and lava flow as flat features on the surface. That may be acceptable, but an interesting option is to give them additional (and artificial) height to increase visibility and to help reveal spatial relationships with the surrounding landscape.

Exaggerating the height of features is known as *extrusion.* Extruded point features are presented as vertical lines, line features as vertical walls, and polygons as 3D blocks. Image 2 presents Volcano Island with the new extruded features. The rivers are unrealistic-looking ribbons, but they are easily visible and can be appreciated in their surrounding. Even the rivers normally out of sight from this view (on the opposite side of the volcano) have some visibility thanks to the additional height from the extrusion.

The lava flow also has an unrealistic scale, but it is interpreted without difficulty and can be seen as a volume feature (which it is). There are various settings for the extrusions, such as extra height or thickness, which present different types of views for analysis. Two different rotations and tilted views are shown. Extrusion of 3D features is an interesting and useful visualization technique that helps one to interpret and understand the landscape.

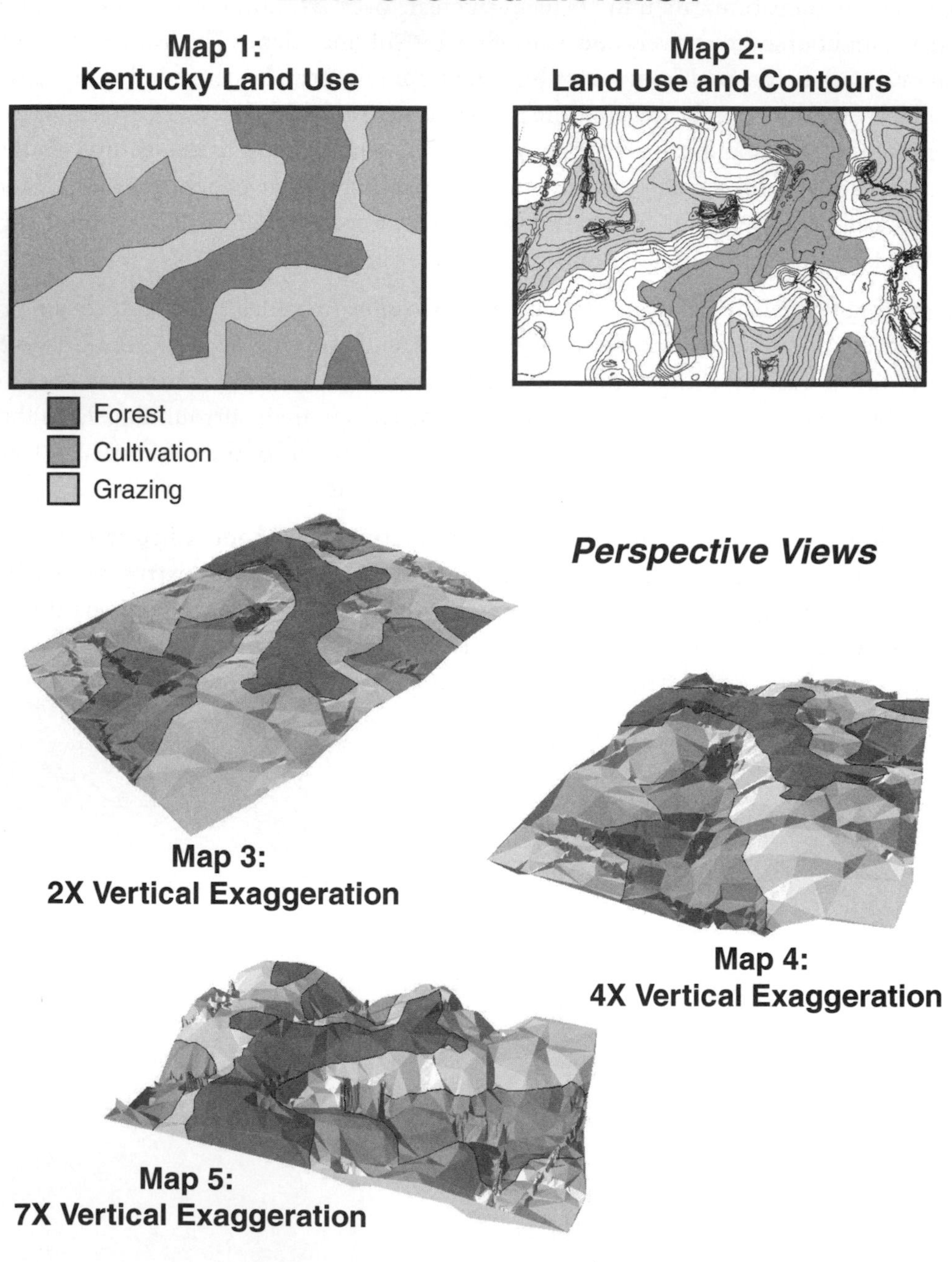

Fig. 9-18: Example of draping land use over topography.

Draping

Overlay of complete data layers onto 3D views is often termed *draping*, with the products called *drapes*. A common need in many applications is to view both the raised elevation presentation and an associated non-elevation theme. Map 1 of figure 9-18 is a land-use map of the Kentucky area, with woodland in the darkest tone, farmland in the medium tone, and grazing in the lightest tone. Map 2 shows the contours overlaying the land use, which is difficult to interpret except by experienced users. Spatial relationships between the various land uses and topography are not easy to interpret. Draping the land use over the TIN will help.

Maps 3 through 5 presents perspective views of the drape, each of a different vertical exaggeration and rotation angle. Interpretation is now much easier. It appears that the woodland is primarily on the higher elevations and farmland is in the valleys, with the steeper slopes being used by grazing cattle. Which perspective is best for an overall appreciation of the relationship between land use and topography? Which one is best for viewing farmland?

The analysis could continue with additional draping; for example, property lines and district zoning boundaries. The ability of GIS to integrate various types of data and to present themes in a variety of views is a significant assistance to a wide range of applications. The data visualization aspect alone is an impressive tool for both general and specialized user.

Perspective Views

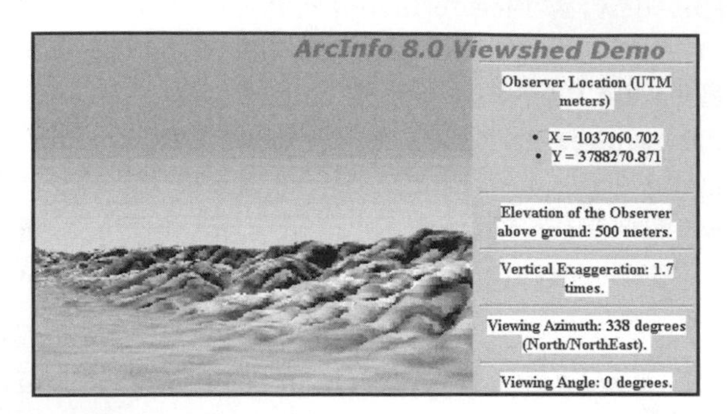

View 1: Hills
500-meter Altitude

View 2: Hills
1,000-meter Altitude

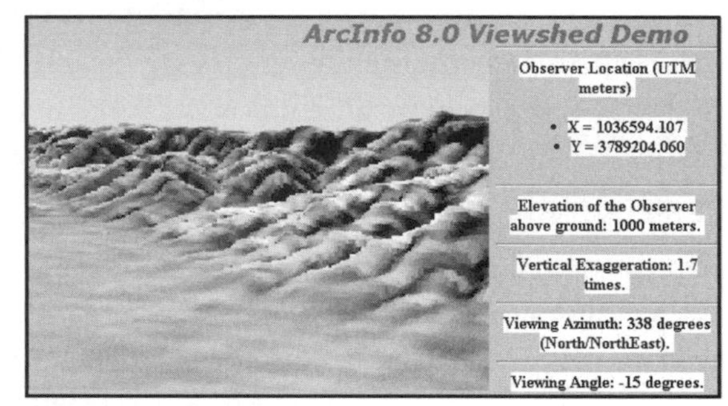

ArcInfo 8.0 Viewshed Demo
Observer Location (UTM meters)
X = 1035660.918
Y = 3786871.086
Elevation of the Observer above ground: 10000 meters.
Vertical Exaggeration: 2.5 times.
Viewing Azimuth: 338 degrees (North/NorthEast).
Viewing Angle: -45 degrees.

View 3: Fault Valley
High View

View 4: Fault Valley
Low View

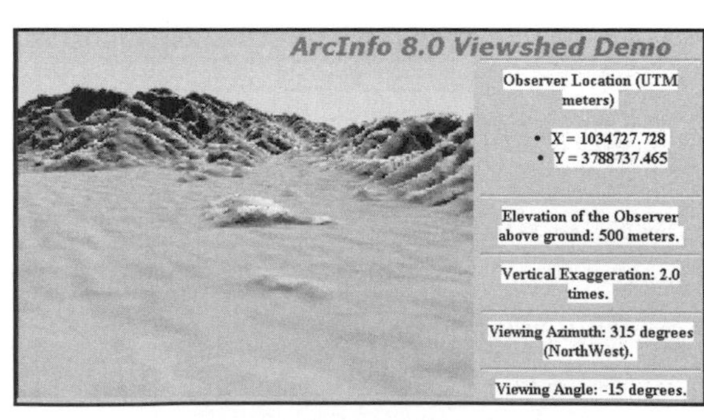

Fig. 9-19: Examples of perspective views.

Perspective Views

Perspective views are effective 3D displays and are valuable for both spatial data visualization and for analysis. They are useful and attractive to nonspecialists as well as to the professional user. In the construction of perspective views, GIS offers interactive control of map rotation, height and angle of the view, and scale of landscape exaggeration. Different features can be interpreted from different perspectives. Often, the user explores many variations to see which one fits the project task. This is a very useful visualization technique that helps to prepare for analysis.

There are several on-line (Internet) GIS web sites that offer interactive perspective views. The interface of the views shown in figure 9-19 allows the user to control the elevation of the observer (altitude), vertical exaggeration of the data, viewing azimuth (rotation), and vertical viewing angle (negative numbers look downward, 0 = horizontal, and positive numbers look upward). Four examples of a hill area in Southern California, USA, are presented that demonstrate the variety of possible views. Also note that these are shaded relief views, projected from the standard vertical viewpoint to lower-angle perspective views.

Views 1 and 2 of figure 9-19 show the hill area from several heights. They use identical settings but the altitude differs from 500 meters (view 1) to 1,000 meters (view 2). Note the differences in visual information that can be obtained. Which one is best for geologic interpretation? What applications could use the other view?

Views 3 and 4 view what appears to be a fault valley, about 90 degrees left of the view 1 site. A geologist needs to observe the area from several perspectives to confirm or reject the fault valley interpretation. The two views differ in elevation, vertical exaggeration, azimuth, and viewing angle.

View 3 is from a high perspective, showing the regional context and most of the valley. Some of the smaller subsidiary valleys, as well as the overall shape of the valley, are easily seen. A synoptic view is gained by the geologist. View 4 is from a low position, from near the ground (or so it seems, even though the altitude is 500 meters), showing some of the geologic detail near the open end of the valley. Each view has advantages and disadvantages, but the interactive capabilities help the geologist explore a variety of perspectives that will help arrive at a satisfactory interpretation.

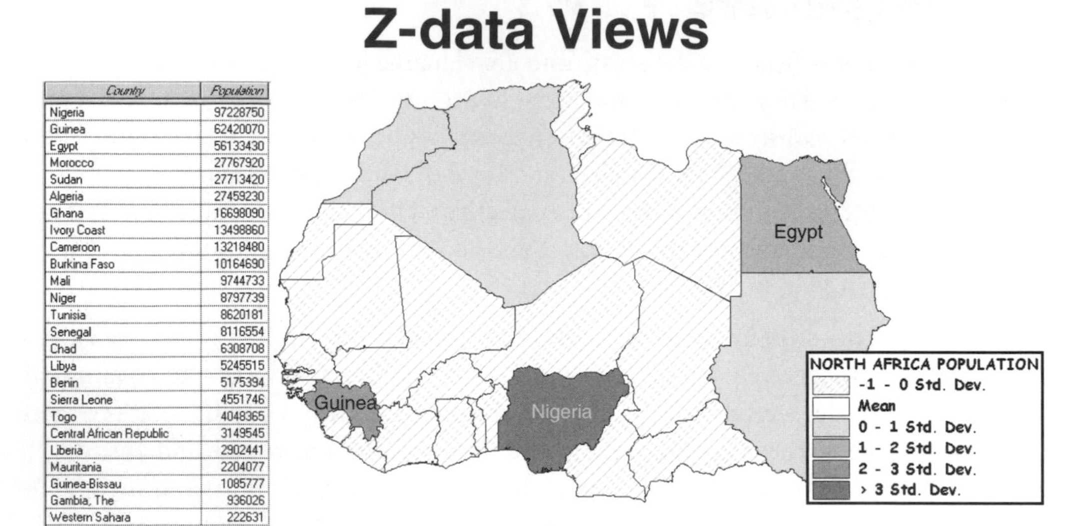

Country	Population
Nigeria	97228750
Guinea	62420070
Egypt	56133430
Morocco	27767920
Sudan	27713420
Algeria	27459230
Ghana	16698090
Ivory Coast	13498860
Cameroon	13218480
Burkina Faso	10164690
Mali	9744733
Niger	8797739
Tunisia	8620181
Senegal	8116554
Chad	6308708
Libya	5245515
Benin	5175394
Sierra Leone	4551746
Togo	4048365
Central African Republic	3149545
Liberia	2902441
Mauritania	2204077
Guinea-Bissau	1085777
Gambia, The	936026
Western Sahara	222631

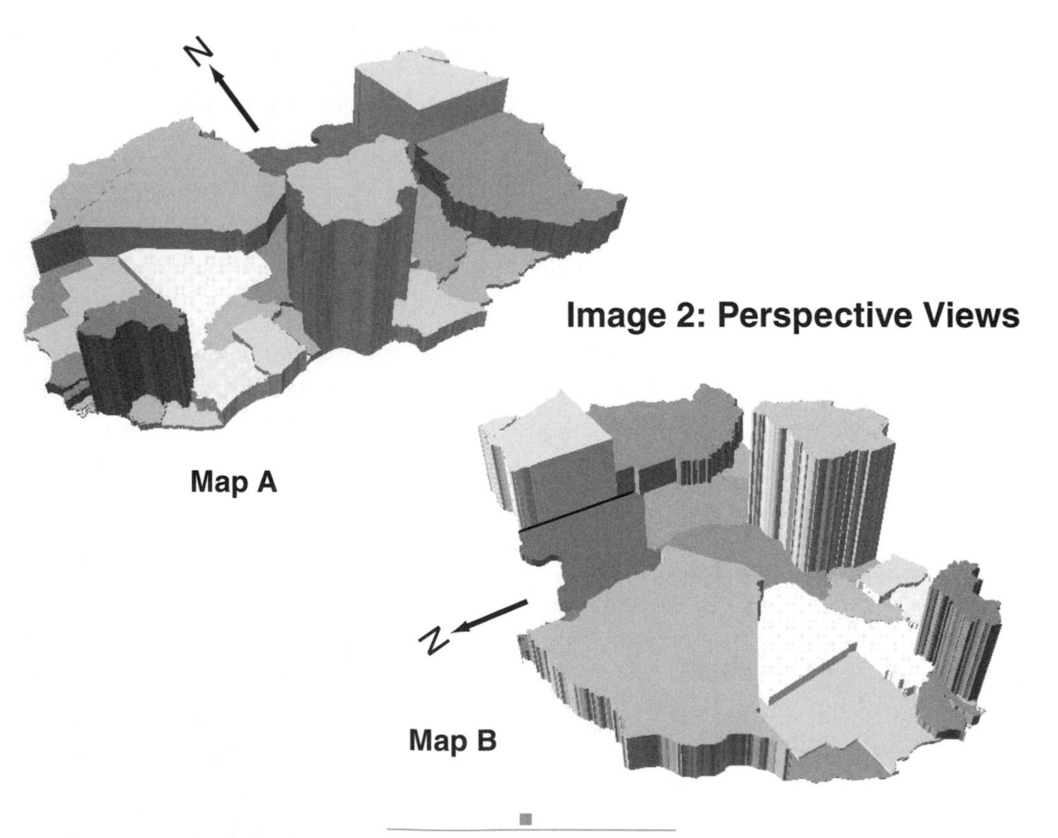

Fig. 9-20: Examples of Z-data views. (Data permission of ESRI, http://www.esri.com)

Z-data Views

Terrain analysis techniques can be applied to other types of data. Any attribute that can have a measure assigned to it can be expressed as Z-data. GIS treats all numeric values the same way, regardless of meaning (after all, the computer is only a very fast, but quite ignorant, number cruncher; it does not care what is being computed). For example, polygons of population data can be elevated to indicate population magnitude in order to show comparisons. This type of data visualization often may be better than normal classification maps.

Perspective Z-data views are presented, with image 1 of figure 9-20 showing North African countries (with only selected names, to avoid clutter). The database at left presents sorted populations, from highest to lowest. Note that Nigeria has the largest population, which is 55 percent larger than the second most populous country, Guinea. Note also that there is a wide range in terms of population among the countries.

Regional population and their comparisons are difficult to interpret on a map displaying a different color or tone for each population number. Even the map here, classified by standard deviation of region populations, still forces the reader to scan and interpret. Nigeria is in a class of its own (+3 standard deviations from the mean), but the real magnitude of population differences is still difficult to appreciate. A better format will help.

Using population as the Z-value, a 3D map raises the polygons according to their magnitude. Image 2 shows two perspective views. Map A has the region slightly tilted and rotated for easy interpretation, with north toward the upper left. Nonspecialists are not disoriented by these views, and are able to understand the polygons. Nigeria's dominance is clear because of its height. One quick glance gives the viewer a comparison of North Africa's populations in both magnitude and distribution.

Map B is a bit more unusual, with north toward the lower left. However, observation of the northern nations is improved, with the high values of Nigeria and Guinea still apparent. Perhaps no single rotation gives a perfect presentation, but the perspective views represent relative populations much better than the 2D planimetric maps.

Almost any quantitative data can be processed as Z data for 3D display. Those views work very well for visual comparisons and dramatic representation. Even classifications (generalized data) can be projected vertically. In fact, with a little effort, the vertical elevation can indicate one measure (population), and color or shading another (income).

Network Analysis
Shortest Route

Map 1: Shortest Route Madrid to Lisboa

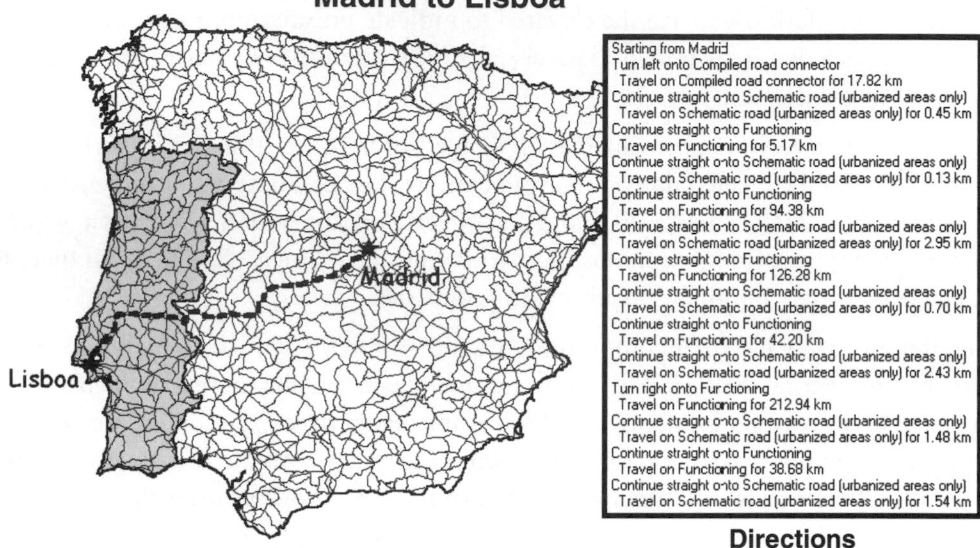

Directions (Partial)

Map 2: Shortest Route Madrid - Oviedo - Lisboa - Madrid

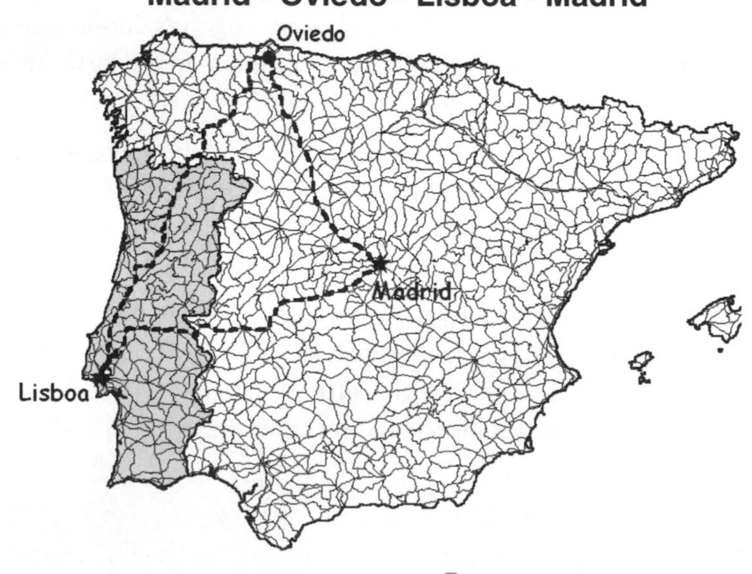

Fig. 9-21: "Shortest route" processing under network analysis.

Network Analysis

The sections that follow discuss specialized applications in regard to analysis of spatial networks. The processes explored are routing, the concept of "nearest facility," and service area parameters.

Routing

In recent years, GIS has branched into specialized fields, with dedicated techniques and software for specific tasks and applications. Network analysis is another field that uses specialized GIS. This and the sections that follow explore network analysis techniques, primarily with an emphasis on distance operations.

Geographic networks consist of multiple, connected lines of a particular theme (typically transportation, rivers, and similar features), though any type of route or linear phenomenon may be included. Basically, network GIS operates on these line features, but it also includes the surrounding area and associated attributes. Distance along roads or other line networks is fundamental information in many applications.

Map 1 of figure 9-21 shows the Iberian Peninsula nations of Portugal and Spain, with a complex road system. A simple assignment is to find the shortest path between Madrid, Spain, and Lisboa (Lisbon), Portugal. Additional information is required, such as directions and distances. The network analysis operation asks for the starting and ending cities (or other points) to be connected.

Finding the shortest path from one location to another involves tracking all possible routes and presenting the one having the shortest distance. Computers are very fast with this type of task (manual measurement would take a while on this high-density road system), and results are presented in just a few seconds.

The heavy dashed line is the shortest distance along the roads between the two cities, covering 617.53 kilometers. Directions are presented, usually in a long list of individual roads, turns, and distances (a partial list is presented in the accompanying box). The travel industry has many applications for this type of information, as do trucking companies and other transportation activities.

Any number of connecting cities can be used. Map 2 shows a route from Madrid to Oviedo (in the north), on to Lisboa, and then returning to Madrid. The return option is normally just a click selection. The total distance of this trip is 1,346 kilometers (directions are too long to present). Any number of stops can be added, either through point and click manual selection or from a list of selected cities.

Distance can be expressed as linear extent (any unit of measurement), cost, or time. Cost is easily calculated using either the price per kilometer or per unit of time. Time may be linked to average speed or even to individual speed limits or conditions on each section of road. For example, construction along one segment slows movement to a certain speed (sometimes called a *friction surface*), which is entered in the database as a network analysis factor. It is possible that the shortest route may not be the fastest.

Efficient routing is a major requirement for many businesses and services, especially for emergency response applications. It is often said that "time is money," and for many enterprises GIS can help save financial resources. Other applications include route planning and vehicle fleet management. In fact, there is a separate subdivision of GIS devoted to transportation, sometimes called transportation GIS or GIST.

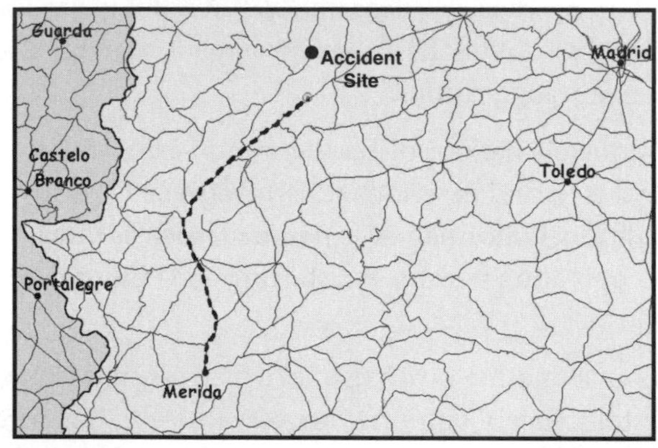

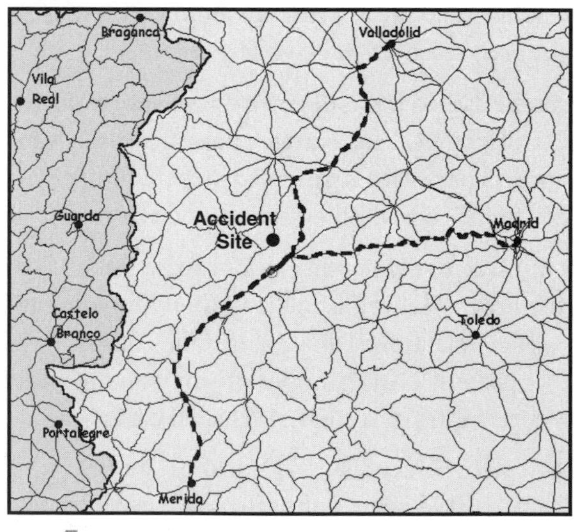

Fig. 9-22: "Nearest facility" processing using network analysis.

Nearest Facility

A primary task of network analysis is to find the shortest path between features, as demonstrated in the previous section. This includes the use of attributes other than simple distance; for example, connecting only those cities with populations over a given number. The concept of *nearest facility* refers to features along the network that are designated as a "facility" in some way. The most common meaning relates to a service, such as a hospital, although it could be defined by any attribute important to the project.

Emergency response applications are major users of nearest facility operations. Map 1 of figure 9-22 shows an accident site in Spain, distant from the major cities where advanced medical facilities exist. It is very easy to assign the accident site as a beginning node simply with a point-and-click action. Then the nearest facility operation is activated, with the program reading the database for the cities with appropriate attributes. The shortest (or fastest) path between the accident site and the city defined as a facility is calculated and marked, as shown. Distance is also given (or time or cost), which in this case is 240 kilometers to Mérida. As usual, directions could be printed for driving instructions.

The nearest facilities (plural) could be calculated, if needed. Perhaps multiple sites are needed, or backup facilities must be known. The operation can easily include the closest three facilities, as demonstrated in map 2. A major road accident occurred at the designated site, resulting in more patients than can be managed by any one or two hospitals, so the nearest three are requested.

It is possible to limit the search distance when needed. For example, under situations when driving time must be restricted, the maximum distance can be designated and the search is then limited to those settings. Although a major medical facility is preferred, a smaller one may be acceptable under certain conditions. The procedure also lets the user decide travel direction, either to the accident or from it. In most cases, the measure may be the same, but of course the directions are dependent on which point begins the path.

In the city, a system of one-way streets may make the outgoing route different from the incoming, thereby changing the distance and time between the accident and the facility. Response time to the accident may be different than return travel time to the medical facility. Further, on certain days the local football game makes the primary road impassable or very slow, thereby creating a significant friction surface for route determination. These types of variables can be included in the database, ready for use in an emergency.

It is apparent that the graphical part of the shortest route and nearest facility operations are fairly simple, but the database must be defined appropriately for various applications and conditions. Being prepared for disaster is very important in many applications, and the flexibility and speed of GIS in incorporating, analyzing, and interpreting an infinite number of variables can make significant contributions.

Network Analysis
Service Areas

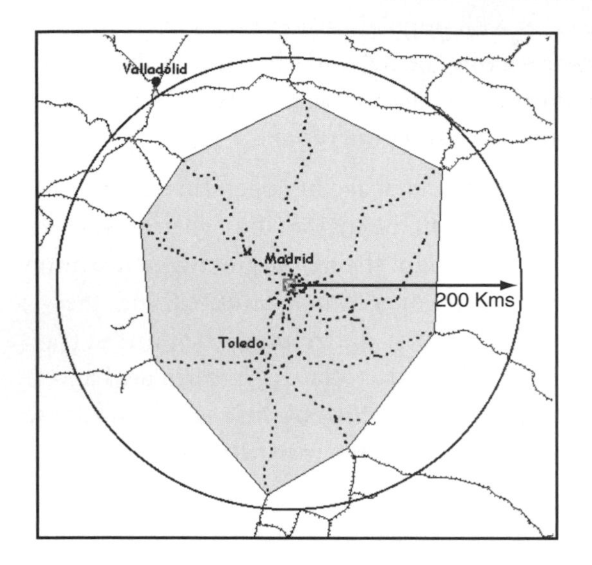

Map 1: 200-kilometer Service Area

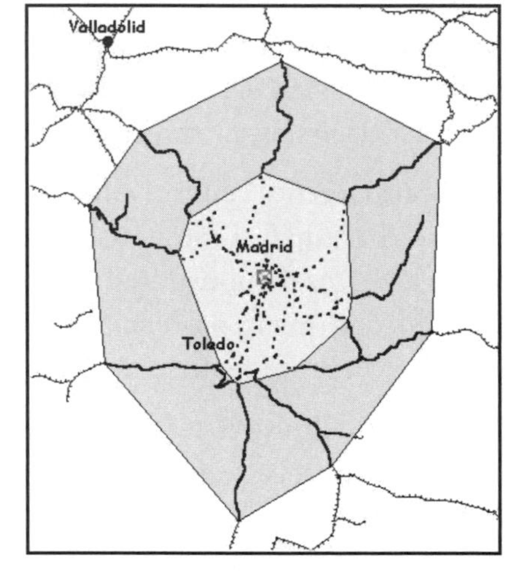

Map 2: 100- and 200-kilometer Service Areas

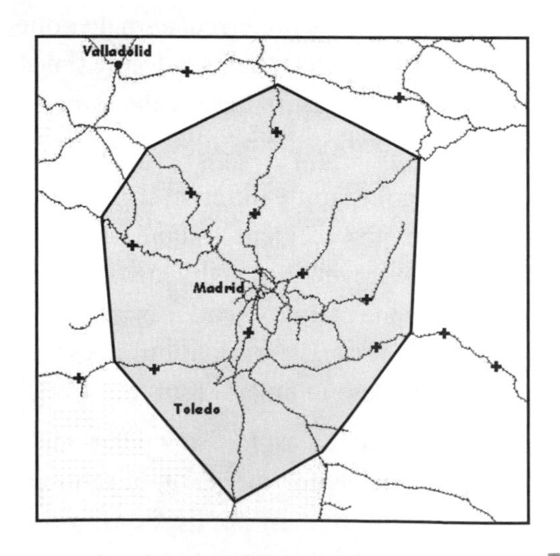

Shape	ID	Units
Point	1	12
Point	2	14
Point	3	5
Point	4	23
Point	5	21
Point	6	15
Point	12	10
Point	13	17
Point	14	11
Point	7	11
Point	8	11
Point	9	21
Point	10	14
Point	11	18

Statistics for Units

Sum: 128
Count: 9
Mean: 14
Maximum: 23
Minimum: 5
Range: 18
Variance: 31
Standard Deviation: 6

Map 3: Service Area Overlay Customer Inventory

Fig. 9-23: "Service area" processing under network analysis.

Service Area

Buffers create zones at given distances from features, normally using straight-line measurements. It is easy to draw a 200-kilometer circle around Madrid, but in many applications the true on-ground travel distance is more appropriate. Most people depend on transportation lines for movement, so it is realistic and logical to measure the actual travel route rather than using a simple circle distance. Network analysis provides techniques to measure and mark such routes and to define the area within the designated range. That area is termed a *service area* because it is the territory served by the measured transportation routes.

Map 1 of figure 9-23 shows the 200-kilometer service area around Madrid as measured along the railroad routes. A heavy machinery business transports its products by rail and needs to know the area to be served in a new customer service plan. The service area is an odd-shaped polygon in the darker tone, centered on Madrid, and the rail system within it is depicted as dotted lines (which continue as railroad line symbols, barely recognizable at this scale, outside the 200-kilometer reach). A 200-kilometer circle centered on Madrid is shown for comparison. As expected, it extends further than the railroad-defined service area.

Service areas can be defined by increments of distance from the center. The machinery company wants to know the section of the 200-kilometer service area that is within 100 kilometers of Madrid. Map 2 shows two service area zones of 100 kilometers each. The railroads of the inner area are dotted lines, and those of the 100- to 200-kilometer zone are solid lines. Other increments could be explored to help refine the new business plan.

Service areas can be used as standard themes for further analysis. The company now needs to know the customers served in the new service area and their projected business. Map 3 presents the service area in a darker tone, and customers as + symbols. The customers within and outside the service area can be seen, but a GIS overlay will identify and select those meeting the company's criteria. The database at right shows the nine service area customers (darker records) and statistics on the number of machinery units purchased in the previous year. The company can make projections based on those numbers and then modify the business plan if necessary.

Service areas are useful tools in many applications. They function as territories defined by travel criteria (e.g., distance, cost, or time), and they can be flexible GIS features for spatial analysis. In some ways, they are a nice bridge between network analysis and standard GIS point-line-polygon operations.

CHAPTER 10

SITE SUITABILITY AND MODELS

Introduction

THIS CHAPTER TAKES GIS ANALYSIS TO THE NEXT LEVEL, namely, applying various operations to site suitability analysis and developing GIS models. The wide range of GIS operations discussed in previous chapters serves many purposes. One of the major applications that uses a mix of procedures is *site selection*, which is finding sites that meet specific conditions. This may include the best, worst, or most sensitive sites. Site selection is both an application and a methodology.

Modeling is a major GIS development. GIS models have several definitions and are used in a variety of applications. For example, models help to simplify an environment and its processes, enabling us to understand it more easily. They also are used to establish multiple-step procedures for completing tasks. Four types of GIS models are discussed in this chapter: descriptive, process, predictive, and procedural.

Site Suitability
Finding the Best Sites
Image 1: Simple Site Suitability

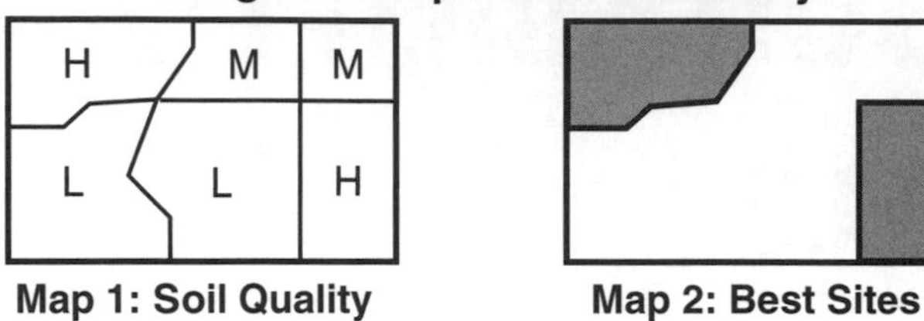

Map 1: Soil Quality **Map 2: Best Sites**

Best site = High soils

Image 2: Site Unsuitability

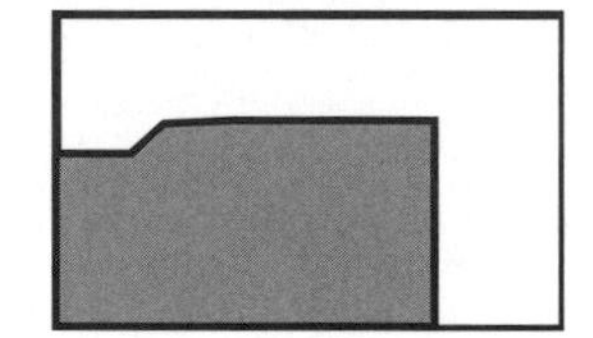

Map 3: Unsuitability Map

Worst site = Low soils

Image 3: Two-theme Site Suitability

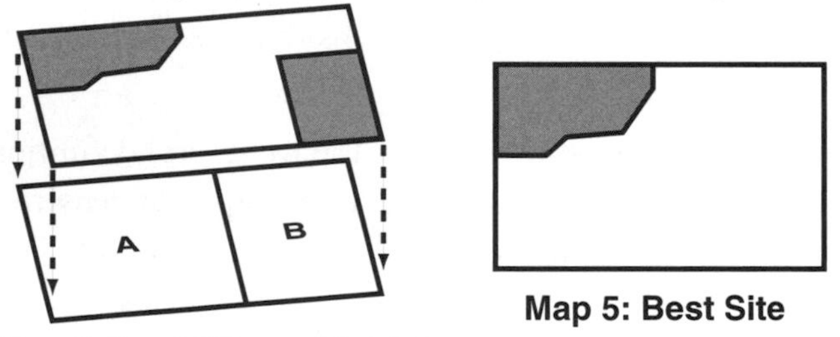

Map 5: Best Site

Map 4: Best Sites + Districts

Best site = High soils in District A

Image 4: Multiple Suitabilities

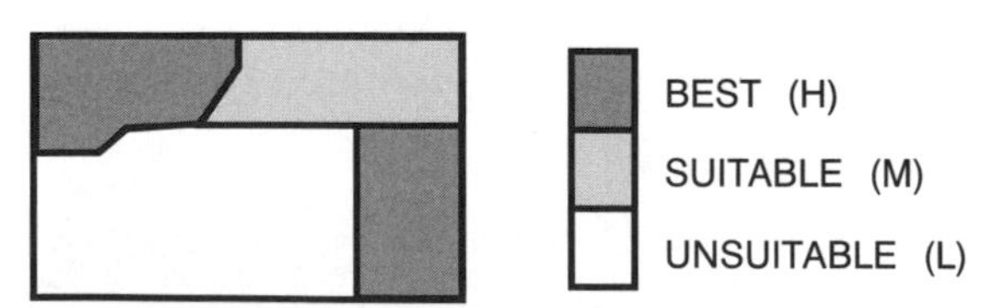

Map 6: Site Suitability Classes

Fig. 10-1: Site suitability and site unsuitability.

Site Selection

Site selection is the generic process of finding locations or features meeting specified conditions or criteria. For various purposes, it is often expressed as *site suitability, site unsuitability,* or *site sensitivity* analysis. Site suitability is the process of finding the most suitable (best) location or locations for a particular purpose. It typically involves sifting through data to select the best attributes defined by specific spatial or descriptive attributes.

Map 1 of image 1 in figure 10-1 shows a rectangular area with soil quality polygons having H (high), M (medium), and L (low) attributes. For a specific application, the best sites (map 2) are defined as those having high soil quality. The site suitability analysis operation searches the database for records having the High attribute and then selects the polygons meeting that condition. Two areas are shown on the Best Sites map. This is standard GIS site suitability analysis. Notice that the important work was performed in the database, using graphics only for display of the results.

Some applications are concerned with sites that are *unsuitable,* such as the worst place to locate a polluting industry or the least favorable location for a tourist facility. The procedure is identical, except that the criteria place the least suitable preference as the highest value and then work the analysis accordingly. In this case, the Low soils are the worst areas to place a proposed agricultural development. Image 2 shows a site unsuitability map, with the Low quality soil polygons selected and presented as the areas to avoid (map 3).

Image 3 presents site suitability on an overlay of the soil map and a district theme (districts A and B of map 4). The best site (map 5) is defined as the high soil quality areas in district A. Therefore, only one H site is selected. Two themes have been combined to find the most suitable site. Using additional attributes from a more detailed database simply involves matching the required conditions.

It is not difficult to define site suitability in stages or levels of acceptability, such as Best, Suitable (acceptable), and Unsuitable. The original soil criteria have been modified for image 4: High soils = Best, Medium = Suitable, and Low = Unsuitable. The site suitability map (map 6) presents soil qualities in terms of suitability classes, ranging from best to unsuitable.

Sensitivity Analysis

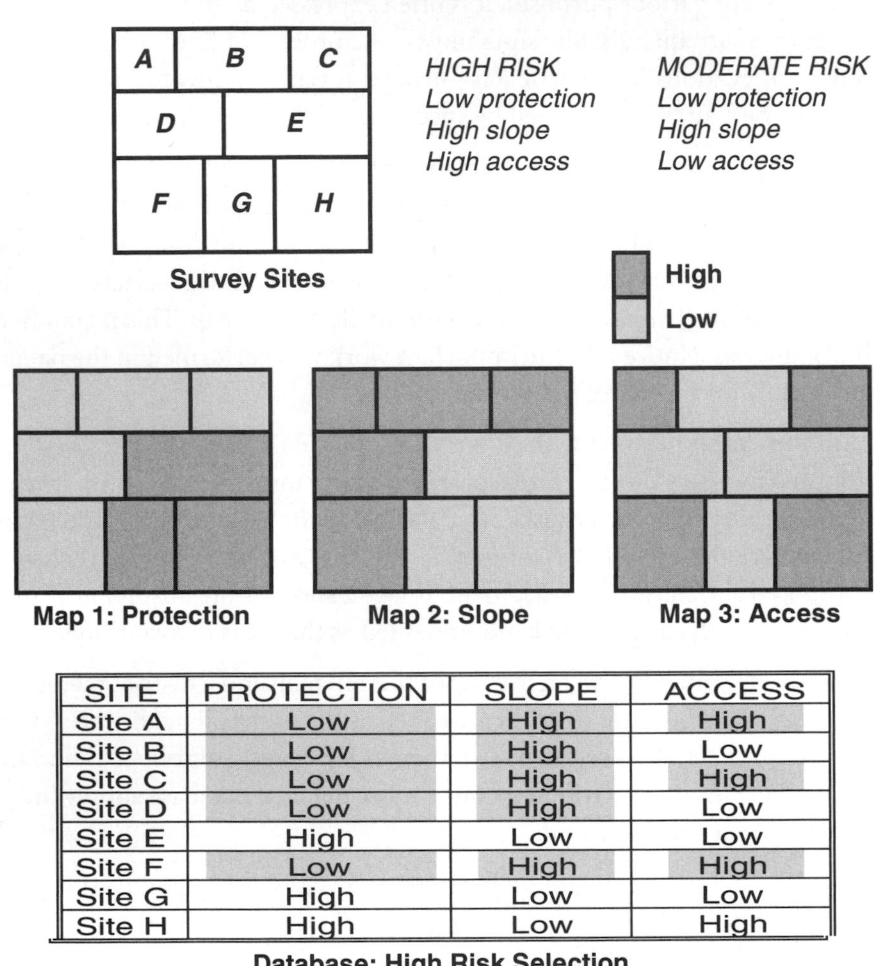

SITE	PROTECTION	SLOPE	ACCESS
Site A	Low	High	High
Site B	Low	High	Low
Site C	Low	High	High
Site D	Low	High	Low
Site E	High	Low	Low
Site F	Low	High	High
Site G	High	Low	Low
Site H	High	Low	High

Database: High Risk Selection

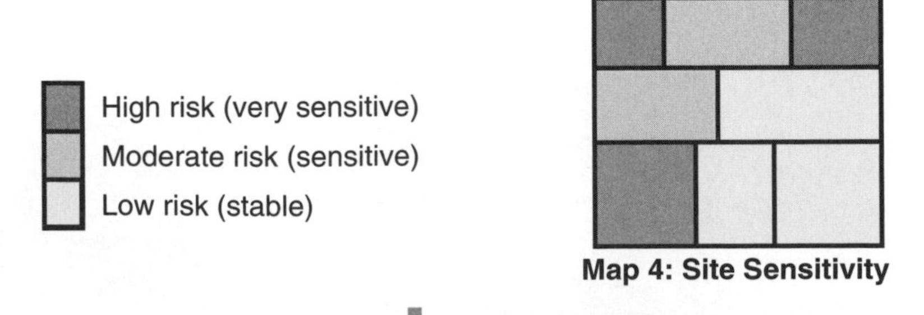

Fig. 10-2: Examples of sensitivity analysis.

Site Sensitivity

A major ecological application of GIS is the protection of sensitive areas; that is, environments at risk for harm or destruction from some proposed feature or process. Locating sites that need protection is a site suitability operation, usually termed *sensitivity analysis.* Figure 10-2 shows an environmental protection application with eight survey sites (A through H). The task is to find the areas having high sensitivity (i.e., a high risk of environmental damage).

Rated as high or low in each site are available protection devices (e.g., fence), the degree of slope (steep slopes are potentially unstable), and accessibility by people. High slopes with low protection and high access are classed as having the highest sensitivity, and thus the highest risk. Sites with low protection and high slopes are still at risk, but low access means there is less sensitivity. All other sites are classed as stable, either with high protection, low slopes, and/or low access.

Maps 1 through 3 show the rating for each theme, with sites marked either high or low. The database, which includes all sites and their attributes, is shown below the three maps.

The sensitivity analysis procedure reads the attributes of each survey site and then assigns the appropriate risk classification. High risk attributes are shaded. In effect, the operation first selects the records meeting the required conditions in the first column (Protection), then moves to the next column to determine which of those selections continue to meet conditions under Slope, eventually ending with one or more records that have all of the sensitivity attributes. Note that all sites except E, G, and H meet Protection and Slope conditions, but only A, C, and F "survive" the Access condition selection. Another iteration (cycle of computing) will be performed for the moderate risk analysis, using the new set of conditions.

Map 4 presents the site sensitivity analysis, showing both high and moderate risk sites, and therefore the stable sites as well. This is an easy task for GIS, either in raster or vector data format. In addition, the operation can be performed using either graphic overlay techniques or the database.

Site Suitability: Map Algebra

Image 1: Binary Overlay, Multiply

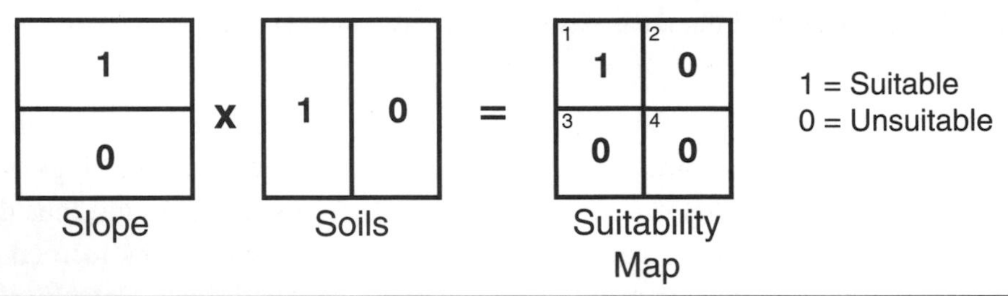

POLYGON	SLOPE	SOILS	CODE (SLOPE X EROSION)	CLASS
1	1	1	1	Suitable
2	1	0	0	Unsuitable
3	0	1	0	Unsuitable
4	0	0	0	Unsuitable

Polygon number in upper left corner of Suitability Map

Image 2: Binary Overlay, Add

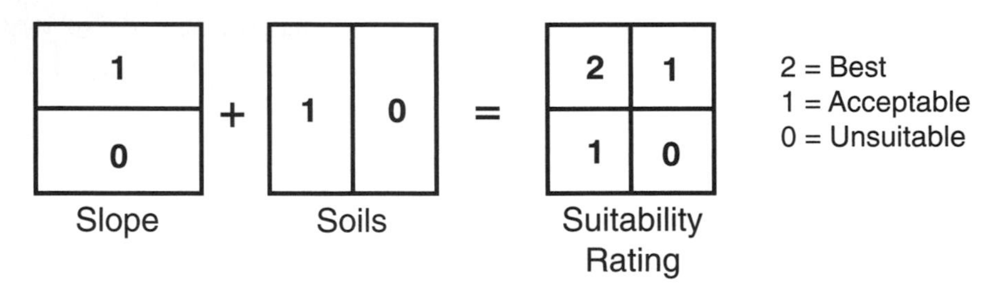

POLYGON	SLOPE	SOILS	CODE (SLOPE + EROSION)	CLASS
1	1	1	2	Best
2	1	0	1	Acceptable
3	0	1	1	Acceptable
4	0	0	0	Unsuitable

Fig. 10-3: Map algebra applied to site suitability.

Map Algebra Technique

In either vector or raster systems it is sometimes advisable to reduce complex data to the simplest classification format for site selection analysis. If each theme can be recoded to suitable or unsuitable classes (a binary classification), the final operation (overlay or database work) can be managed with the least amount of confusion. For example, if the suitable areas on two or more combined themes match, obviously that area is suitable (and the same for unsuitable measures). Techniques for resolving mixed attributes are described in the following sections.

Using raster map algebra techniques of planning overlays can be useful for either data format. Image 1 of figure 10-3 shows two themes, Slope and Soils, presented as binary classifications: 1 = Suitable, 0 = Unsuitable areas (recoded from standard themes). Raster overlay or vector database analysis using multiply will result in the class 1 Suitable (where 1 and 1 occur) and Unsuitable code 0 with any other combination (1 x 0, 0 x 0, 0 x 1). This produces a simple but effective Suitability map. The merged database, from each theme, shows the four new polygons and how a new field named Code uses the numbers to produce a suitability classification. Study the database and see how easy the concept and operation can be.

To get a suitability *rating model*, the themes are combined using the map algebra operation Add instead of Multiply (image 2). The Best sites are where two Suitables occur (1 + 1 = 2), Acceptable is where one Suitable exists (1 + 0 or 0 + 1 = 1), and two Unsuitables are clearly unacceptable (0 + 0). This database is almost the same as the previous one, except that Code adds the numbers, and a more detailed classification is produced.

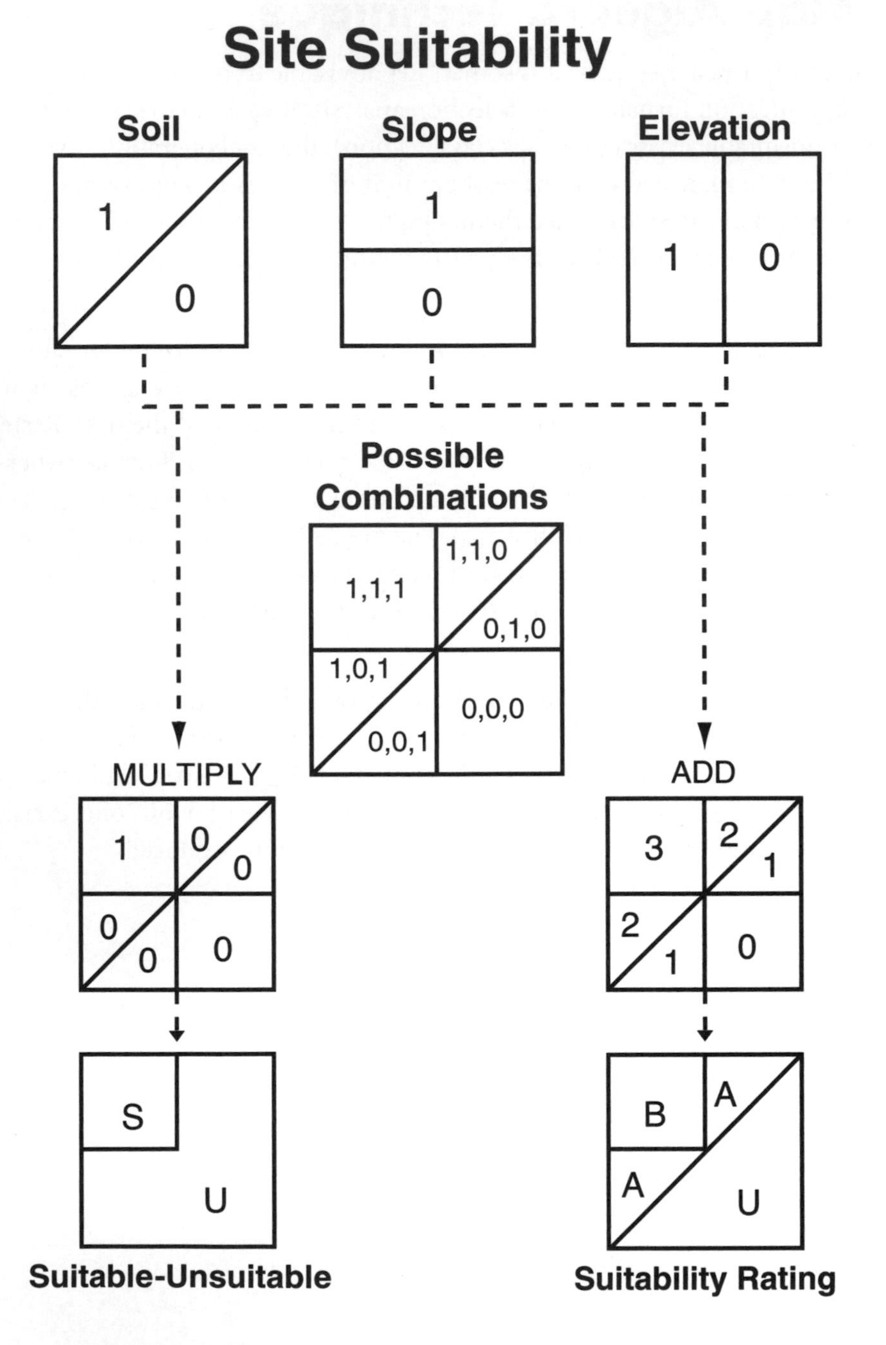

Fig. 10-4: A three-theme overlay for site suitability calculation.

The map algebra approach, whether in raster or vector, can work with more than two themes. The binary multiply procedure will result in values of 1 or 0, regardless of the number of themes (any 0 in the math will always produce 0, of course). If a rating is needed (say to help offset the "veto power" influence of one 0-value polygon or to test middle values between the suitable-unsuitable extremes), the procedure using Add can be used.

Figure 10-4 presents three themes in the site suitability operation: Soil, Slope, and Elevation. As in the previous illustration, each theme has been reduced to a binary code, with 1 = Suitable and 0 = Unsuitable. The polygons are arranged here for demonstration to show the possible combined values (more of a diagram than a realistic map). The Possible Combinations map shows the input codes from each theme in each overlay polygon.

Under the condition that any 0 value from at least one of the input themes will create an unsuitable situation, the Multiply approach can be used. This means that all polygons with a 0 in the overlay will be classed Unsuitable. Only polygons having all 1 values will be designated Suitable (in this case, there is only one). This is a restrictive instruction for determining suitability, but there may be applications where it is warranted. An example is the evaluation of potential housing sites, where all of the environmental factors must be suitable. Otherwise, the site is declared unsuitable.

If there is some flexibility, a rating system will be better. An overall evaluation of the environment is needed. Therefore, the three-theme combination is made using the add operation. Follow the values in each polygon from the Possible Combinations map and note the sums in the Add map. The user then makes a decision on how to classify the values: 3 = Best (B), 2 = Acceptable (A), and 0 or 1 = Unsuitable (U). The concept and operations are not difficult, but they do require logic and organization.

Matrix Planning

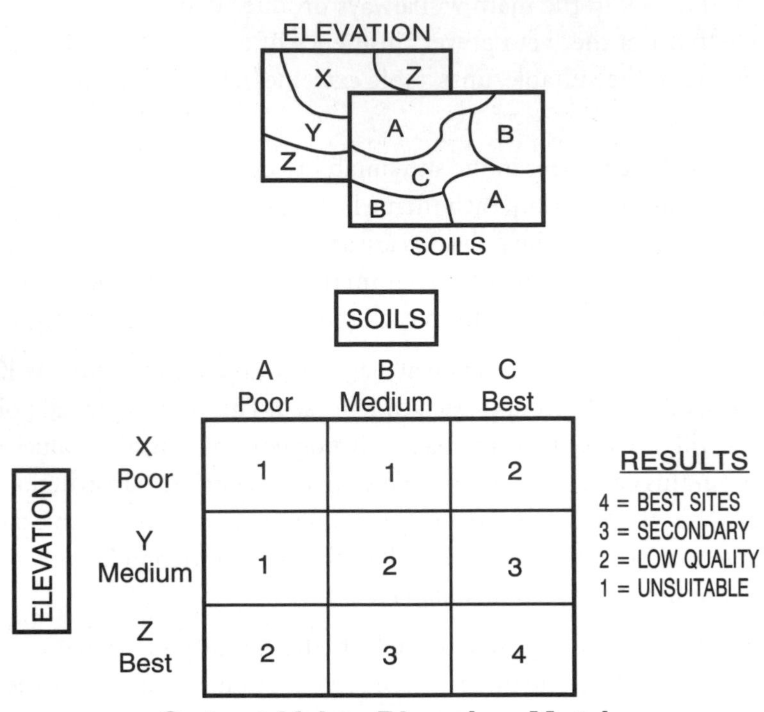

Output Value Planning Matrix

INPUT CODING

Soils Code _ and Elevation Code _ = **?**

Soils Code A and Elevation Code X = **1**
Soils Code B and Elevation Code X = **1**
Soils Code C and Elevation Code X = **2**
Soils Code A and Elevation Code Y = **1**
Soils Code B and Elevation Code Y = **2**
Soils Code C and Elevation Code Y = **3**
Soils Code A and Elevation Code Z = **2**
Soils Code B and Elevation Code Z = **3**
Soils Code C and Elevation Code Z = **4**

Fig. 10-5: The concept of matrix planning.

Matrix Planning

A similar approach to obtaining suitability ratings is to use a matrix to calculate results (often a hand-written scribble for planning). This is used for GISs that permit user control of output values, rather than depending on map algebra (discussed in chapter 3). The small maps at the top of figure 10-5 are Elevation and Soils of a region. The themes are entered as rows and columns on the matrix. Both themes have been recoded to Best, Medium, and Poor attributes, and the matrix is used to plan the output values for each combination. Four rating levels are desired in this project: Best (4), Secondary (3), Low Quality (2), and Unsuitable (1).

Read the matrix to understand the planned output values. Best sites (output value 4) consist of only Best Soil (class C) and Best Elevation (class Z). Secondary sites, coded 3, have one Best class from either the Elevation or Soil theme and a Medium (soil B or elevation Y). Low Quality sites (2) consist of either a Best and Poor or a Medium-Medium pairing. All other combinations are classed as completely Unsuitable. It may be that the Low class is also unsuitable, but initially it is a separate, slightly higher-value category. It can be recoded easily when needed.

The output coding scheme is presented in the text box at the bottom of the illustration. Recall that the GIS presents the input features and the user fills in the desired code. For example, the program displays "Soils Code A and Elevation Code X = ?" and the user enters the number 1. Data entry is easy and relatively fast; the more difficult task is to plan the combinations correctly.

Matrix planning is a convenient, often necessary, method of combining themes with more than simple input and output binary codes. When various classes are needed (particularly those that cannot be derived easily with map algebra), careful planning and user input can provide direct results without multiple recodes and overlays.

The use of numbers in site suitability analysis has advantages, whether used in raster or vector systems. For example, a variety of formulas can be used in the overlay process to test "what-if" scenarios or theories of influence, such as weighing Elevation over Soils and Slope, as demonstrated in Chapter 8. Map algebra is a valuable concept for analyzing GIS data, particularly in the modeling process (to be discussed in this chapter).

Site Selection Applications

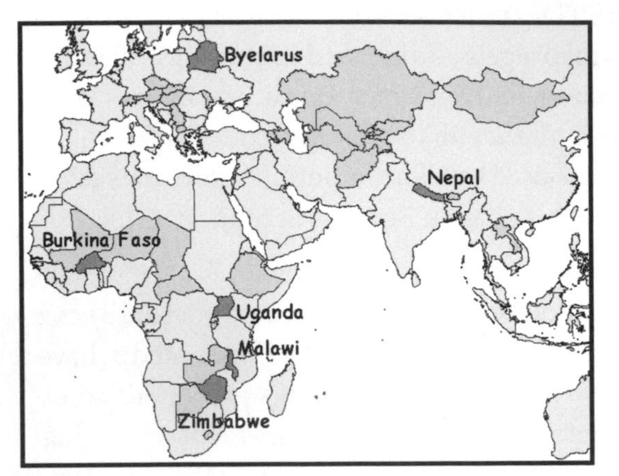

Map 1: Selected Nations

Landlocked
Area 100,000 – 500,000 sq. km
Population 10,000,000 – 20,000,000

Map 2: China Buffers-Airports

Airports within:
150 km of major river
100 km of border (inside)

- Selected Airports

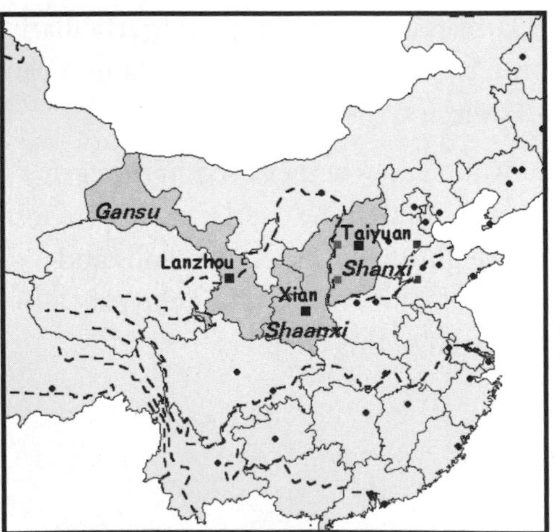

Map 3: China Province-City

Province population: 11,000,000 – 40,000,000
Province area: 100,000 – 400,000 sq. km
City population: 1,500,000 – 2,500,000
Within 200 km of a major river

Fig. 10-6: Examples of site selection application

Site Selection Applications

Finding locations or features that meet certain criteria is a common objective of many GIS operations and applications. Whether expressed as suitability analysis or merely site selection, the basic approaches are the same. Three relatively simple examples are presented here to show how the process works, although there are no operations that have not been discussed in previous chapters. It is interesting to note how these types of examples and operations integrate with procedures discussed elsewhere in this book. Rarely are GIS applications and procedures exclusive.

Map 1 of figure 10-6 shows a special U.N. development application for landlocked nations. It involves uncomplicated database selections, beginning with all landlocked nations in the world and then finding those with areas between 100,000 and 500,000 square kilometers. From those, selecting populations between 10,000,000 and 20,000,000 finishes the process. The map displays the six nations meeting the required conditions (dark tones with labels; lighter tones are other landlocked countries). It is easy to understand how fast this can be. It is also easy to see how GIS facilitates the "what-if" method of testing various conditions (e.g., simple changes of the area and population criteria to see if other nations qualify). The work was performed in the database, but the next example is primarily a graphics overlay process.

Map 2 shows a Chinese transportation development application, in which a query asks for the major airports within 150 kilometers of a major river and within 100 kilometers of the national border (inside China, of course). The first operation is to construct the two buffers and overlay them to derive the common areas. River buffers are shown by the wide zones (300 kilometers wide, 150 kilometers on each side of each major river). The border buffer is light gray along the rim (an inside-only buffer operation), and the common buffer areas (dark gray) are made by the overlay. Then an overlay of airports with the common areas produces the required features. As in the previous example, testing various buffers can be a fast and easy process.

Map 3 has the selection of provinces and cities in China as its objective. The criteria are that province populations must be between 11,000,000 and 40,000,000 and that their areas are between 100,000 and 400,000 square kilometers. Within these provinces, city populations must be between 1,500,000 and 2,500,000 and located within 200 kilometers of a major river. These are not complex criteria; in fact, most of them are simple database selections, followed by an overlay with a river buffer. The map displays the provinces and cities meeting the stated conditions.

Site Suitability Procedure

Criteria:

Major Roads
Best: 0 – 2 km
Good: 2 – 4 km
Railroad: < 30 km
In city limits
Proper zoning
On level land

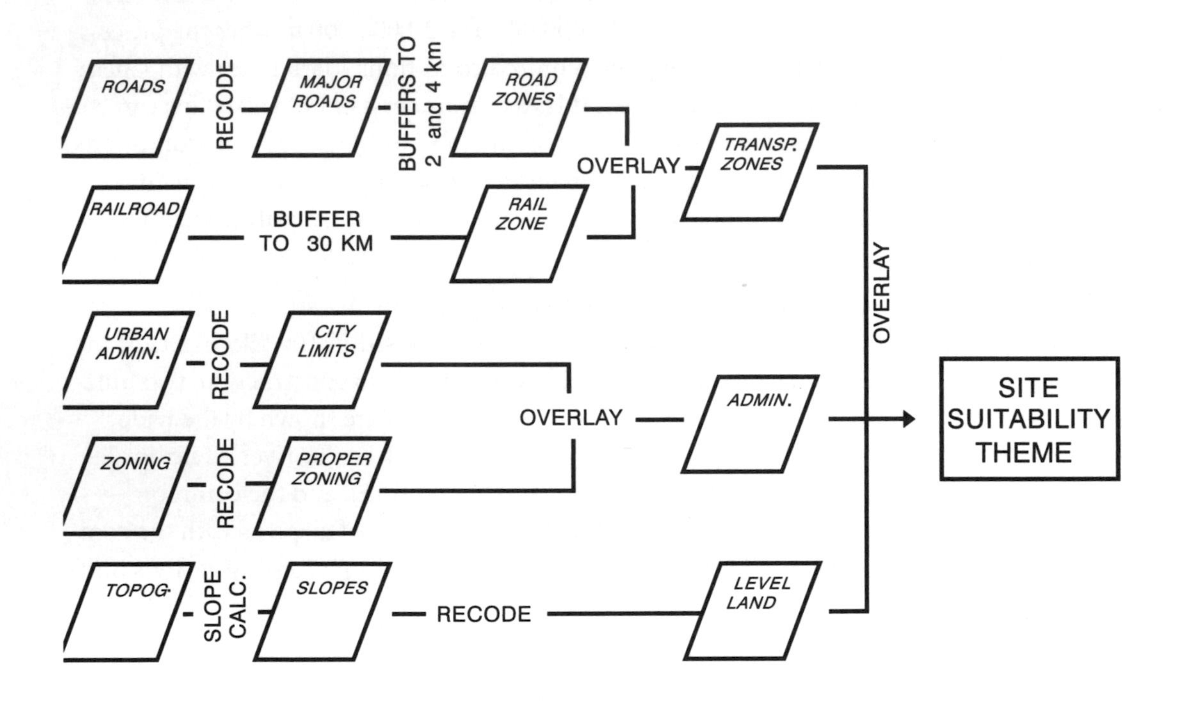

Fig. 10-7: Flow chart for a site suitability evaluation procedure.

Site Suitability Procedure

Although the concept is easy, sometimes site suitability can be a complicated process and a bit confusing, particularly when multiple themes and different types of criteria are used. Therefore, good preparation and organization are needed. Figure 10-7 presents the criteria for one task and shows a diagram (or flow chart) of the steps that can be taken to prepare input themes and to achieve a final product. The list of criteria is slightly complicated, asking for spatial selections as well as attribute conditions. Buffers, recodes, and overlays are needed in the analysis.

Note the organization and the steps. The linear network themes (Roads and Railroad) are worked initially, first individually (making buffers) and then in an overlay giving transportation zones. The administration themes are treated together, first as recodes and then in overlay. Slope is calculated from topography, and then recoded into level land. The final overlay combines the three major themes in order to achieve the site suitability analysis. Follow each step to understand the process.

A logical structure makes the process more efficient, reducing time, workload, and potential error. This type of organization is recommended for a complex set of data and needs.

This presentation uses generic symbols for data and text to describe the processing. As will be seen in the modeling section of this chapter, there are more conventional symbols that relate meaning through their shapes, and the process can be programmed simply by defining the data to be included.

GIS Models

In GIS applications, work and results can be for a single goal or a set of data, but we often hope they are useful for more than one site or task. When the procedures are found successful, it is nice when they can be applied to other sites involving similar applications. If successful, the process can be helpful for other work and sometimes a model can be constructed.

There are various types of GIS models, each with several definitions, depending on the task or objective. The four basic model types described in this chapter are descriptive, process, predictive, and procedural (although other names are used throughout the literature and by individual software vendors). Often there are no firm differences between these types. For example, a prediction model can include both description and process; process models typically also include procedures. The following section discusses descriptive models.

Descriptive Models 1

Image 1: Data Modeling Contour Maps

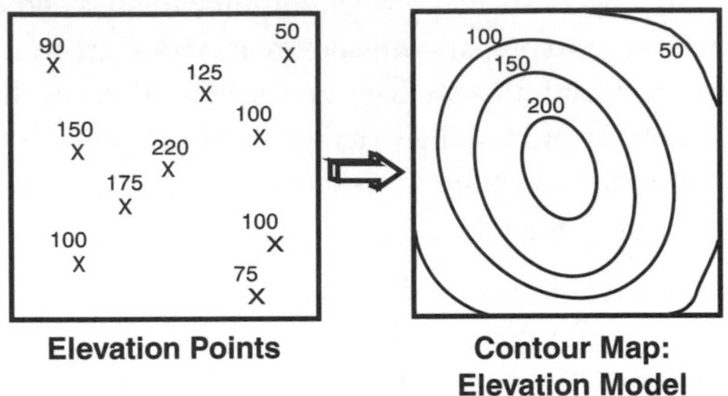

Image 2: Bus Route Model

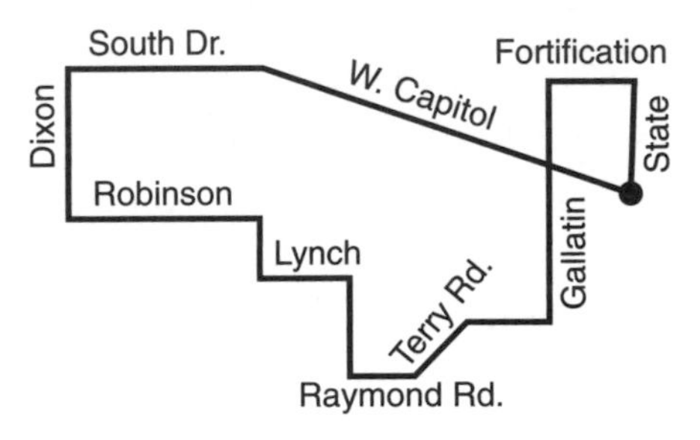

Bus Route, Jackson, Mississippi

Image 3: South America Area Chart

Spatial but not location

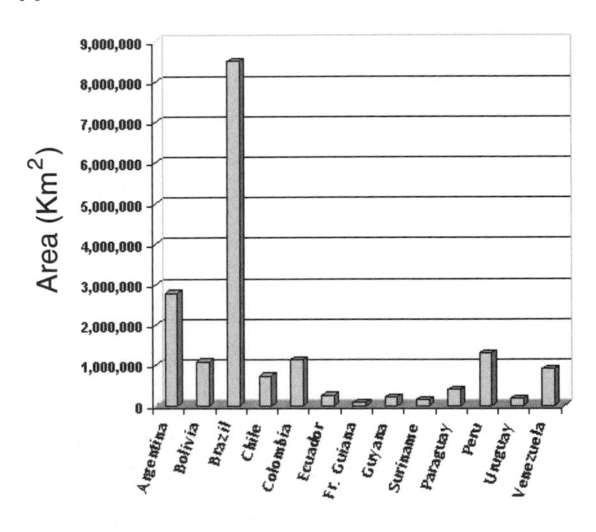

Fig. 10-8: Examples of types and applications of descriptive models.

Descriptive Models

Descriptive models illustrate the region, conditions, or circumstances under study, in a generalized way, to make the area or themes readily understandable. Maps are traditional and efficient models of areas and their features. A descriptive model is a representation of reality, an analogy or a simplification of the real world. It does not include everything on the landscape but uses only those items important to the task or goal at hand. The model reduces complexity and confusion.

A simple model can take available data and arrive at generalizations of an area. Without additional information, the model assumes that what is seen or known is representative of the area overall. For example, elevations can be determined at specific data sites by surveying, and because the points are believed to be representative of the entire area, *data modeling* will build an elevation map.

Image 1 of figure 10-8 shows an example of an elevation model composed of survey points that are constructed into a contour map. A contouring program can expand the survey measurements to the rest of the area using contours (lines indicating specific elevations). Some inaccuracies probably exist on the lines; the only precise elevation remains at the surveyed points. A contour map is a standard way of depicting the topography of an area, and GIS extends the data to the types of 3D models discussed in Chapter 9. A contour map is only a generalization of the real world; it is a model.

Bus and subway maps are typically simplified for better understanding by the general public. Image 2 is an example of bus route map for Jackson, Mississippi, USA. Because there is little need to depict irrelevant details (such as highly accurate distances or every street), the maps are used only for basic passenger information. The maps are models of reality and appropriately serve a particular application. The detailed map on the left shows the route of a given bus, starting downtown at West Capitol and State streets and winding through the suburbs, eventually returning to the starting point. The model on the right is the version posted for the public on the bus and at stops. Note that it is simple, with reduced detail, but is effective as passenger information.

There are other ways of expressing a map (though probably not as efficiently), such as a list of features, a diagram of features and how they relate, descriptive statistics of the important spatial variable, or even text description. These could be models because they are generalizations of the features or processes under study; they are descriptive models in that they describe the environment.

Image 3 is a basic bar graph of the sizes of geographical area covered by South American nations. This is one type of useful, explanatory presentation of geographic information concerning regions. Although the graph does not contain location information (except country name), it includes better quantitative data about the area of each country than can be gained from visual comparison of countries outlined on a standard map.

Descriptive Models 2

Image 1: Image to Map Model

Satellite Image
(Reduced Reality)

Map 1: Major Roads and Lakes
(Selected Modeling)

Image 2: Generalization Model

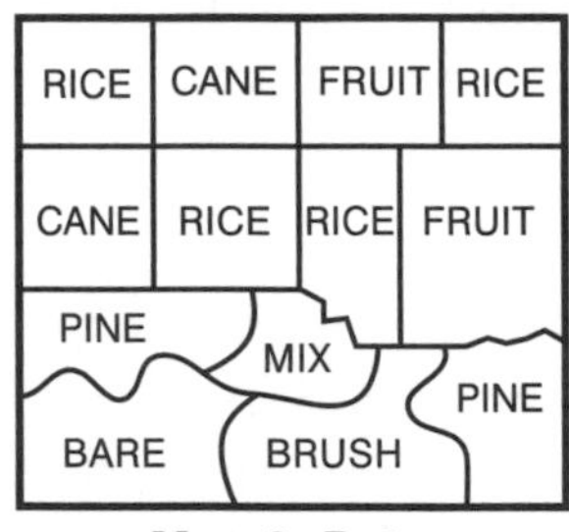

Map 2: Data

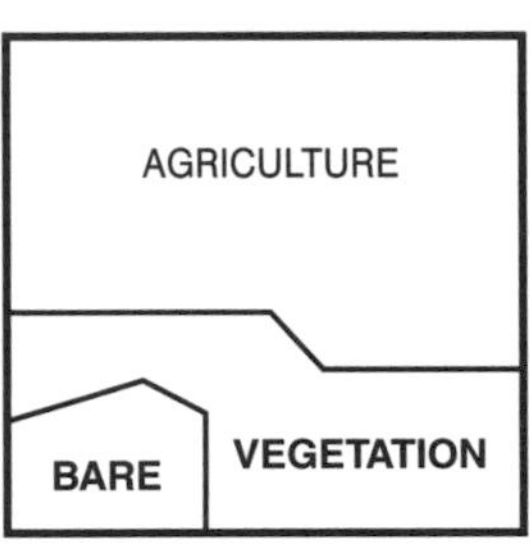

Map 3: Model

Image 3: Landscape Model

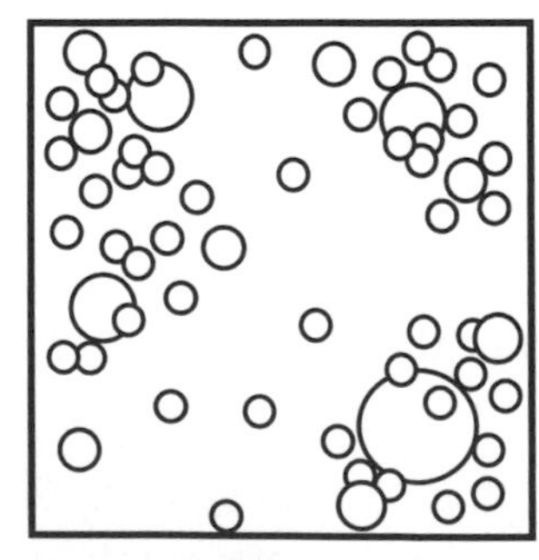

Map 4:
Lunar Landscape Map

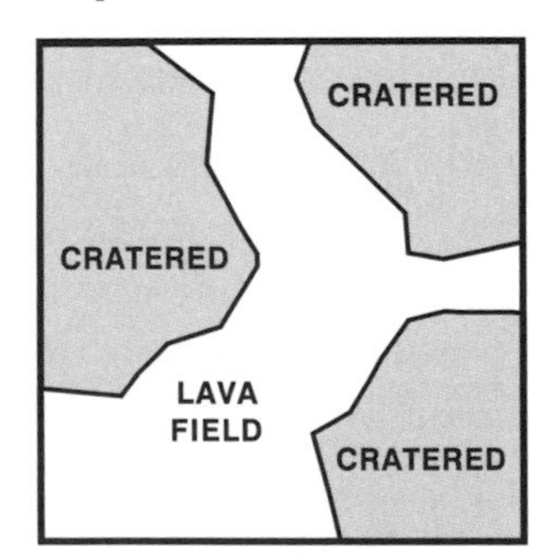

Map 5:
Descriptive Model

Fig. 10-9: Descriptive models, continued.

Image 1 of figure 10-9 shows a GIS model in the form of a simple thematic map (map 1) derived from reducing the satellite image to roads and lakes (lines and polygons), even though there is much more land information available. The satellite picture presents all of the geographic features on the ground (within its scale and electromagnetic setting), but the map presents only selected geography, which is a descriptive model. In this case, there are more roads and a number of smaller lakes than are depicted on the map.

Image 2 is another simple descriptive model. It reduces the data (map 2, original field or image data) from specific crops and vegetation to a land cover theme (map 3). This model generalizes data to a thematic theme and generalizes the spatial details of the polygons. It may be useful for some applications, such as regional planning. Thematic maps are generalized models of the real world.

Image 3 shows a cratered lunar landscape (map 4, presented in Chapter 9), with the clustered generalization (map 5) serving as a model for the terrain. It presents reduced detail to facilitate understanding, but is still a logical representation for basic geologic research and mapping. Both versions of the data are useful. Again, even map 4 is a model because it does not present everything that can be seen. Rather, it functions as a first-generation map that can be converted to thematic models as needed by the project.

Process Models

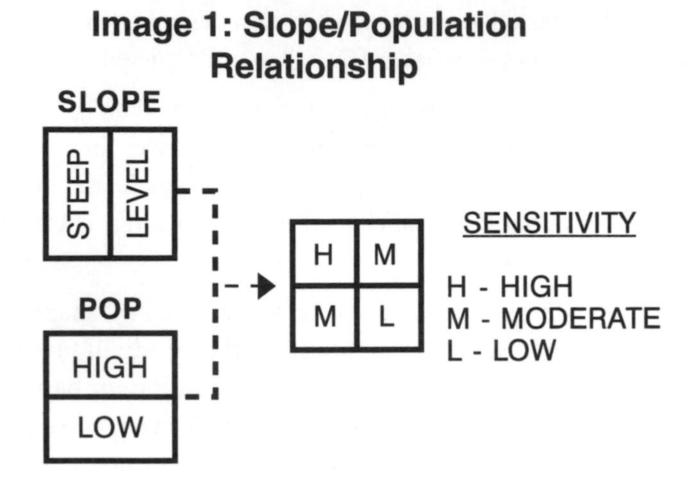

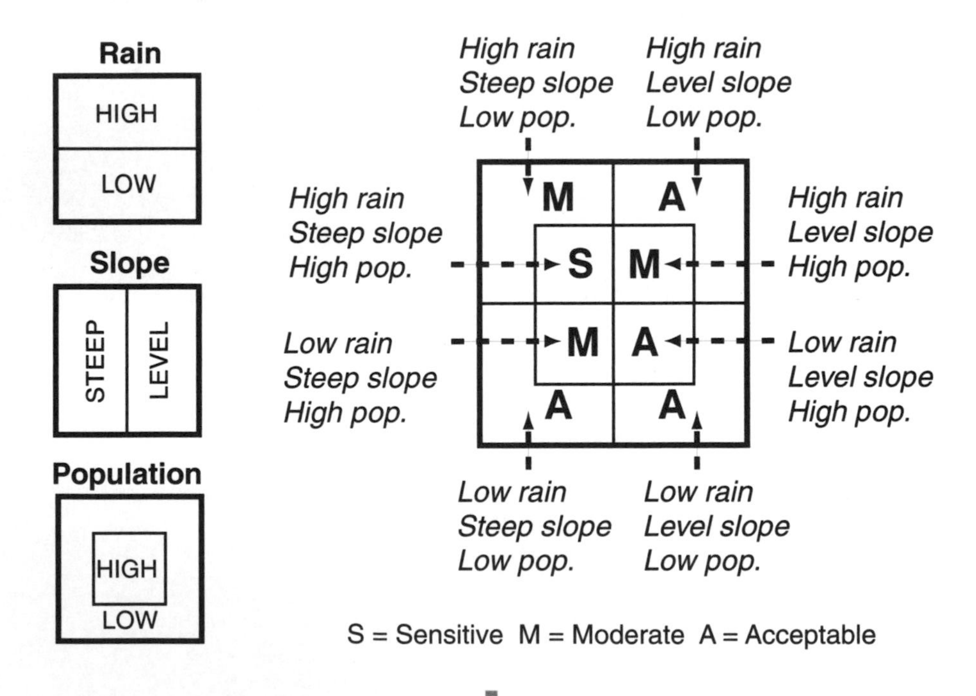

Fig. 10-10: Process models imitate natural processes.

Process Models

Process models attempt to duplicate or describe natural or human processes in order to simulate or provide new information about the features and processes under study. In other words, a process model illustrates how natural or human processes work. Image 1 of figure 10-10 shows how GIS uses the association between slope and population density to produce a sensitivity model. This is a good model because it explains the land-people relationship in a simple diagram, but does not show the actual landscape. Process models can help us understand *cause and effect*, or the factors that cause specific effects. They also can help us predict what can occur under selected conditions.

Understanding process is often a major goal of GIS analysis. Once the process has been understood and can be described, rational responses can be planned and further problems either avoided or at least decreased.

Image 2 depicts *sensitivity modeling* as a means of understanding the relationships between variables that determine environmental sensitivity. Sensitivity models show what can happen under given data or circumstances. Before applying data and GIS operations in environmental analysis, the user must understand potential results and their meanings or effects (part of the cause-and-effect process). The model is a diagram showing the resulting sensitivity impact when merging the three themes of rain, slope, and population density. Each theme has a high and low measure (steep and level for slope), which create environmental sensitivity variables.

When an overlay of the three themes is produced, the model shows that the most sensitive sites are those having high rainfall, steep slopes, and high population density. Moderate areas have two high measures and a low or level one. Read the other combined measures and note how the model shows all possibilities. Reality may be different, of course, but this type of model can be a good initial guide. This will be applied in prediction modeling (described in the next section).

Mathematical Models

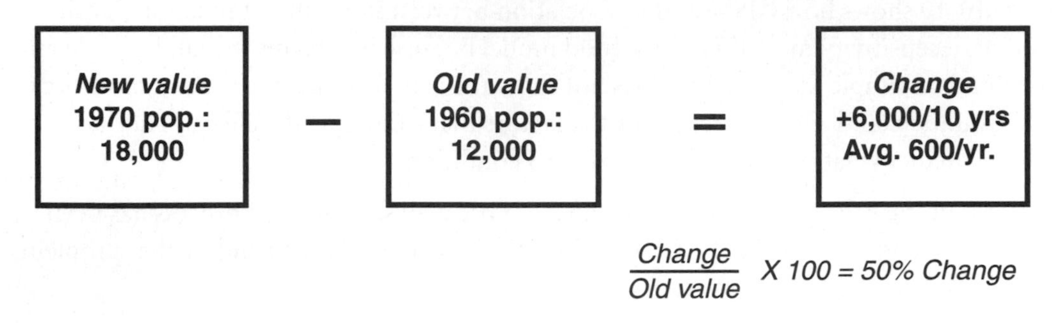

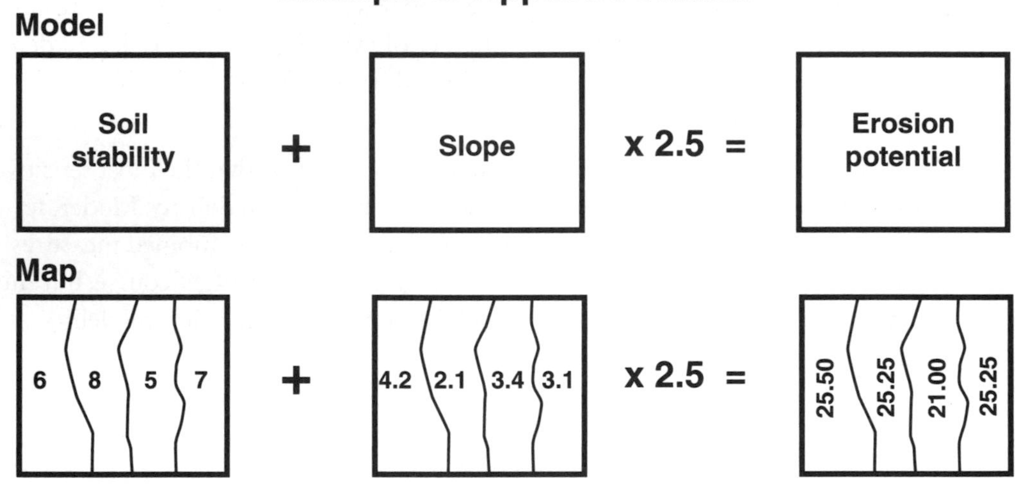

Fig. 10-11: Examples of mathematical models.

Mathematical models attempt to describe or explain processes using numeric analysis (or formulas) to express given environmental or landscape dynamics. Example 1 of figure 10-11 is a simple formula of a very basic model for determining change: Change = New Value – Old Value. This means that the attribute value of a feature or site in the oldest data is subtracted from the value of the newest data to show the amount of change that has occurred.

As applied, this formula shows, for example, that a population in 1960 at 12,000 subtracted from 1970 at 18,000 indicates an obvious growth of 6,000 people over 10 years (an average of 600 people per year). This describes the demographic dynamics of the region and hopefully, may be a first step in explaining the reasons. A simple addition to the formula can determine the percentage of change: ((New Value – Old Value)/Old Value) x 100.

In example 2, the erosion potential of a test site has been calculated as (soil stability factor + a slope measurement) x 2.5, a fairly simple ecological model. This calculation can be applied to other regional sites by GIS to make quick regional assessments of erosion hazards. It is easy to apply simply by overlaying the soil stability theme with the slope theme (using the slope measurement as the working attribute) and then multiplying the resulting values by 2.5. A regional erosion potential map can be constructed, as displayed under the model. Planners will use the model in hazard assessment.

These are very simple examples. Mathematical models can be complicated, employing many variables and advanced analytical techniques. GIS has been applied to sophisticated ecological processes, hydrologic dynamics, and many other applications.

Statistical Models

Example 1: Land Use Growth

Graph 1: Land Use Growth 1960 – 1990

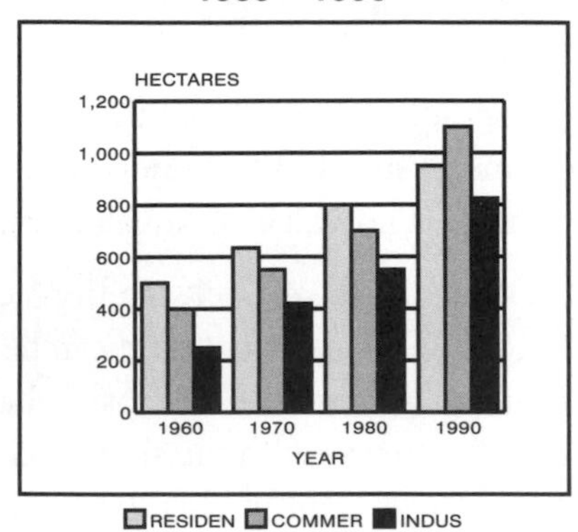

Graph 2: Land Use Growth 1960 – 2000 (Projected)

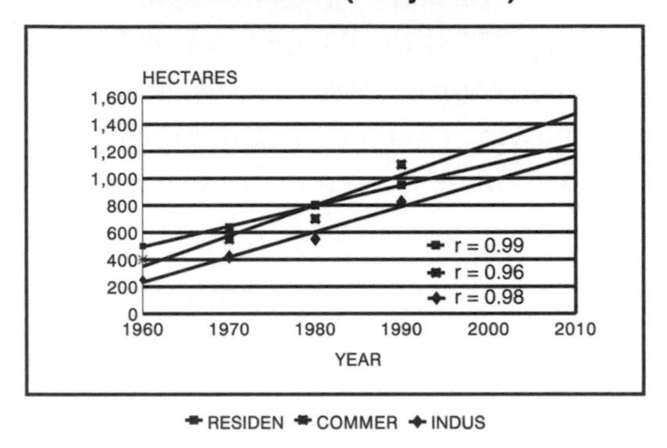

Example 2: Population Growth Southeast U.S. States

16000000
14000000
12000000
10000000
8000000
6000000
4000000
2000000

North Carolina | Tennessee | Alabama | Mississippi | Georgia | South Carolina | Florida

Pop1990
Pop1999

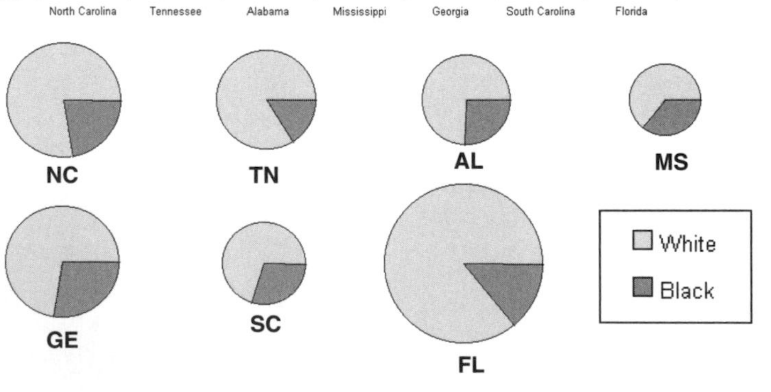

Fig. 10-12: Examples of graphs as models.

Statistical techniques model processes by presenting descriptive statistics over time or space. Figure 10-12 presents ways of representing GIS data other than as themes or databases. The bar chart in graph 1 of example 1 shows the area of three types of land use (residential, commercial, industrial) from 1960 to 1990. Comparisons of land use within a given year are easy to see with the three different colored bars, and growth from one period to another can be distinguished by comparing each color of bar between years. The graph is both a general comparative view of land use for a given year and a record of growth over 30 years, indicating trends.

Graph 2 is another method of describing that growth. The point symbols represent each land use measure at each time period, but a line connecting each point will show no more than the bar graph. However, by calculating the point value for each land use in a regression formula (a statistical technique for analyzing relationships between X and Y variables), a general trend line showing growth can be presented. It is a "line of best fit" (properly termed a *regression line*) that represents a continuum from 1960 to 1990 and beyond.

The r-value, shown for each land use, expresses how well the line represents the trend. Numbers close to 1.0 are very good, indicating a strong correlation between the X and Y value at any point. A regression line also gives an indication of what will happen if current trends persist. In this case, land use sizes will continue to increase and the predicted number can be interpreted for years up to 2010.

As in any statistics, reality may be different at any given point, especially for the predictions into the future. Nonetheless, this is a good start at understanding the area's land use change, but it is only a start. These types of graphs often accompany maps and data as valuable complements.

Example 2 presents two graphs of the population of seven states in the U.S. Southeast. The bar chart displays the total population of each state for 1990 and 1999, permitting a comparison of growth for individual records, but showing a comparison of regional distribution of population for each year. The set of pie charts presents a comparison of racial proportions for each state. In addition, each pie is sized according to the total population. Both graphs offer several types of information and can easily be included on maps and reports.

Predictive Models 1

Image 1: Time Series Model

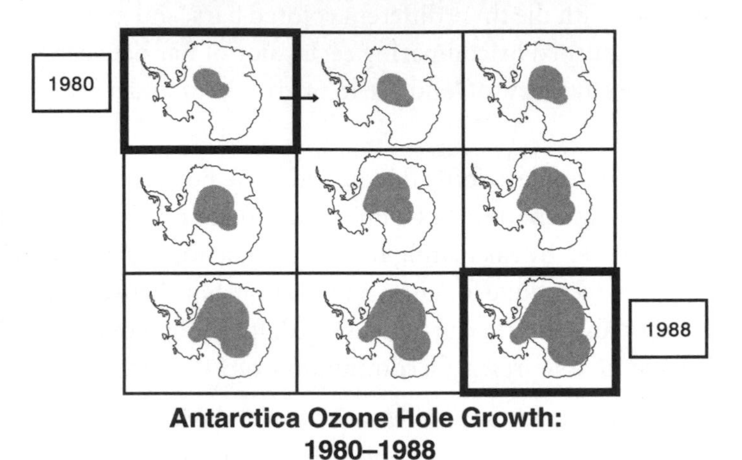

Antarctica Ozone Hole Growth: 1980–1988

Image 2: Sensitivity Prediction

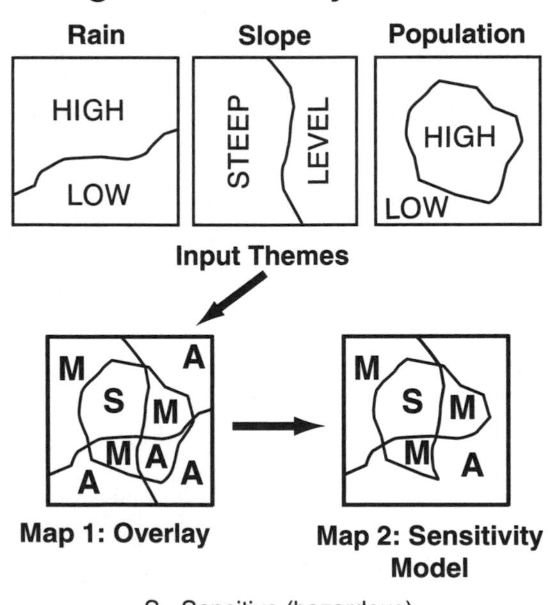

S - Sensitive (hazardous)
M - Moderate
A - Acceptable

Fig. 10-13: Time series and sensitivity predictive models.

Predictive Models

When applied to time, models can help us understand what is happening over a period of years. Predictive models can estimate what probably happened or will happen based on data provided for other times. These are called *time series* models when they represent a sequence. The regression graph of the previous section demonstrated how observed data over a few decades can calculate forecasts for future changes. Weather forecast maps are predictive models that obviously are not always accurate, but they do serve as guides to expected weather.

When conditions and spatial results for at least two times are given (more is better), a model may project backward, forward, or between two times. Image 1 of figure 10-13 shows a hypothetical case in which the 1980 and 1988 Antarctica ozone hole configurations are provided (based on observed data), and the time series model estimates changes during the intervening years (1981 to 1987). The sequence can be based on simple graphical "morphing" (computer-generated graphics changing from one shape to another) or from analysis and modeling of sophisticated environmental processes that create the various shapes.

These conditions may not have actually occurred, and the model should not be considered authoritative; it does not provide specific, conclusive data, but presents a credible set of scenes from which an idea of the changes can be considered. A model can do no more than that.

Prediction of *spatial* characteristics is a major function of GIS. Actually, this type of application has been discussed earlier in the chapter under site selection, and in the previous section in regard to sensitivity models, where the conditions promoting possible hazards are identified. That is, to predict the sites most likely to experience environmental problems due to rainfall, slope, and population density characteristics, an overlay function merges the attributes for the region and gives sensitivity values that permit easy interpretation and location of the candidate areas. They are predictions based on potential conditions.

Image 2 applies the sensitivity model (see figure 10-10) as a predictor or indicator of possible hazardous sites. The Rain, Slope, and Population overlay is displayed in map 1, showing each polygon coded according to the legend under it. Recall that accumulated high values make a sensitive area and lower values are more acceptable. Map 2 presents the final results (where common borders are dissolved into single polygons). This will guide the user or applications to the sensitive areas for further analysis.

Predictive Models 2

Image 1: Spatial Prediction

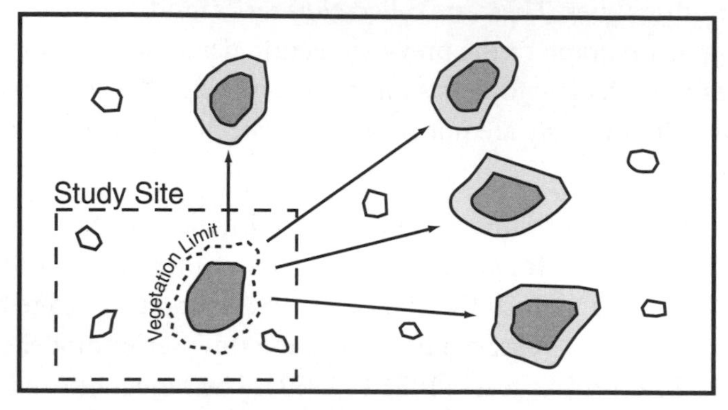

Lake Vegetation within 100 meters of Large Lakes

Image 2: "What-if?" Scenario Model

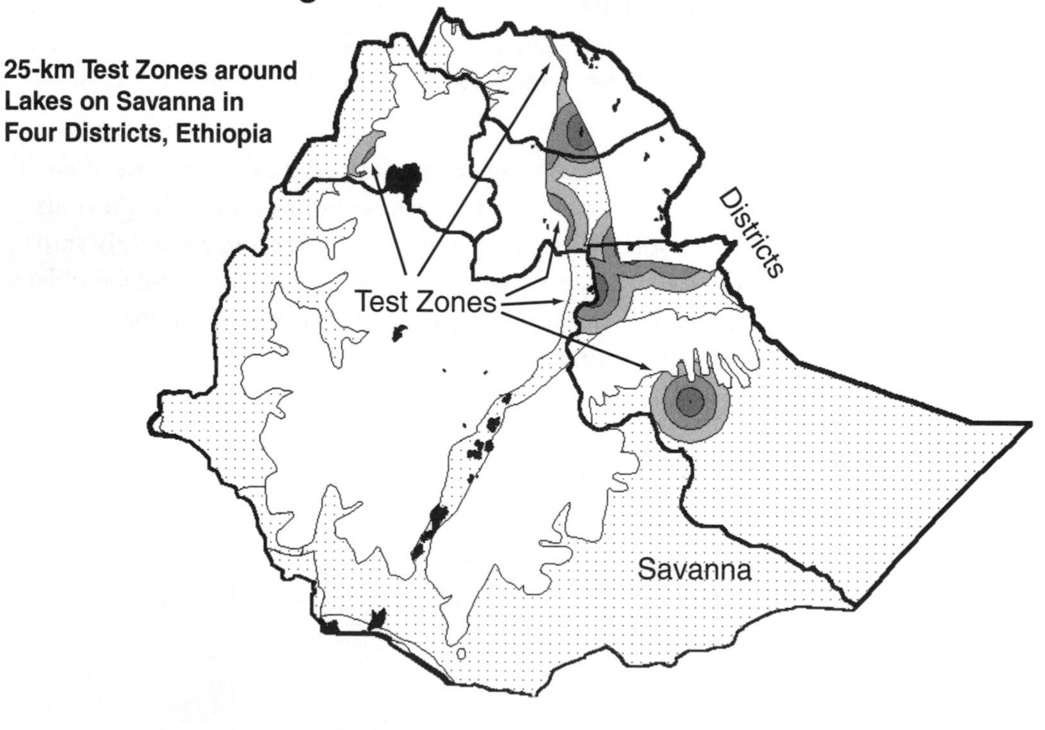

Fig. 10-14: Spatial and "What-if?"-scenario predictive models.

Another aspect of predictive models demonstrates that what is happening here (or in one place) will probably happen there (or another place) if conditions are similar. Image 1 of figure 10-14 shows a study area of vegetation near lakes, where a specific plant has been found only within about 100 meters of a lake larger than approximately 8,000 square meters (not the smaller ponds).

Given the environmental factors of this distribution, it is predicted that the same vegetation will be found within a 100-meter zone around other large lakes (light-shaded buffers around the dark-shaded polygons). When mapped, the field crew will know which lakes to investigate (pointed to from the study lake), as well as the potential distribution around those lakes.

GIS models can be used to test various scenarios ("what-if" cases) of proposed spatially related projects or processes. In Ethiopia, a hypothetical dangerous insect swarm begins in water bodies and spreads across grassland/savanna areas. Authorities are unsure of the maximum extent and plan to test within three 25-kilometer increments in the four north and eastern districts. If successful, they will then be able to target responses in the rest of the country.

Image 2 shows the planning map, with the circular test zones in dark to light gray tones. Lakes are in black (most are small), savannas are the dot-fill polygons, and the four districts are outlined at top and right. Basically, these are "what-if?" scenario areas that will show the spatial impact of three swarm distributions. If conditions or ideas change, GIS can respond very quickly to make a new set of themes and maps.

Procedural Models

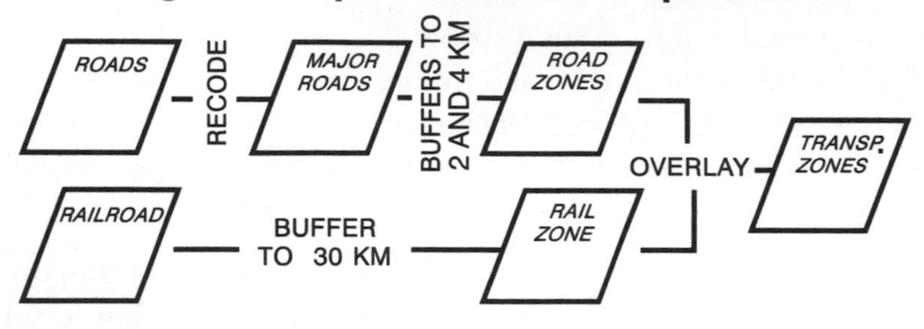

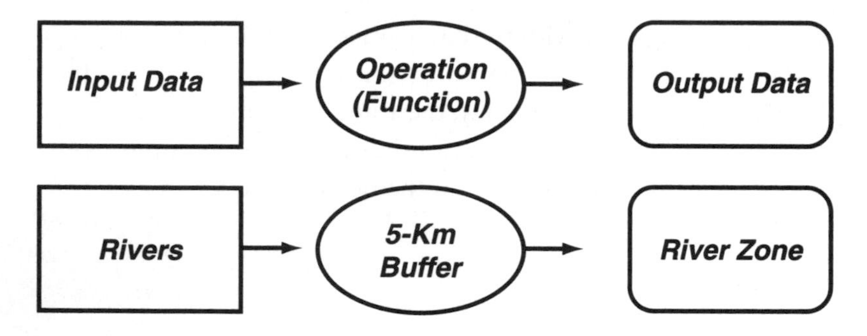

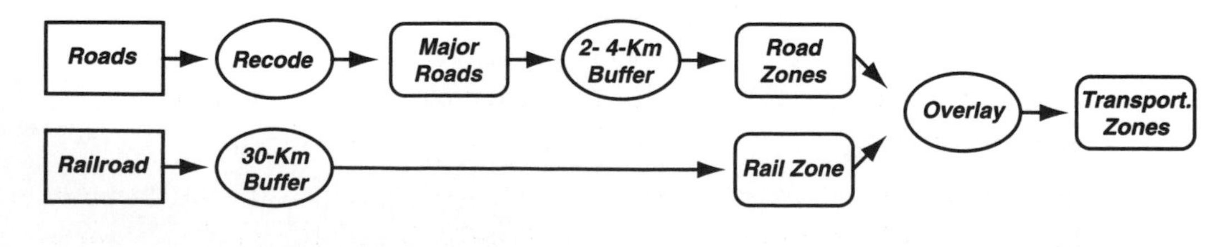

Fig. 10-15: Examples of procedural model application.

Procedural Models

A *procedural model* represents the steps needed to accomplish a specific task or goal. It can be a diagram or flow chart that shows the GIS steps needed to complete a general or a specific objective. Image 1 of figure 10-15 is part of the site suitability outline of operations discussed earlier, in regard to figure 10-7. It is a graphic model of site selection processes that can be applied to different data sets and in different parts of the project study area.

This type of model, usually a simple project document, has been referred to as the "cookbook" approach in that it is a "recipe" of steps and techniques to use. It is particularly useful when a test site has been worked on (a pilot study) and the procedures can be applied to the rest of the region consistently. This helps reduce error because the same techniques are being applied to all data sets, providing greater consistency to output, which supports regional comparisons.

In recent years, the GIS community has developed the procedural model approach into software that aids in making and running operational models. An exciting and valuable paradigm is now available to users in the form of modeling programs, or model building. The basic principles are easy to understand. Each type of input, operation, and output has a functioning graphic symbol that controls the program, linking each part of the process so that the user does not have to perform every task. Once the model is constructed, the tasks are activated and the program runs through each step, entering data, performing the operation, and producing derived data (a GIS user's dream!).

Image 2 shows the three major symbols: rectangles are input data, ovals indicate the operation (function), and rounded rectangles are the output (derived) data. The example below is the construction of a protective river zone: input the river data, make a 5-kilometer buffer around it (operation), and produce the output river zone theme. In effect, the user can "program the program" to perform much of the labor. Because most GIS processes can be expressed with the three model symbols, the procedure here is little more than linking relatively simple and logical steps. (These examples are from ESRI ArcView's ModelBuilder; other vendors have similar capabilities.)

If needed, the model can be changed simply by redefining any of the steps. That is, in this case, enter Highways into the input data symbol, change the buffer to 10 kilometers, and change the name of the output to Highway Zone. In addition, each part can be run separately to check individual steps before the entire model is completed.

Image 3 applies model building to the generic steps in image 1. Compare the two diagrams and note the differences. Whereas the original model (image 1) is only a graphical outline for organization, the model-building version (image 3) is a functional programming structure that actually runs the GIS.

Model-building programs offer great assistance to GIS users, particularly when the same sequences of data and operations are needed for different parts of a study area, or when there is need for testing small changes in procedures. Although many GIS operations and functions are not difficult, they require time and effort. When operations and functions are programmed, there is a significant decrease in time demands on the user and less chance for error. Model building is one of the most significant developments for GIS in the past decade

CHAPTER 11

GIS OUTPUT

Introduction

ONCE ALL OF THE DATA COLLECTION AND ANALYSIS HAS BEEN ACCOMPLISHED, the GIS project must convey the information in some manner, whether on a traditional map or in a new digital format. This chapter discusses some of the basic means of presenting data, with an emphasis on GIS cartography. Good maps communicate efficiently and effectively on the computer monitor, Internet, CD-ROM/DVD, paper, or publication. The GIS user must be proficient in the respective communications medium.

As in previous chapters, there is much more to say about the topic than can be delivered in these few pages. This is not a tutorial on mapmaking (which would require an entire book), but an attempt to offer a few essential guidelines on GIS presentation.

GIS Output

Image 1: Displays

25-inch (63.5 cm) Monitor Display

Stereo Monitors

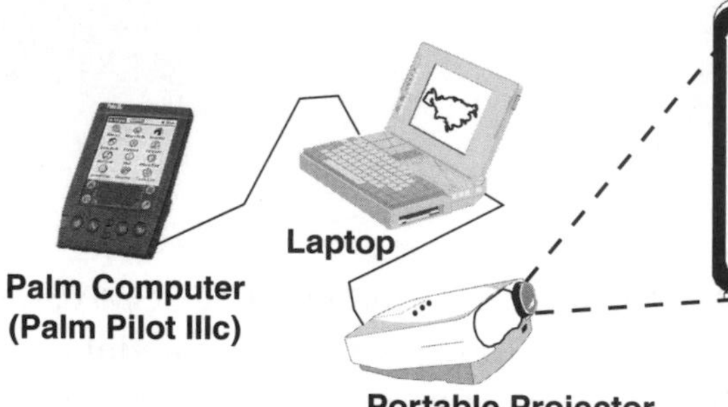

Palm Computer (Palm Pilot IIIc)

Laptop

Portable Projector

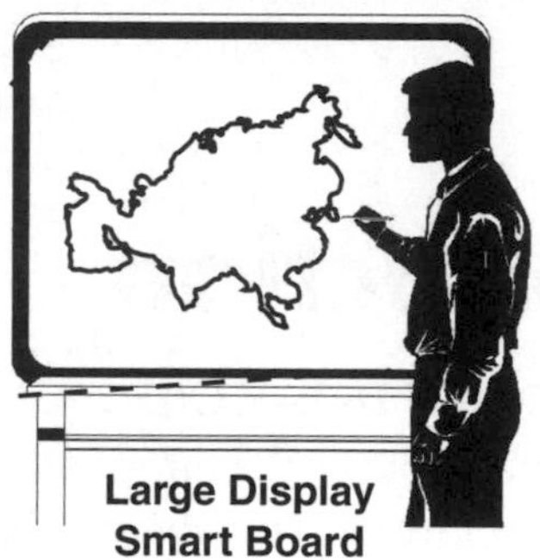

Large Display Smart Board

Image 2: Text and Graphics

Reports

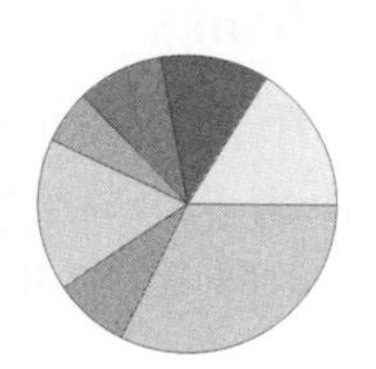

Graphs

Central American Database

Country	Population	Area Km^2	Pop Density	Currency
Belize	207586	22175	9	Dollar
Costa Rica	3319138	51608	64	Colon
El Salvador	5752470	20697	278	Colon
Guatemala	13021270	109502	94	Quetzal
Honduras	5367067	112852	48	Lempira
Mexico	92380850	1962939	47	Peso
Nicaragua	4275103	129047	33	Cordoba
Panama	2562045	74697	34	Balboa

Statistics

Image 3: Digital Media

CD/DVDs

High-density Disks

Local Networks

Internet World Wide Web

Fig. 11-1: Types of GIS output.

Types of Output

In GIS, there are numerous ways of presenting data and information. The computer monitor is the most common medium. Because of the flexibility GIS offers in making different versions of results for projects, it is practical to view initial output on a screen rather than printing maps for every change. Viewing data in the "old days" of GIS involved either sitting (or crowding) in front of a 14-inch (35.6-cm) monitor with limited color capability (and before that, even smaller non-graphic monochrome displays and poor printouts).

Today, however, computer displays are available in a range of sizes and formats, from handheld "palm" computers (ideal for field work) to wall-size projections using interactive "smart boards" for audiences, and even including stereo viewing with special software, monitors, and glasses, as shown in image 1 of figure 11-1. The future holds even more exciting promise (see the next chapter).

GIS typically produces statistics and text as part of the reporting. These are standard information formats, such as printed text, various types of graphs, and tables of data (image 2). Census tables, for example, can provide an important frame of reference, as well as detailed information for an area. Although these products are also "going digital," printed versions will continue to be used.

Many GIS projects are now presenting data and results in digital format (image 3). The media for this technology include computer disks, CD-ROMs, DVDs, high-density disks, local networks, and the Internet. In the last few years, the Internet/World Wide Web has become a major medium for dissemination of data, and its importance is growing very rapidly. Certainly increased application of digital products will continue, and although hardcopy maps will undoubtedly be used less in the future, they will probably never become obsolete.

GIS Maps
South America
General Reference

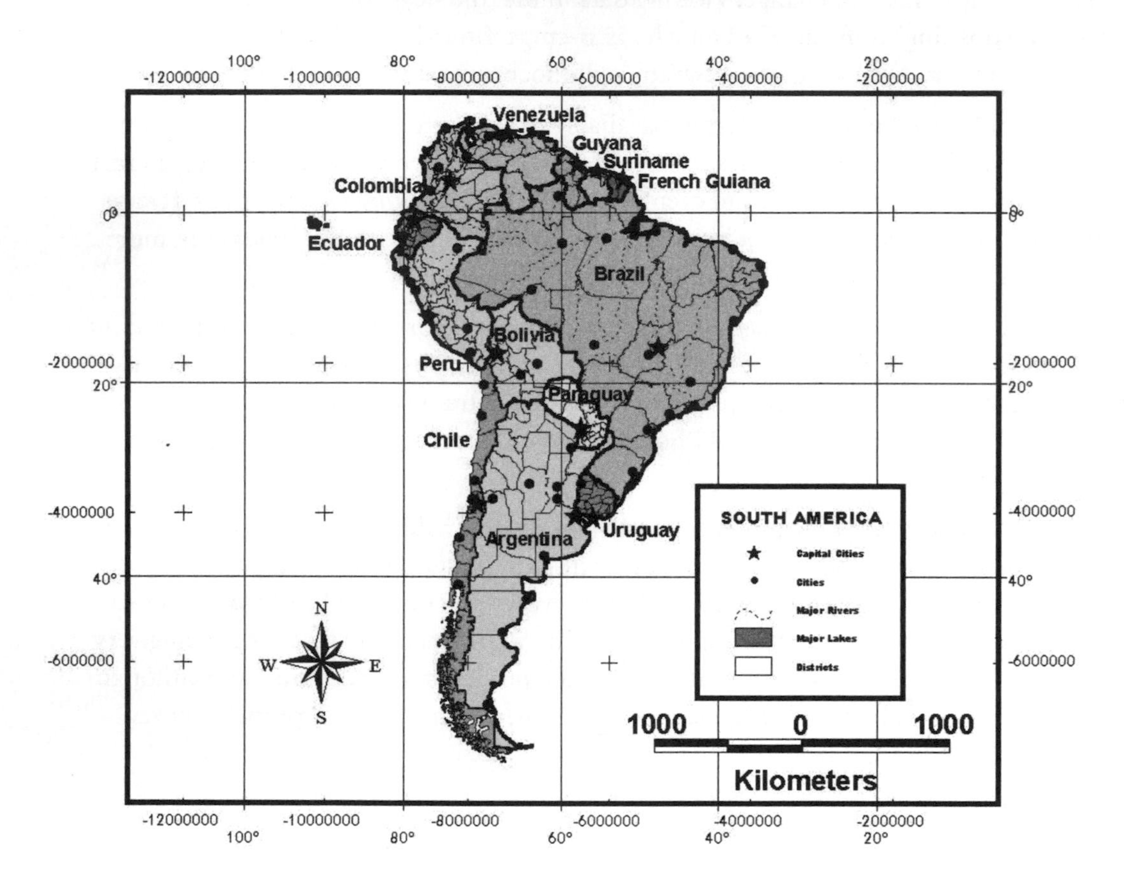

Fig. 11-2: Example of a general reference map.

GIS Maps

GIS is a powerful research tool, but it is also a data visualization system, with map production as a core medium. "Maps" in this sense include digital and paper formats. Each type of product has individual considerations for presentation, but the general concepts of cartography and visual management are valid for maps of any type. Although "quick-look" project maps can be produced without much regard for cartographic standards (they are typically for temporary in-house use), most GISs are capable of making excellent publication-quality map and graphic presentations.

GIS data can be presented in numerous ways, using a variety of map types and visualization options. Previous chapters use a variety of presentations, including selected feature maps, color-coded data, 3D relief maps showing slope and aspect, contour maps, diagrams and charts, and others. GIS is a very flexible presentation technology.

Two basic types of maps are usually made in geography: general reference and thematic. General reference maps present the basic geography of an area and are not specialized for a particular theme. These include project base maps, standard atlas and road maps, and topographic maps (which have elevation as a central theme but also show the general geography).

Figure 11-2 is a general reference map of South America, including a few key features, a title, a legend, a scale, and two reference grids (latitude/longitude and a system using meters from Equator and Prime Meridian). More general reference features can be added, but the small presentation size limits the number of elements. This type of map can serve as a project base map; for example, in association with this chapter's focus on South America as a regional example for mapping.

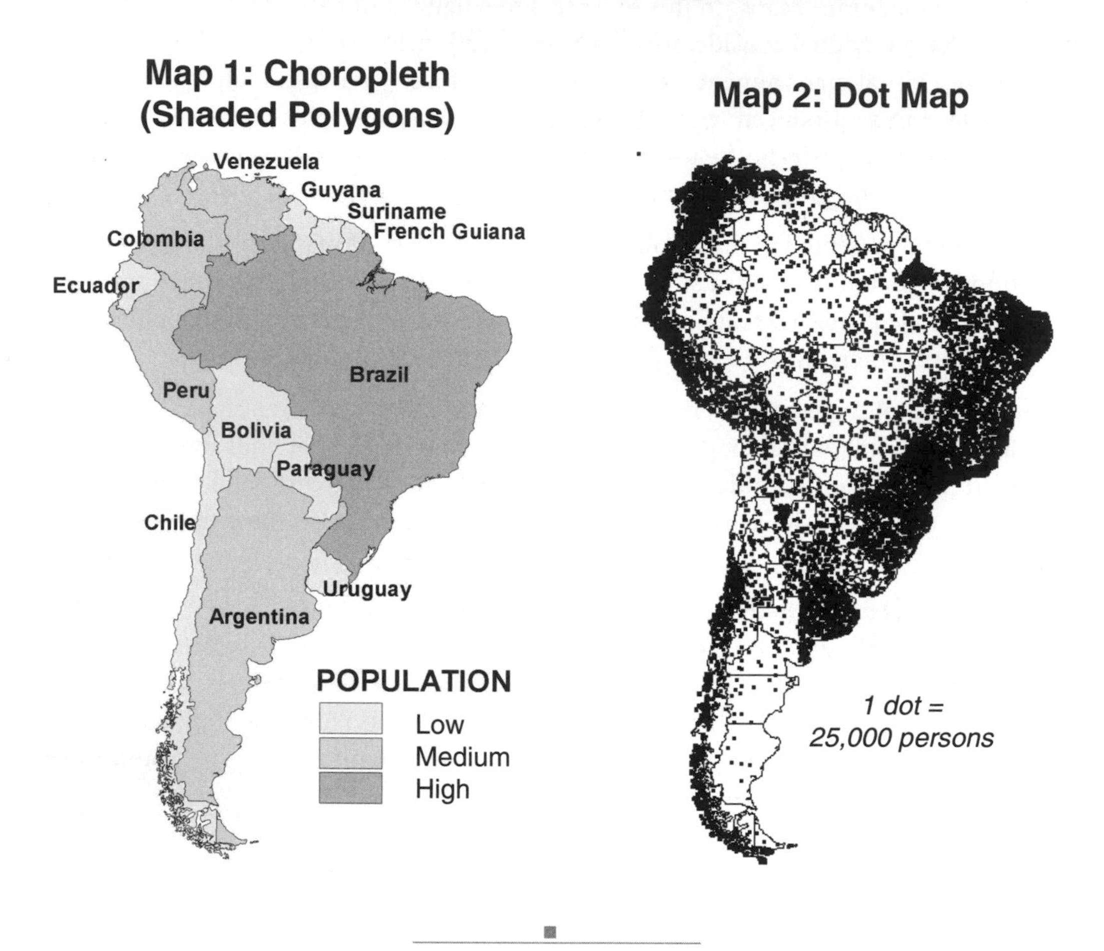

Fig. 11-3: Examples of thematic maps.

Thematic maps present selected themes, such as population or land use. Most GIS projects focus on specific themes, and associated maps are mostly thematic. Thematic maps can be presented in various formats, depending on the features and information to be shown. Point and line data use symbols, size, and color to show identification, attribute measures, and classification. Polygons are typically presented as shaded areas (colors or patterns).

Shaded area, or choropleth, thematic maps are the most common types in GIS. Map 1 of figure 11-3 shows South America population presented as shades of high, medium, and low. The map gives a good comparison of national populations, but is limited in terms of distribution and actual numbers (although the range of each population class can be included in the legend, if needed). This is not a complete population map of South America.

Map 2 uses a dot distribution format to present a better idea of how and where people are dispersed or spread out on the continent. It shows where the high and low concentrations are and gives a different type of demographic information from the choropleth map. Each dot represents 25,000 persons in each district. Districts are smaller units than nations and offer a better sense of distribution. Figure 11-10 depicts other dot map options.

This chapter will explore other types of thematic maps, some with very specific features and information. To be discussed are presentation strategies, such as the method for dot distribution, polygon shading, use of symbols, and other considerations. Thematic maps are the heart of GIS presentation and the user must have the cartographic skills to produce them as primary project information.

GIS Cartography

Purpose

Clarksville City Planning

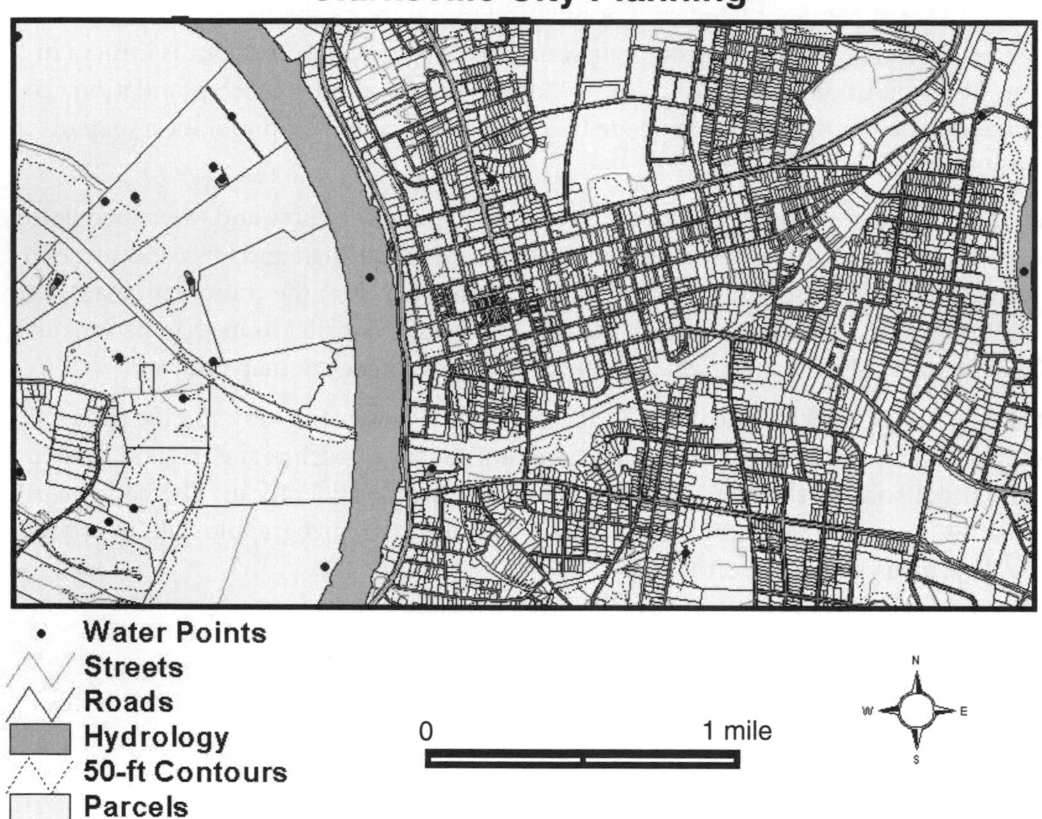

Base map for planning project:
Visually busy but displays
important data components.

Fig. 11-4: A map's purpose is the primary design influence. (Data courtesy of Bob Brundage, Austin Peay University GIS Center.)

GIS Cartography

Maps are still the most common form of GIS output product, whether on paper, in a book or report, or displayed electronically. Most applications eventually make use of hardcopy paper or printed maps because they are inexpensive, portable, and widely useful. Cartography, therefore, will continue to be an essential skill for GIS professionals.

The following sections are a brief tutorial in making maps for GIS projects. Full instructions and experience are the best learning tool, and this presentation is only a brief look at a few fundamentals in order to get started. South America population is used as an example for mapping and presenting data.

Actually, there is little difference between GIS cartography and standard geographic, paper-map cartography; the principles are basically the same, and there is no real separation between the two. Whereas paper maps are produced at various sizes and scales, GIS must also regard the medium of output (e.g., monitor, paper, CD-ROM, or Internet). Each format has particular design considerations, and some of the guidelines discussed here have to be modified as needed.

The most important factor for map design is *purpose*. Because GIS is a primary data visualization medium, a map's purpose is the chief managing cartographic component. With clear focus on the essential reason for the project and its maps, standards can be maintained on what is to be shown and how it should be presented. For example, a city planning project involves diverse urban features, from properties to transportation, and a base map should include the important components.

Figure 11-4 is a project base map for city planning of Clarksville, Tennessee, USA, with a visually "busy" display (at this scale especially), but one that is appropriate to presenting the study area and complexity of features that must be used. If the purpose shifts, say to city transportation planning, the features and design will be changed accordingly.

GIS Cartography
Audience

Map 1: Specialist Map

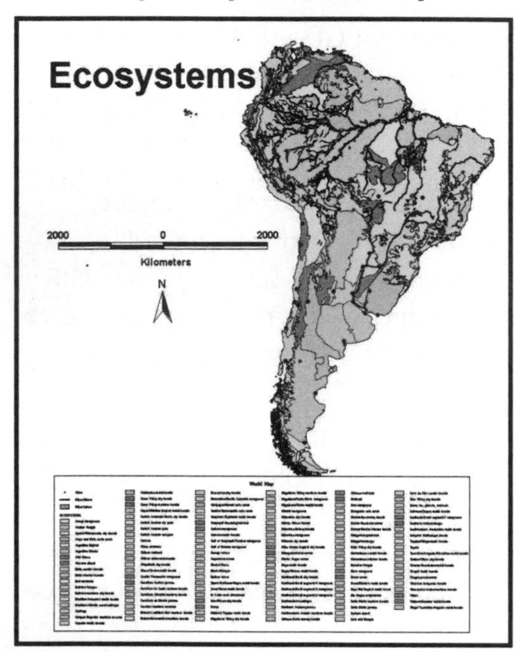

Map 2: Public Map

Fig. 11-5: A map's audience is a primary design consideration.

The second most important design factor, perhaps equal to purpose in many cases, is the intended *audience.* In GIS, there can be several audiences. For example, a map for specialists will be more detailed and complicated than one for students or the general public. It is not unusual for a project to produce several versions of maps in order to accommodate different audiences. Sometimes *who* will use the map can be as important as *why* it is presented.

Figure 11-5 presents a pair of South America environmental maps. Map 1 is for an academic audience who can make sense of the details and numerous classes (the color version is easier to interpret). The list of ecosystems is too large for a standard legend (which probably would be included on a separate page for better reading), and the size of the continent had to be reduced, but these are acceptable modifications to specialists. Map 2 is for the general public and is greatly simplified, with only the more basic ecoregions (habitats) and no other confusing elements. Both maps are useful, but their audiences are different. Because they are thematic maps, some of the common features of general reference maps are not included, such as a coordinate system.

GIS Cartography

Balance

Image 1: Cartography: Art and Science

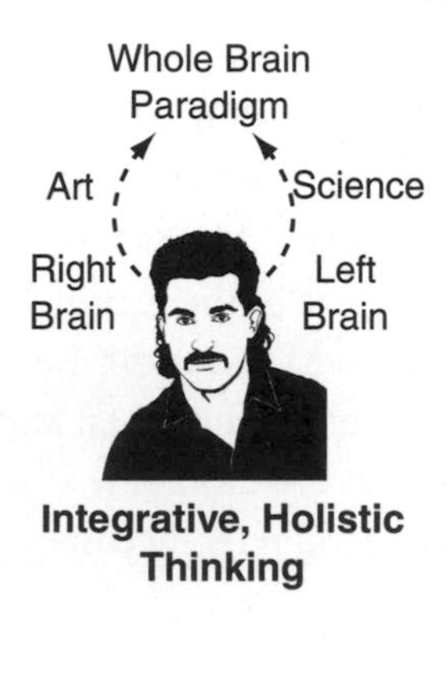

Map 1

Map 2

Image 2: Cartographic Balance

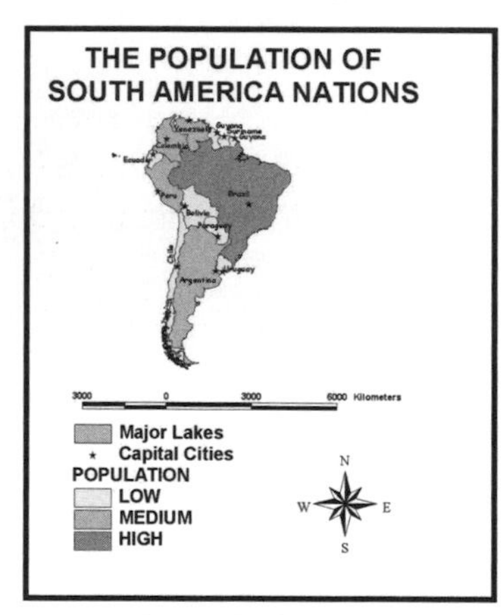

Poor Balance

Title: Too wordy; large.
Central Map: Too small.
Scale: Too long.
North Arrow: Too large.
Poor arrangement.
Off center.

Acceptable Balance

Fig. 11-6: Aspects of decision and balance in cartography.

There are numerous design considerations in cartography, and the following pages discuss the various mapmaking principles. The following four general principles should always be regarded.

- *Mapmaking is both an art and a science.* Data visualization should conform to scientific standards and proper conventions, which are necessary for unambiguous communication. However, because maps are visual, there is considerable need for "artistic" judgment. Several options or valid alternatives for presenting information visually are usually available, and the user must make choices. Right-brain (artistic) thinking can be as important as analytical left-brain in cartography, thereby making GIS cartography a "whole-brain approach" paradigm (image 1 of figure 11-6). For example, each nation of South America presents a particular shape and size (maps 1 and 2). Surrounding supporting elements must fit within the spaces around major features. Notice the name labels and how they must be carefully placed according to the shape and size of each nation. Arrangement, therefore, is a matter of decisions made by the designer. Look at the overall shape of South America and decide where the legend, scale bar, and other cartographic elements can or should be placed. Both maps are probably acceptable for most applications. Decisions can be difficult when several options are available.
- *Visual balance is important for all maps, regardless of purpose or audience.* Balance is the organization of all cartographic components within the overall composition of a map. It involves visual harmony and symmetry among the various elements, proper sizing of components, colors that do not clash, and other variables that make for comfortable viewing. A nice balance is a major requirement in effective presentation. The reader is comfortable with a balanced map, and does not have to labor at interpretation or appreciation. Image 2 shows two versions of the South America map, with poor and acceptable (but could be better) balance. Some of the elements that will help to achieve acceptable balance in this case are discussed in material to follow.

GIS Cartography
Communications

Map 1: Ineffective Communications

Incomplete:
No legend or scale.
What is the message?

Map 2: Old Style Cartography

Ornate and fancy artwork.
No longer appropriate.

Fig. 11-7: Aspects of decision and balance in cartography.

- *A map should stand alone; that is, it should present its information without the need for extraneous (outside) explanation or support.* If a map can do that, it has achieved its primary objective of effective *communication*, which is the fundamental purpose of any map. This means that the map should contain all information needed to communicate the central ideas, and it should be presented in a manner that is effective and efficient. Map 1 of figure 11-7 shows how incomplete information presents confusion about the map, which therefore makes for poor communication.
- *GIS maps typically avoid the fancy artwork of older-style maps.* No longer appropriate are dragons, ships, blowing cloud-faces, and multicolored scrollwork. GIS is, perhaps sadly, a utilitarian medium (practical, sometimes plain) that does not incorporate the artistic flourishes of yesterday (map 2). This is not to say that GIS output should be boring and staid, but its maps are usually meant to convey information and not art. However, as noted, there is considerable need for artistic judgment, and maps should be pleasing as well as informative. Perhaps modern cartography is as artistic in its own way as its hand-drawn forms of the past.

There are other considerations in making maps, many of which come from experience rather than tutoring. The following pages discuss other important factors, but they offer only initial information on achieving cartographic skills. Start with the principles in this book, but keep going to develop your own style and expertise.

Classification 1

Image 1: Classification - Data Reduction

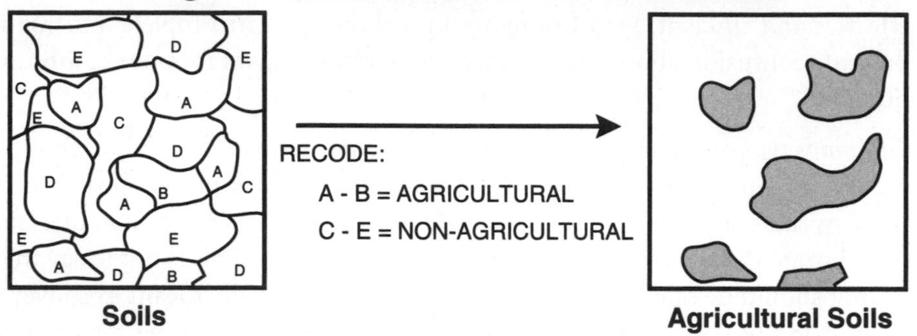

Image 2: Unique Classes

Original	*Option 1*	*Option 2*	*Option 3*
1 - 2	1.0 - 2.0	1.0 - 1.9	< 2
2 - 3	2.1 - 3.0	2.0 - 2.9	2 - 3
3 - 4	3.1 - 4.0	3.0 - 4.0	> 3

Values 2 and 3 are in which classes?

< > at each end only; middle values need Option 1 or 2 when more than 3 classes are used.

Image 3: No Gaps

Data	*Invalid*	*Suggested*
1,3,4	1 - 4	0 - 5
6,8,9,10	6 - 10	6 - 10
13,14	13 - 14	11 - 15

- All values included
- No gaps
- Even increments
- Comparable with other data

Fig. 11-8: Examples of the use of classification.

Classification

Thematic maps offer meaning to a theme's many features. Presentations with unique symbols or colors for every feature and value are not usually effective maps; making map data interpretable is the primary objective of cartographic communications. Classification is a data reduction structure, reducing a list of numbers or descriptions to a manageable few.

Whereas a general reference map may show each nation of South America in a different color or tone, it is the purpose of thematic maps to present how each country compares to the others regarding the topic under investigation (e.g., population classes). In addition, classification is an analytical procedure, and one of the first considered in trying to understand the theme better.

Instructions on the theory, principles, types, and uses of classification are beyond the scope of this chapter, but a few points on the cartography of classification are appropriate. Purpose and audience are the primary guides to the type and nature of classification, as indicated, and decisions have to be made on which type to use and how to present it.

Basically, classifying data helps to locate and describe where some type of homogeneity (similarity) exists, or conversely, where dissimilarities exist. Image 1 of figure 11-8 is the recoded soil map from Chapter 8, showing that when numerous confusing polygons are reduced to a few classes, distribution is more evident. The following are four basic principles of classification.

- *Each class interval is unique; that is, no data value can go into more than one category.* Although it is common, and perhaps understandable, to express values in easy increments of 1–2, 2–3, 3–4 (usually meaning *from 1 up to 2*), this creates the problem that the value 2 seems to belong to two classes. It is a bit awkward, but correct, to use classes such as 1.0–1.9, 2.0–2.9, and so on. Note that even the 1 must be designed as 1.0 in order to keep the same number of decimal places. There may be sets of data where the greater-than and less-than signs (> <) can be used, but obviously they must be employed in the proper context, especially only as the beginning and ending values. It is improper to use < 1, < 2, < 3. Image 2 demonstrates the problem and offers a few possible solutions (none of which is perfect, but they are improvements).
- *No gaps should exist between class intervals, even when there are no intervening values.* A data set, as shown in image 3, includes three groups of values in the first column, with separating gaps. It is invalid to use convenient classes of, for example, 1–4, 6–10, 13–14. The suggested reclassification can be 0–5, 6–10, 11–15. This covers the gaps, which may be important for comparison with other data sets that have values not in the first set. In addition, this scheme groups numbers in increments of 5, which is more logical, intuitive, and consistent.

Classification 2

Image 1: 0 Is Valid Value

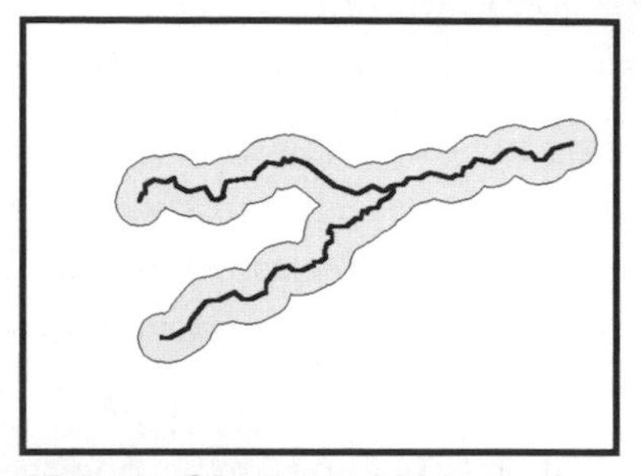

Map 1: Observed Vegetation
In a zone along main channels

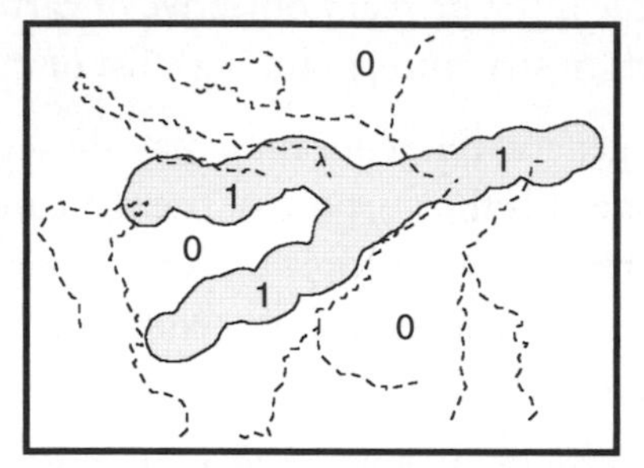

Map 2: 0 - 1 Classification
Overlay with Features
0 - No Vegetation
1 - Vegetation
Question: Why not along other channels?

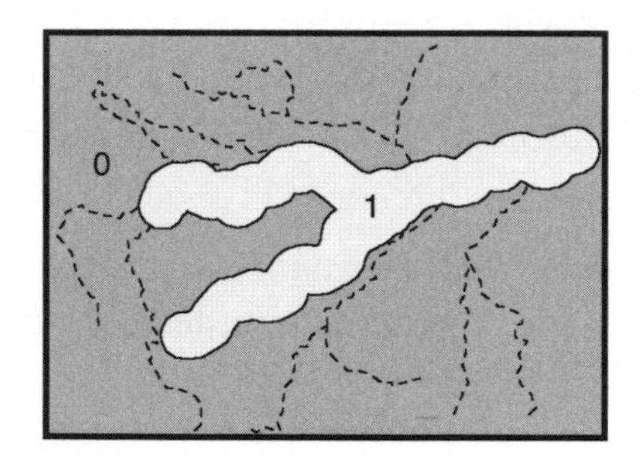

Map 3: 0-Value
Spatial analysis of "No Vegetation"

Image 2: Rational Classification Scheme

Data Groups	*Original Classification*	*Suggested Classification*
2,10,11,12,13	1 - 13	0 - 15
16,17,22,25,26	14 - 26	16 - 30
31,34,36,37,39	27 - 39	31 - 45

Fig. 11-9: Classification values and schemes.

- *The number 0 is a valid value.* The absence of something can be as meaningful as its presence. Therefore, when possible, include 0 as a value, whether or not the number exists in the current data set. Image 1 of figure 11-9 shows the distribution of vegetation around two main stream channels (map 1). When values of 0 (no vegetation) and 1 (vegetation) are used (map 2), the classification speaks as much to where vegetation does not occur (around other channels) as to where it does (along the two main channels only). This leads to questions in search of the reasons for the distribution. The 0 value can be mapped as a valid feature (map 3) for better analysis, and possibly can be combined with other factors (e.g., the other streams, geology, soils, or slope). 0 has become a value equal in importance to the vegetation value 1.
- *The classification scheme must make sense; that is, it should be consistent and use understandable categories and intervals.* Even though the numbers in the image 2 data set can be grouped into intervals of 13, it is best to use even increments and common classes when possible (e.g., 15 in this case). Intervals of 5, 10, 25, 50, 100, and so on are understandable to the reader; using classes of 13 or 33 may be statistically valid, but not very clear. However, if there are special groups of numbers that need to be identified, various statistical classifications may be available, such as standard deviation or quantiles (see the following section).

Correct classification is critical to project analysis and objectives, yet it is a difficult operation for many users. Statistics can be misused, of course, and it is a major responsibility to present reliable and honest classification. The next section offers a few pointers on classification mapping.

Classification Mapping

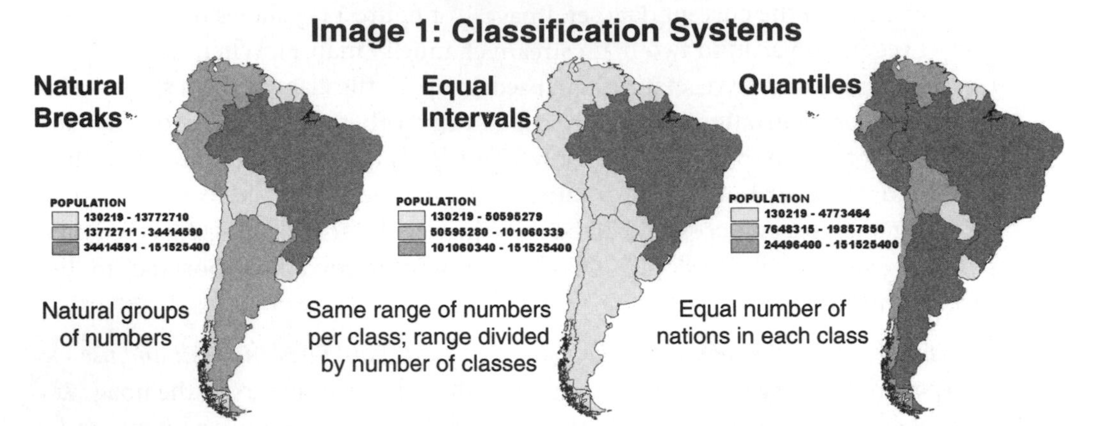

Image 2: Dot Map Densities

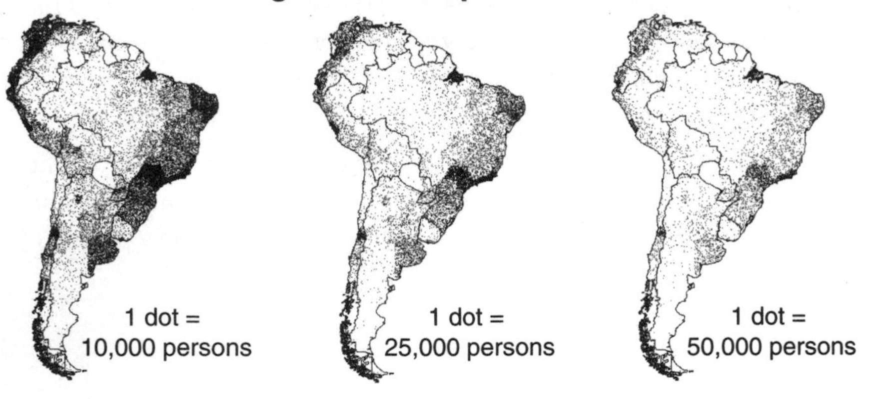

Image 3: Dot Map Options

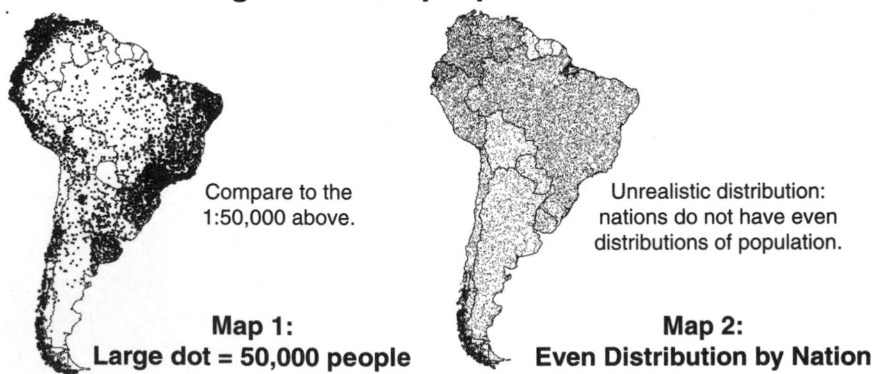

Fig. 11-10: Examples of classification schemes used in mapping.

Classification Mapping

Once the classification scheme has been properly devised, it must pass the cartography test. Statistics and cartography can be synergistic (both contributing to a presentation that is much better than if they are displayed separately), but they are not always a natural pair. The user must work at making effective and communicative maps. Discussed in the following sections are thoughts and hints on the visual elements of cartography.

Using the principles from the previous section, an initial classification scheme is devised, followed by editing to keep the number of classes to five or so (for visual convenience). Although sometimes many features must be identified (as in a complicated ecosystem map), thematic maps try to communicate a sense of comparison and therefore should not be visually complex.

Many classes on a map are difficult to interpret, and reduction to fewer can make the map more effective. However, forcing the classification scheme into an artificially abbreviated format may have consequences on the statistics and analysis. Sometimes it may be necessary to show only a generalized map of the more complicated classifications.

Most GISs offer at least several methods of data classification, each suitable for different types of data and applications. Sometimes it is difficult to know which method is best, so the user must understand the data, classification techniques, and the intended application.

Image 1 of figure 11-10 shows three classifications of South American population. Each may be appropriate for a specific application. All are statistically correct, but they communicate different messages. Changing the number of classes on the three maps will complicate the dilemma even more. Deciding how to classify data is a difficult task.

Dot maps can be particularly troublesome. Classification considerations relate to the number of persons represented by each dot (population-dot ratio) and how the dots are distributed (dot density). Obviously, too many dots will result in black regions, with too little sense of regional differences (low population per dot = more dots on the map). Too few dots (high population per dot) will not convey realistic densities or distributions. Experimentation may be needed to decide the appropriate balance.

Image 2 shows several versions of a South American population map, using different population-dot numbers based on district populations (but with national borders displayed). Each map is useful, but the 1:50,000 map seems sparse and does not give a good sense of population distribution. The 1:25,000 seems preferable (or does it?).

Image 3 demonstrates two dot map options. Map 1 shows a larger dot symbol size than image 2, at 1:50,000 display. Note the difference in appearance between it and its counterpart in image 2. They seem to be different data. Is it as good as the 1:25,000 dot map?

Map 2 is based on nations as the unit of measurement rather than the smaller districts, which is basically a different mapping resolution. Note how the dots are evenly distributed over an entire

Colors and Cosmetics

Symbols, Tones, and Patterns

Diverse Map Features

Glasgow, Scotland

General Reference

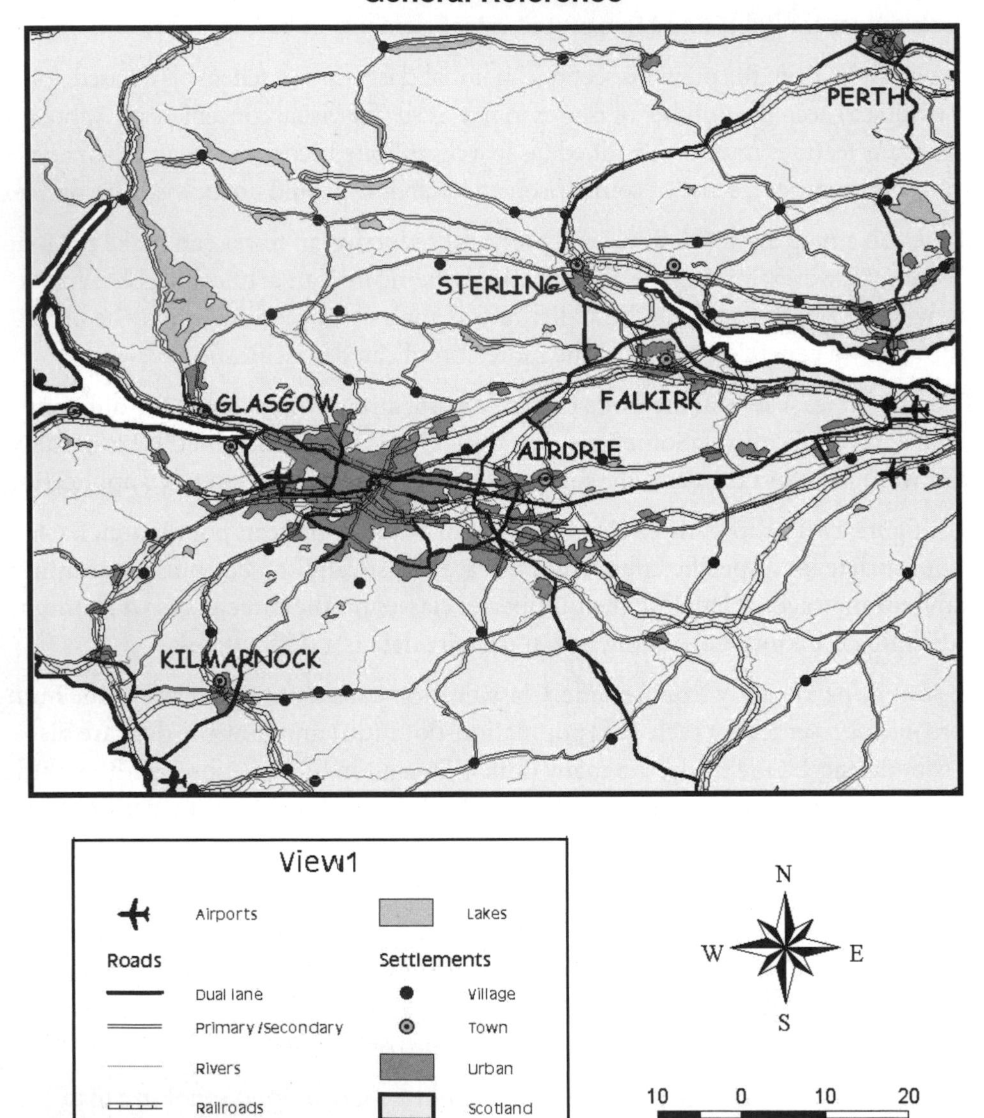

Fig. 11-11: The visual properties of map content.

country. This may be statistically valid, but it is visually deceptive. Certainly the Amazon region does not have the same population distribution as the Brazilian coastal area. Districts offer more spatial definition and, when possible, even smaller units (census tracts) are preferred.

Colors and Cosmetics

Effective classification and communication depend on the visual properties used to convey information. The cosmetics (appearance) can be highly contributive or disastrously destructive to the message being delivered. As discussed, good GIS presentation is both right-brain design and left-brain objective analysis. The following are a few pointers on using cosmetics in a GIS project. Like classification, common sense is paramount, but experimentation is useful.

Figure 11-11 is a map of the region around Glasgow, Scotland. Maps with point and line features use symbols depicting identifications or classes, such as different colors and sizes to show magnitude. For example, roads are normally shown in various line weights (thickness), from thin (minor roads) to thick (major highways). Color or tone can add information (e.g., type of road surface). Points can be depicted as different symbols, such as dots, circles, or an icon suggesting identity (e.g., an airplane for airports). They can also make use of color or size (such as small to large, meaning low to high traffic) to indicate further distinctions.

Symbols should be intuitive and clear to the reader, so that interpretation and understanding of the information is not difficult. Towns can be dots or circles rather than an X or a triangle. For example, stars are usually reserved for capital cities. A hierarchy of dot/circle sizes usually indicates settlement populations or types (village-town-city). The map shows a variety of symbols and shades (but color and a larger size would be much easier to interpret). There is room for improvement, but this map demonstrates how the careful use of point symbols, line types and weights, and polygon shades can present diverse information, even in a small monochrome map.

Polygons normally use shaded areas, colors/tones, or patterns. Like symbols, the colors and patterns should be organized for easy reading. For example, arrangement of low-to-high classification in a light-to-dark sequence is logical and visually sensible, especially using a single color or tone rather than separate primary colors. That is, sequences in shades of a single color or tone are more easily interpreted than are three or four different individual colors; light pink to dark red is better than pink-blue-green-purple.

Classification is understood much better when presented for comfortable visual interpretation. The population maps in this chapter are organized from light to darker tones, indicating Low-Medium-High populations.

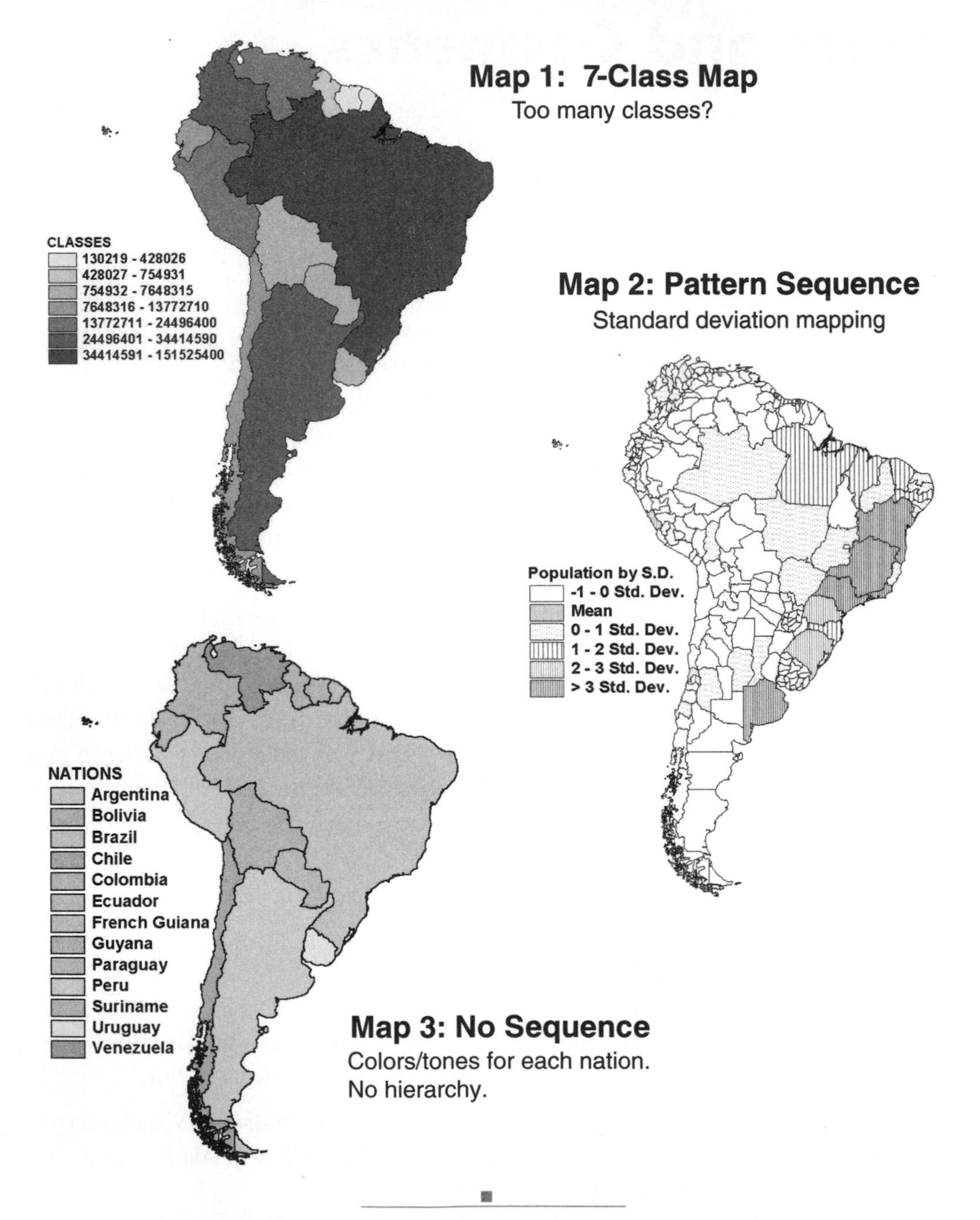

Fig. 11-12: Patterns and shading used in various types of maps.

The eye can understand and track only about six colors; more than that can lead to visual confusion. It is a good idea to limit the number of classes for a given variable. Map 1 of figure 11-12 shows a seven-class map of South American population. Seven is too many for easy interpretation, as well as too many for a 13-nation region. There is probably little need for seven classes of population, anyway.

Color or shade choice can be tricky. Six distinct shades of a given color, such as red, may not be possible. Using patterns on top of the color can help, but care is needed. Two colors can be used, each with a sequence of shades, but it is difficult to make the entire range visually progressive. For example, the darker tone of one color may be more dense than the next higher color's lighter shade, and the eye will not perceive the sequence easily. However, two color ranges can show two different aspects of a classification (e.g., low population group and higher population group).

Map 2 shows South American population mapped by standard deviation, using symbols and shades. Positive deviations (higher than the mean) progress from dots through increasing vertical bar densities. The negative deviation is a distinctly "weaker" tone (white), giving the eye easy recognition of the hierarchy. The mean is neutral gray, but there does not appear to be any district with that measure.

Sometimes, more classification colors are required than the eye can discriminate, such as separate colors for each South American nation (map 3). In this case, colors serve only to separate and help identify the nations, and no hierarchy or sequence is inferred. Distinctive borders can enhance separation of the nations (e.g., lines that are heavier than the default thin ones). However, heavier lines do not work very well on detailed borders. Notice the difference in definition of the islands of Southern Chile between maps 1 and 3.

There are many more hints and suggestions to cartographic cosmetics. Perhaps a useful strategy for learning is to examine and critique maps from many sources, observing the good, the bad, and the ugly.

Map Framework

Border and Neatline

Image 1: Map Framework

Major features.
Other elements include:

- Inset map
- Text
- Data sources and dates
- Map publishing date
- Credits

Large Title

Central Map Figure

North Arrow

POPULATION
LOW
MEDIUM
HIGH

500 0 500 1000 1500
Kilometers

Scale Legend Chart

Image 2: Layout Formats

Landscape

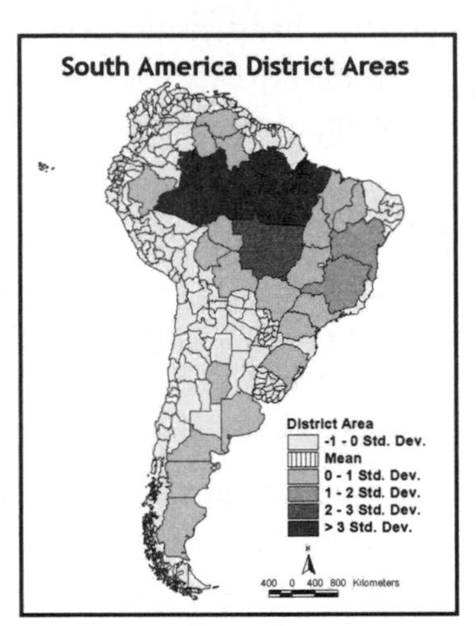

Portrait

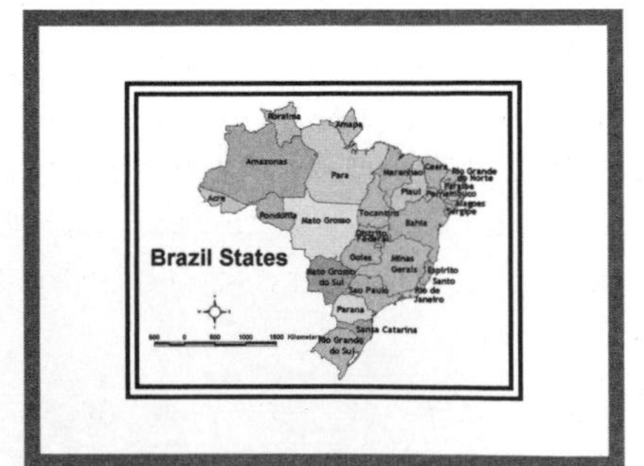

Image 3: Double Neatlines

Large page format, with heavy outside border.
Double neatline or border and neatline inside.

Fig. 11-13: Types of map frames.

Map Framework

There are various approaches to proper cartography. The one used here begins with the outside framework, then proceeds to the central map figure, and then to the support elements. However, personal preference may be used. Many GIS technicians choose to insert the central map first, then build the framework around it. Image 1 of figure 11-13 presents the basic map layout features discussed in the following sections.

Most GISs offer a few options to help the cartographic process, such as a ruler display for precise measuring and placement, grid snaps for automatic placement, and other devices. Once the basic configuration has been selected, map construction can begin.

The first task is to complete the page setup. This includes the page orientation, either landscape (horizontal format) or portrait (vertical). Image 2 shows both displays, with an appropriate map in each (although Brazil could be presented in a portrait layout). It is possible to place vertical maps into landscape frames (or the reverse) if there are other map elements (such as text, charts, and so on) that will fill the mapping area.

Maps normally have a frame of some type, typically a border. However, there are situations in which the map "floats" without a frame (a format used in this book as a design preference). A border is a visual support structure, so it should visually encompass (surround) the primary map parts.

A common and effective border is a heavy squared line (though rounded corners can work at times). Because its purpose is to frame the map, it should not be visually distracting (i.e., plain and not fancy). The "weight" or thickness is a matter of judgment, and usually a balance between thin and thick is best. Of course, this depends on the planned size of the printed map. Large maps can use fairly thick lines, but the actual weight number is a matter of trial, error, and judgment. Simply drawing a box is usually sufficient to make a border. The weight and color can be adjusted at any time. Color is typically black or dark. Bright, eye-catching reds or pastels are not normally appropriate.

Neatlines are used in two ways: as complements to the border and as special borders for the map. In the first case, they are thin lines just inside the border (image 1). Neatlines supposedly help to draw or focus attention from the border into the center, but the decision to use one is a judgment to be made by the cartographer. The line should be notably thinner than the border, and placed a small increment within it. (Thin outside and thicker inside lines have been used, but that is not recommended normally.) The neatline should be the same color and style as the border (e.g., dark and squared).

Neatlines are useful where heavy borders are well outside the central map and act as page or larger borders (image 3). The neatline, often a double neatline, serves to frame the map components like a standard border. The choice of models depends on the application, format of presentation, and other considerations.

Central Map Figure

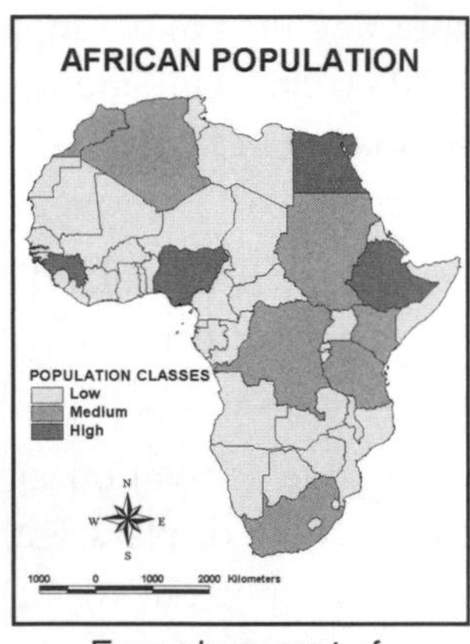

Easy placement of support items

Image 1: Shape Influence

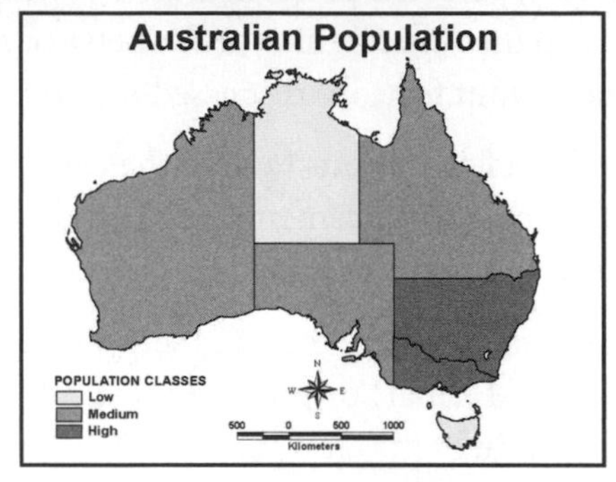

Limited space for support items

Image 2: Colors, Patterns, and Tones

Map A: Primary Colors to Gray Tones

Map B: Fill Patterns

Map C: Balanced Gray Tones

Fig. 11-14: Considerations for the central map figure.

Once the framework has been established, the size and placement of other components has some guidance. Crowding the line with other elements is not recommended; some empty "white space" is needed to help focus the central map feature.

Central Map Figure

Because a presentation map consists of all of the features discussed in this chapter, including the map at the center, the term *central map figure* is used here to denote the region or place being mapped. It is the primary component of the output, and can be constructed next, although the order of parts to be entered is not particularly important. Because the map figure is the defining element and the most visually obvious, it should be the dominant item and as large as possible within the framework (while leaving sufficient white space).

Usually a click-and-drag box is used to define the area for the map figure (depending on the software). Normally, the area is centered in the frame, but the actual shape of the central map figure will define the best placement. Begin with the center position and then rework placement as the overall presentation develops. Africa's shape suggests locating the legend and other elements in the Atlantic Ocean space, but the rectangular Australia leaves only a small area beneath it for these items (image 1 of figure 11-14). Each nation and mapped area has particular spatial design factors, and the GIS user must make special decisions. Examine South America and note its possibilities.

The project and presentation purpose will determine what is shown on the South America example. This is a population project, and a simple thematic map is to be produced, showing three population classes. Presumably, classification choices have been made prior to the mapping, but it is usually possible to make changes at any time. The original view is in color, but if the map is to be printed in black and white, modifications are often required.

Map A of image 2 shows how bright, primary colors translate into monochrome. The map is too dark and lacks contrast; adjustment is needed. Map B uses fill patterns only. There are many types of patterns that can be used, from dots to line and other designs. Are these satisfactory? Probably not. Even though the low, medium, and high nations can be interpreted, the mix of patterns is unappealing. Better aesthetics is needed. Map C has a balanced set of a few gray tones, which seems to work well. Some trial and error is normal in deciding which tones work best. Perhaps the map still needs some work, possibly tones and patterns. How would you improve it?

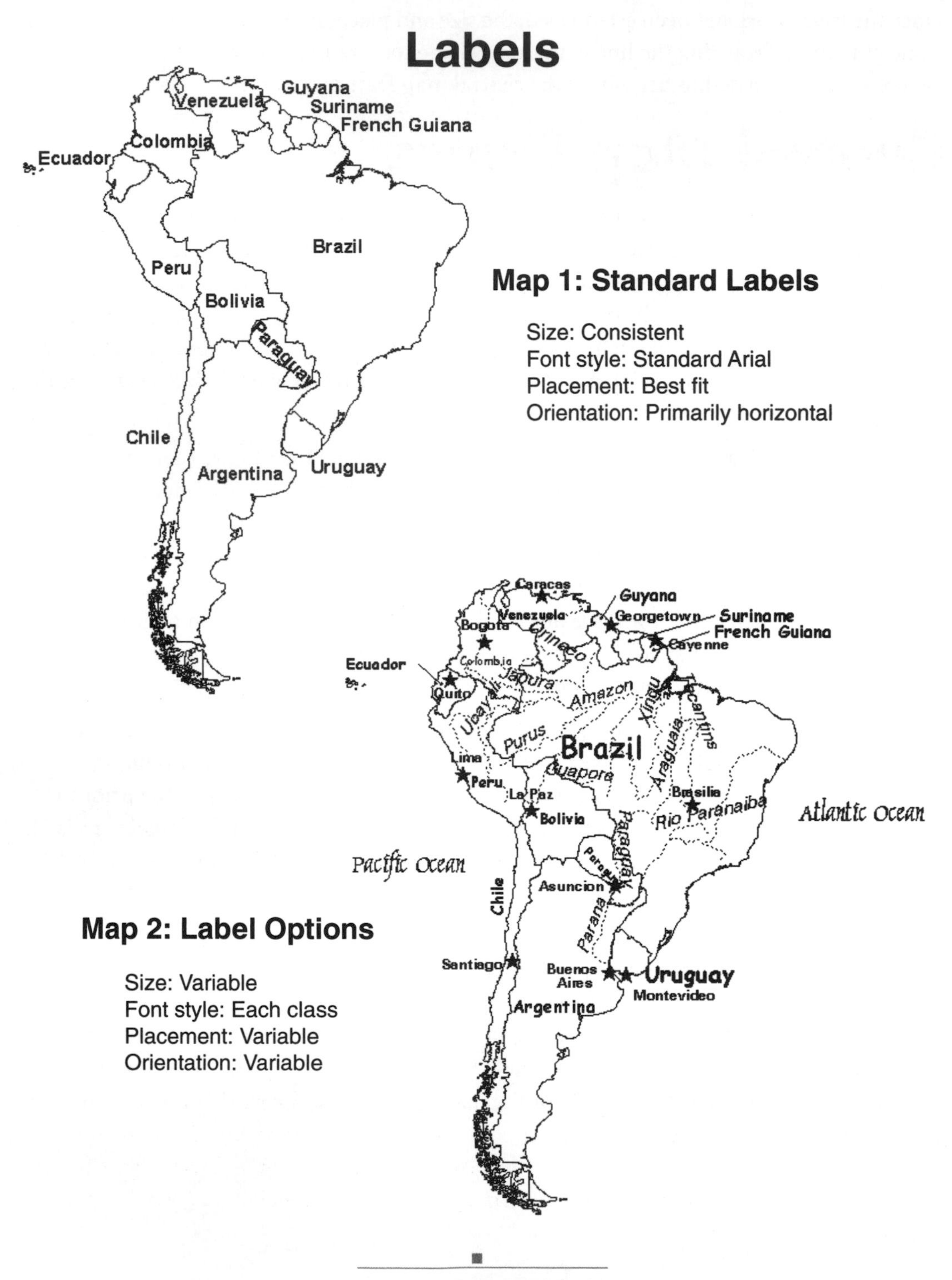

Fig. 11-15: Using labels

Labels on central figures in maps can be useful to many readers, especially when numerous countries or polygons are mapped. There are four basic considerations in the application of labels; namely, size, font style, placement on the map, and orientation of the text. Two maps of South America with various label situations are presented in figure 11-15. Map 1 has country labels typically in the center of the nation (as in Brazil), but off-center is acceptable (Peru), or even rotated when necessary (Paraguay). Because of Chile's narrow shape, its label is placed mostly outside (and it can be rotated vertically, if needed).

Ecuador and Uruguay are too small to contain a name, and therefore these labels are also placed outside the countries' boundaries. The three small northeast nations present a label placement problem. Names must be carefully stacked and offset, but even then each label does not line up perfectly with its nation. Fortunately, readers usually understand and can easily make the visual adjustment.

Note that labels have the same size and font. Even though there are problems of fit, it is best to remain consistent in font type, size, and color for labels that identify features of the same classification. Different fonts help to separate identification of different types of features, as demonstrated in map 2.

Of course, the names should be large enough for easy reading, but when possible, they should fit inside the nations without touching the borders ("line bleeding" can result). Venezuela and Colombia barely fit. The font style can be changed after some experimentation and altered to fit a particular audience, if necessary. Some people prefer formal, standard fonts, such as the Arial bold used here, whereas informal and unusual fonts are preferred at times (used in map 2).

Map 2 shows more complicated problems and overall it is not a good map. Because country names, capitals, and major rivers are presented, the map is more crowded and more difficult to read. Several labeling alternatives are offered, some of which seem to work in this map, whereas others do not. Different label size, placement, and orientations are presented.

Different country name sizes are used, mainly to help fit within borders. Varying sizes seem to confuse more than assist. As noted, in maps with various classes of features, label consistency is expected. The informal Comic Sans font is used, sometimes perceived as too odd and inappropriate for maps. The capitals have a smaller size and font, but they interfere with the country names and visual crowding results.

Water features, including the oceans, typically use an oblique or slanted font (in blue when color is used) in order to distinguish the names from the country labels. The major rivers use a unique font, which makes their names easy to recognize, even on busy maps such as this one. The oceans use a more elaborate font, perhaps too ornate for this type of map, but they are distinctive.

Support Features

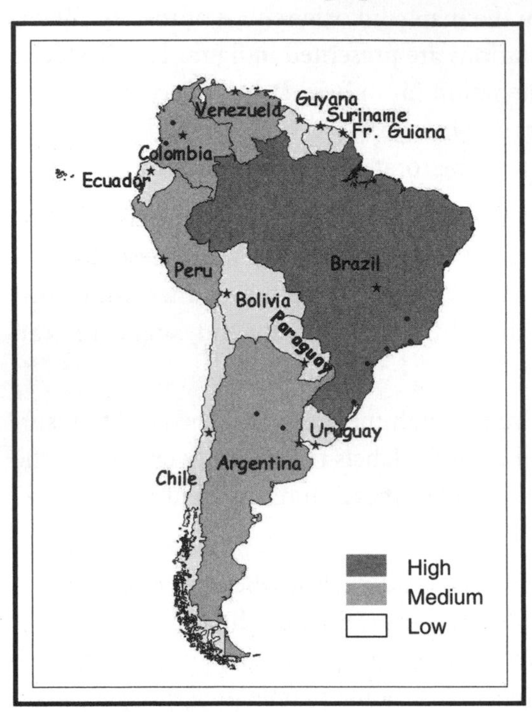

Map 1: Population with Selected Cities

Cities > 1,000,000 population
Label place adjustment

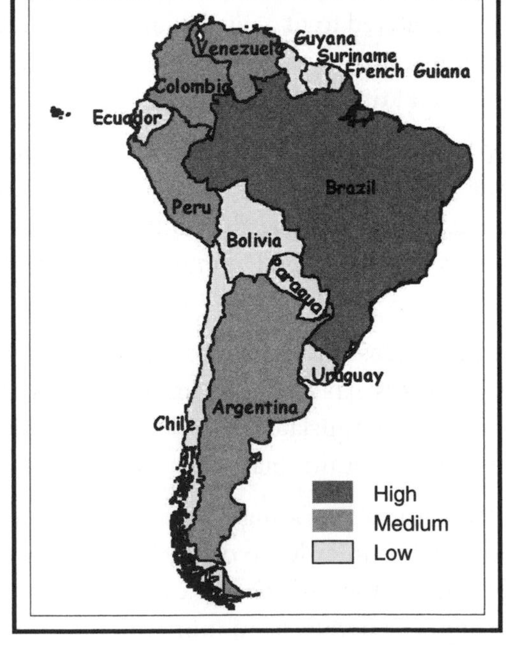

Map 2: Added Contrast Distinct Boundaries

Trade-off of southern Chile island and coastline details

Fig. 11-16: Support features applied to a central map figure.

It is useful for the GIS cartographer to understand the major font types and various ways of presenting them. Some GISs offer ways of changing spacing between letters and lines of text, which can help in troublesome situations. If the label for Colombia does not quite fit, but the font style and size is preferred, a slight adjustment of letter spacing may achieve the fit without sacrificing consistency of size and style.

Adding cities, rivers, and other features can be helpful or harmful, depending on the basic theme and the nature of the additional elements. This is where visual balance is very important. Because the purpose of the maps in figure 11-16 is to show population classes of each country, the secondary elements should not be prominent; they are intended to lend support to the theme display. Decisions must be made about what to show and how the display it.

Map 1 displays the population map with cities. The database contains 66 cities, too many to show on a basic thematic map of this size without making visual clutter. Thus, only the major cities (with populations over 1,000,000) are represented. Avoiding visual confusion is important. The cities are dark dots, easily visible when printed at the darkest tone.

Capitals are included as a separate category, and are shown as star symbols. In addition, note how some of the country labels have to be shifted after the cities and capitals are added. Fortunately, relocation is usually a simple click-and-drag operation.

All of the needed elements are in place at this stage of the map construction, but South America still seems to be lacking visually in some way. It does not seem sufficiently "crisp" or contrasting, so heavier borders are added to give contrast to each nation, especially where identical classifications are adjacent (map 2). However, note the ill effects on southern Chile's coastline and islands. There is a trade-off, as usual, and decisions have to be made. Which is better: nicely defined national borders but a degraded coastal region, or weak borders but better definition of the detailed coast? It is possible, though difficult, to reformat the spatial data so that the islands do not become heavier when the other borders are increased, but that adds more work. Decisions have to be made.

For many tasks dealing with cosmetic changes, however, GIS offers easy operations. Trial and error is rather painless and even fun. Because GIS cartography is full of small and large decisions, the ability to try different views without difficulty is a significant benefit.

Legends 1

Map 1: Map Layout

Map 2: Standard Legend

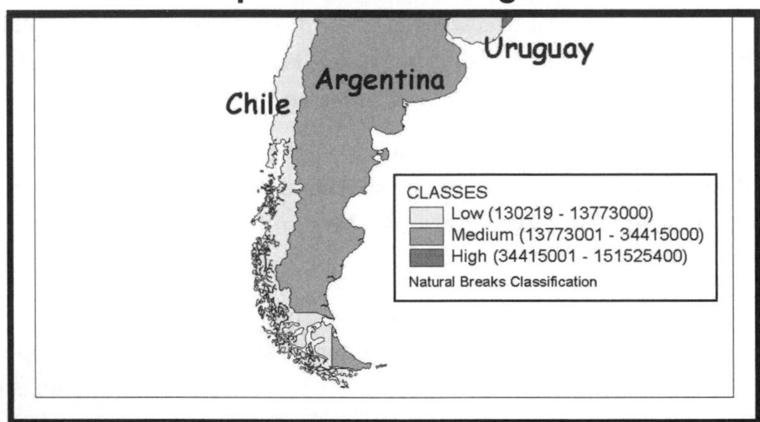

Map 3: Alternatives

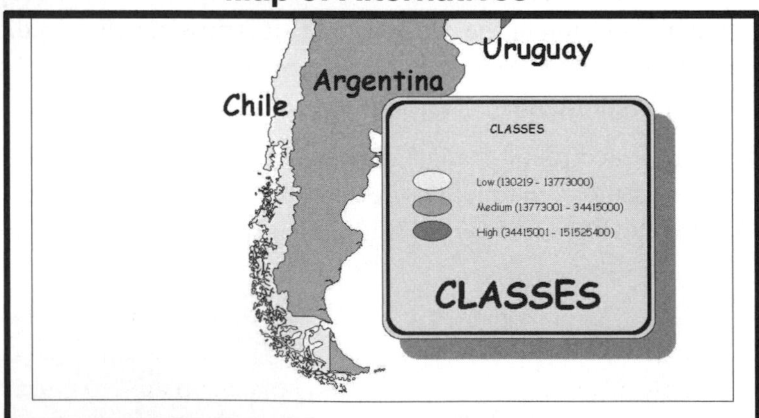

Bad alternatives? How many design problems can you see?

Fig. 11-17: Examples of map legends.

The Map Legend

The legend is probably the second most important component of a map, next to the central map figure itself. Its purpose is to explain, in a visually convenient manner, what is seen on the central map, such as identification of the features and the meaning of classification colors. There are several considerations in the construction of a legend, including its frame; content, style, and placement of text; and style, size, and placement of symbols and classification boxes. Although the design of legends can vary considerably, two elements of all successful legends are an effective style (visual balance) and effective substance (clear explanation).

The size and placement of the legend is a major decision. As noted, the shape of the central map feature greatly influences the nature of the surrounding map components. In these South America illustrations, the legend is placed at the lower right, although it probably could fit at the middle left as well. Map 1 at the top of figure 11-17 shows the overall layout, whereas the legend examples in maps 2 and 3 are illustrated with only the bottom portion for clarity; the scale of the presentation on these pages is too small for reading the details when the entire map is displayed.

An initial task is deciding how to present the representative symbols and classification boxes. The boxes can be displayed as squares, rectangles, circles, ovals, or other shapes. They are normally filled with the classification colors, followed by a label, either the class name (High, Medium, Low) or the actual numbers. Sometimes both are used as a convenient way of presenting both qualitative and quantitative measures in one line (map 1).

Colors and tones in the classification boxes must be the same as shown on the map. Otherwise, the reader will be confused. It is best to use distinctive shades in order to avoid misinterpretation, but sometimes shades can appear definite in the legend but less clear on the map; care is needed. The legend must duplicate the map to maintain consistency and readability. Some experimentation may be necessary.

The symbols in map 1 are neatly aligned along the left side, spaced apart for easy reading, and are basically centered within the legend box. These are standard formats, but the user can make other decisions based on artistic merit or special considerations when space is restricted. Most GISs offer plenty of options in construction of the legend, and there may be numerous versions; decisions are not always easy.

Map 2 shows a few of the legend design possibilities, most of which are poor choices, demonstrating what to avoid. The legend frame has been changed to a rounded double line (an interesting option that offers a nice touch), but it is too large in relation to the continent. It is a "drop-shadow" format with gray background inside, which is nice for some presentations but too ornamental for this simple thematic map.

Legends 2

Map 1: Mixed Symbols and Design

Map 2: Large Floating Legend

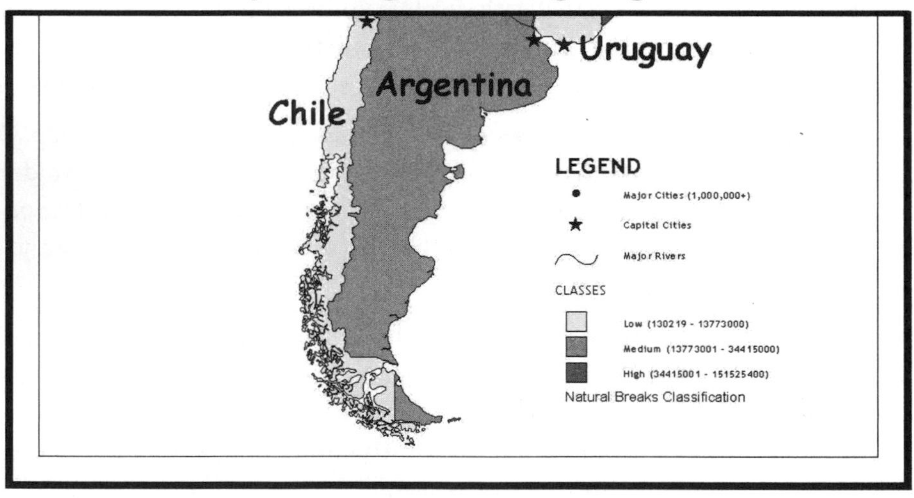

Fig. 11-18: Further examples of map legends.

The legend covers part of Argentina, which in some cases can be permitted; for example, where there is insufficient room for the legend, a bit of overlap onto the central map figure is acceptable if important features are not concealed. There are two "Classes" labels or titles in the legend, with the bottom one being too large and redundant (unneeded duplication), and probably in an inappropriate font. The ovals are interesting, but not visually balanced in this case. In addition, the legend appears to have too much empty or unused space, especially given the limited room available for it. Larger print in the legend box would be helpful.

Other features can be presented on the map, such as cities. In map 1 of figure 11-18, the capital cities are designated with a small star and all other major non-capital cities (over 1,000,000 population) are shown with dots (none are present on the southern part of the continent). Note that major cities are defined in the legend so that the reader understands the term *major.* Because capitals are in a separate theme, they are also noted in the legend (although the star symbol is a common indicator of capital cities, and some general reference maps do not put them in the legend). The double-line frame is used here. Does it work?

Some features do not need to be shown in the legend, such as a box indicating the country polygon. Sometimes rivers and mountain ranges do not need to be explained because they are well recognized and are labeled on the map, but when in doubt, they should be included in the legend. River names are normally displayed as blue lines and blue italic text when color is used.

Note that the classification scheme (natural breaks) is shown in small text at the bottom of the legend. Because there are many ways to classify this type of data, an explanation gives the reader better information. Note also that *Legend* is the dominant text because it is a title; either larger, bolder, all capital letters, or a different font may be used.

Another title for the legend may be applied (e.g., "Classes" in the first two illustrations), but it should not be worded the same as the map title. The legend title can be positioned anywhere around the boxes and labels (e.g., centered above, along the left margin, or below). In addition, the title may be placed outside or inside the legend box. Note that the "Classes" title of the classification boxes is larger than the label text but smaller than the legend title.

Legends can float or be boxed. Floating legends often look very nice, but sometimes a border is needed. This is primarily an artistic decision, but limited space may be used more efficiently without the border, such as in map 2, where there are numerous items to be represented but restricted space to present them. Also notice the shallow curve symbol representing major rivers. Most GISs offer some flexibility in symbols.

Legend size should be large enough to read the smallest text easily, but it should not visually compete with the main map. The text here is almost too small.

There are numerous options and alternative design possibilities for legends. The user must be both right-brained and left-brained. However, the legend should be no more ornate or decorative than the map it is explaining.

Map Titles
Wording, Size, Subtitles

Map 1: Balanced

Map 2: Wordy, Long

Map 3: Poor Design

Fig. 11-19: Making an effective map title.

The Map Title

Maps have elements that support the presented theme, such as title, scale bar, north arrow, and margin information. These are essential components that help make a presentation good or bad. There is no required order for entering them, but the title is logically added next.

Sometimes titles and subtitles (together called "titles") are difficult to word. The user knows what the map is trying to convey, but the titles must be informative in a very few words. Avoiding words such as "the" and "it" is recommended, as are verbs when possible. Instead of "The Population of South America," a better title would be "South American Population." Readers understand that titles need to be brief and are worded accordingly, using correct spelling and proper use of words. Figure 11-19 presents three versions of South American population titles. Map 1 shows an acceptable title, demonstrating a nice balance of words, size, font, and placement.

Is "Population" a sufficient title or should a better descriptor be used, such as "Population Classes" or some other wording? This can be a difficult decision at times, and there are no hard rules directing what to include. However, in this map the reader will understand very quickly that classes of population are presented; consequently, no further subtitle may be needed. Still, if this title and subtitle seem boring and uninformative, further thought may be productive.

Map 2 is a poor example of attempting a better description. It tries to include "Classes" but adds too many unneeded words, making a three-line title. In addition, spelling is incorrect, minor words are capitalized, and it is off center. By deleting some words (such as *of*, *the*, and *Countries*), perhaps the title can be rescued.

Title size and placement can be tricky. It is a primary element of the presentation, but should not be dominant over the central map. Titles can be too large; a careful balance with the other elements is needed. The subtitle is subordinate (secondary) to the title, and should be in a smaller font size. Achieving the proper balance between the title and subtitle size also needs careful attention.

Several iterations of experimenting with size, font, and boldness are normal. Map 3 is another poor example of title design. It is too large, stretching the entire width of the map frame, which is out of proportion with the continent. In addition, the smaller-case subtitle ("Population") is inappropriate, off center, and too close to the map and labels. Complete redesign is necessary.

There is also a question in map 3 of whether "South America" or "Population" should be the main title. This depends on the purpose and overall context of the map. If it is one in a series of South American maps, then *Population* will be the logical main word. If maps of the world are being presented, "South America" will be first. Sometimes it really does not matter, but the user must make decisions.

Map Titles
Shapes and Placement

Map 1: Linear

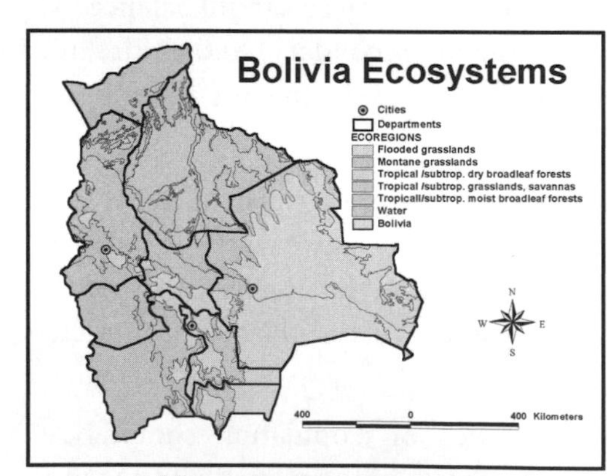

Map 2: Asymmetrical

Map 3: Asymmetrical

Fig. 11-20: Examples of title placement.

Title color is normally dark because bright colors detract from the central map. Although the title is usually centered, sometimes an offset position can be appropriate and effective, especially with an odd-shaped map feature. Chile, for example, is long and narrow. It fits along the side of a map presentation, leaving a large space for other elements, as in map 1 of figure 11-20.

It is not advisable to break Chile into two or three parts unless necessary. (Note that the title font gives the illusion of long and narrow, which seems appropriate.) Bolivia (map 2) offers a space to the upper right for an offset title and legend, whereas Brazil (map 3) can be either at the top or in the lower left (an unusual but possibly acceptable position in this case). Each country and administrative unit has particular shapes that influence mapping design; some are easy and others are more difficult. Where will the title and legend be placed on maps of countries such as Japan, USA, Indonesia, French Polynesia, and Norway (to name a few)?

Support Elements
Scale, Arrow, and Text

Map 1: Acceptable Design

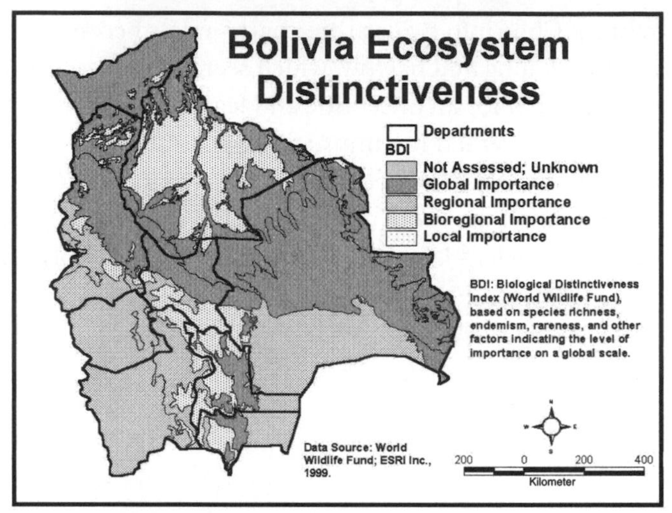

Balanced text, scale bar, and north arrow.

Map 2: Unacceptable Design

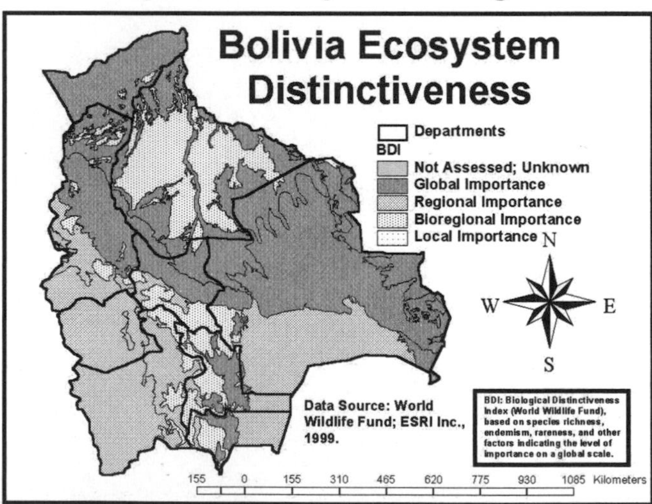

Oversized north arrow with poor placement.
Scale bar too long and poor design.
Text too small and box is unnecessary.

Fig. 11-21: Arrangement, sizing, and placement of a map's support elements.

Support Elements

The scale bar is deceptive and not necessarily an easy item to create. The tendency is to make a long and detailed bar, inferring more accuracy than the map actually supports. Some maps need a detailed scale bar to assist in measuring features or distances (usually general reference maps), whereas others use a relatively short scale bar, for basic reference of distance and size (thematic maps, normally). Striking a balance between too long and too short is essential, though there are no absolute criteria, and experimentation may be needed. Map 1 of figure 11-21 shows acceptable cartographic support elements, and map 2 depicts unacceptable design. Compare them in the following discussion.

The GIS technician typically controls the number of increments and style of the bar. Therefore, choices must be made according to various factors (purpose, audience, art, and so on). Standard distances are the best increments; for example, 50 or 100 kilometers rather than 63 or 155 (map 2).

The horizontal alternating bands of map 1 give the scale a bit more style and visual contrast than the plain open bar of map 2. The word *Kilometers* may go after the last measure on the right or centered under the bar. Placement of the scale is normally at the bottom of the map somewhere, in a visually convenient location; the top part of the map is not recommended, although design considerations may place it in unusual locations.

The north arrow is a cartographic convention that may not be needed on all thematic maps. Its primary purpose is to show cardinal directions, and for GIS maps, simplicity is best. Like the scale bar, its size should not be dominant. A plain style and small dimension are used in map 1. Placement near the scale bar is logical because the spatial support elements are located in one group for visual convenience; scattered elements force the reader to search, distracting attention from the more important parts of the map. Centered over the scale bar is a good idea, but decisions can be based on the available space and appearance.

Additional information (usually text) can include anything the reader needs to know (e.g., the date of the data and date of map production). Recall the importance of *metadata* (Chapter 2), and that it includes all of the data and production information that may be needed by other projects. The legend contains some metadata, but production information (e.g., dates, data sources, and names) can be included in the margins. These maps include text explaining the theme and data source information.

Explanatory text is inserted in a convenient place. For example, the Bolivia shape offers several convenient locations, but the margin (close to the border) is often used for basic reference information, particularly if it is short text. Text may be boxed or floating (a style choice). Fonts are usually small and dark and should therefore be more prominent.

Other elements can be included (e.g., inset maps for location, or charts), depending on the application and nature of the data. Some organizations have standard "templates" (models or guides) for mapping design and layout, but the user still must make whole-brain decisions. GIS cartography can be difficult, yet fun and highly productive.

What Is Wrong with This Map?

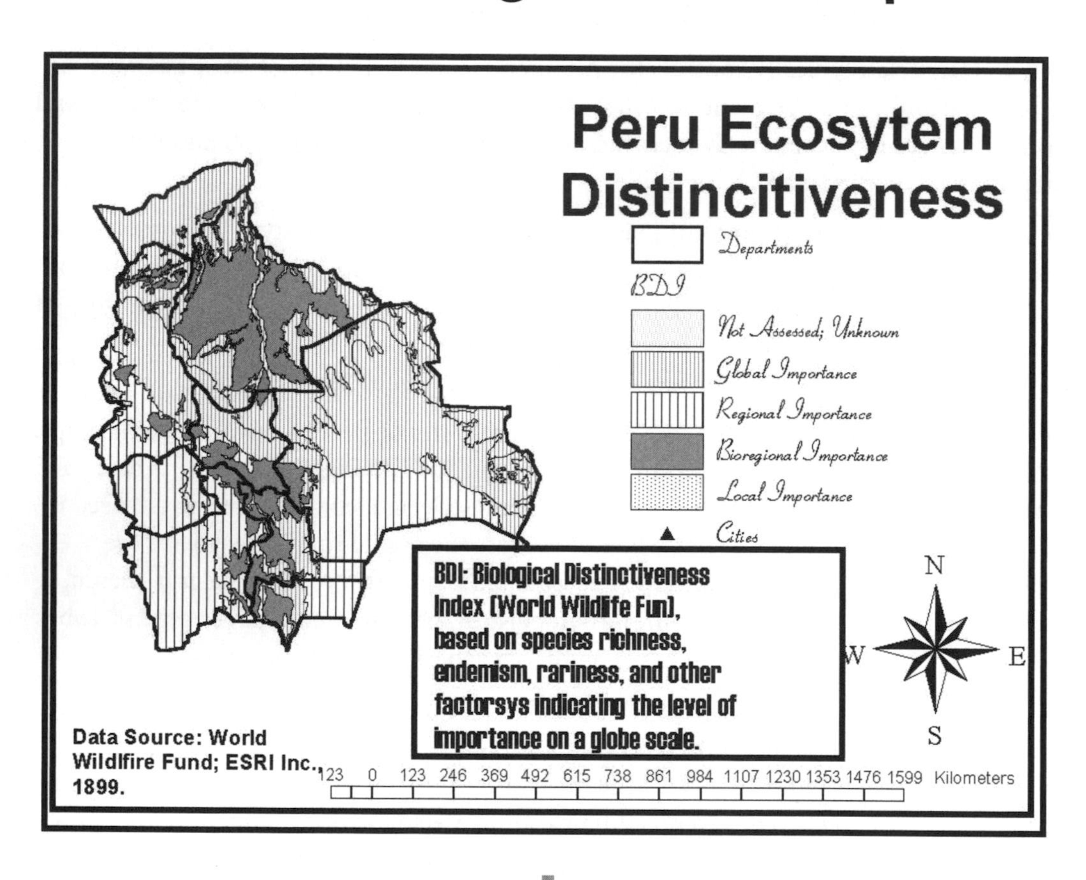

Fig. 11-22: How many types of errors can you find in this map?

Spotting Map Layout Problems

NOTE: *Read and study the material on this page before looking at the answer map.*

GIS cartography is an involved process, requiring balance of numerous items that have varying levels of importance. Remember that purpose and audience are the primary guides, but other considerations can be essential. As an exercise to help grasp the many points in this chapter, examine the map in figure 11-22 and try to discover what is wrong with it and how to make corrections.

Perhaps it is best to explore the map in the order used in this chapter. A good approach is to first make a general exploration of the entire map, trying to understand what is being communicated, and then move to the specific items for evaluation. Then examine the map, list the problems, and check your answers with the answer map (figure 11-23). Remember that good cartography is a whole-brain process. If you score well, you are on your way to becoming an excellent GIS person!

What is wrong with the map shown in figure 11-22? As the old joke says, it is better to ask what is *right* with it? The first question the reader usually has, regardless of the individual elements, deals with what is being communicated. Were you able to tell immediately what the nature of the map is and what it is trying to state? If so, the map is at least partially successful; if not, it is a failure. If the map does not communicate easily, it is not successful. Specific design problems are pointed out. Perhaps there are others. What else can you find?

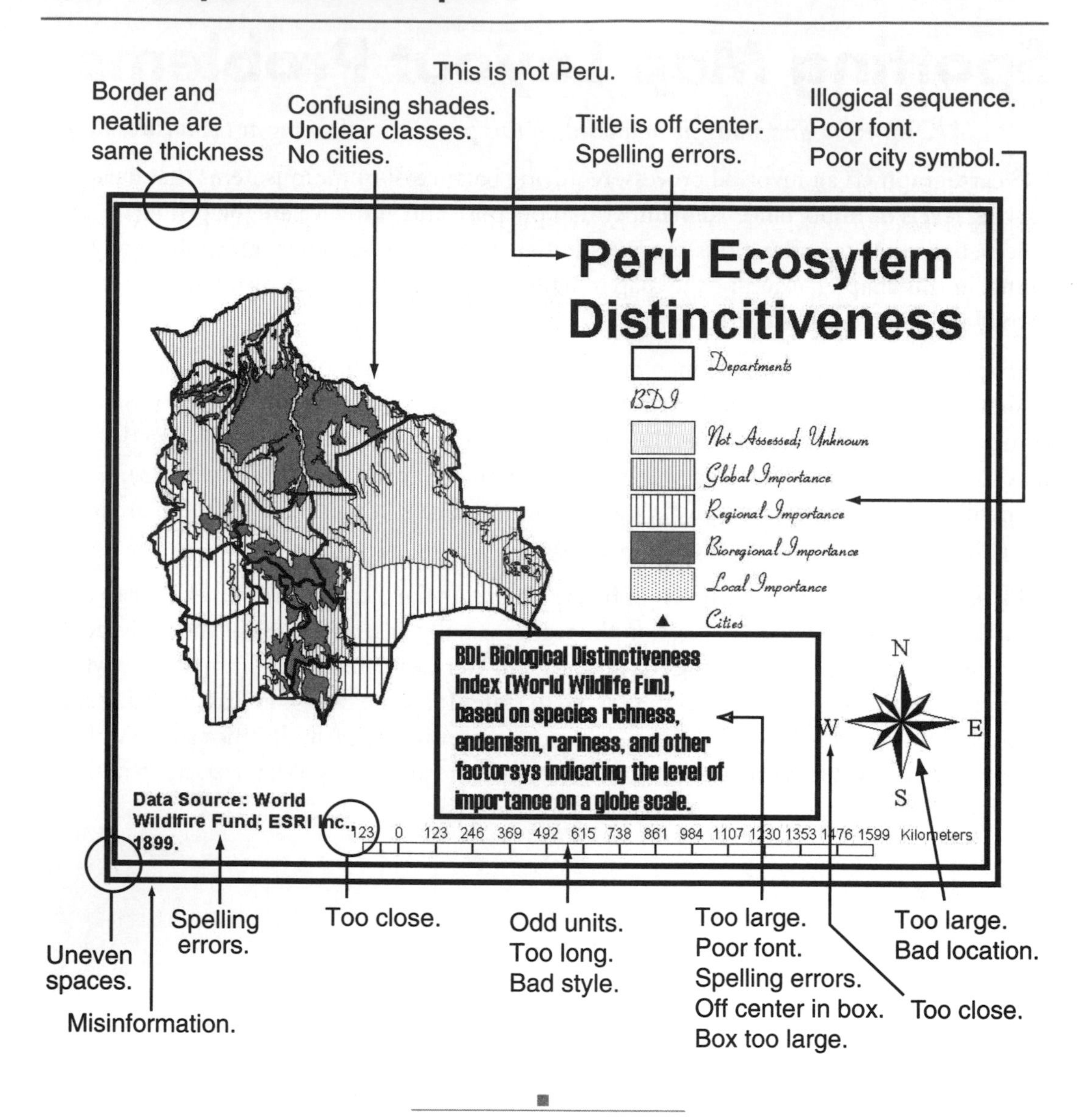

Fig. 11-23: Answer key to figure 11-22.

CHAPTER 12

THE FUTURE OF GIS AND GIS IN THE FUTURE

Introduction

GIS is a powerful information technology, methodology, business, and profession. It will continue to have a very important role in the evolving Information Age. As GIS becomes more widely used and easier to operate and apply, undoubtedly it will become more integrated into the mainstream of science and society. GIS will become a more common component of everyday life and work. It has been said that GIS will be the next "word processing" because it will be a common part of everyday software and applications.

This chapter looks at the nature of technologic change first, and then discusses some of the upcoming hardware and software advances. Their impact on GIS data is examined, followed by what it all means for GIS.

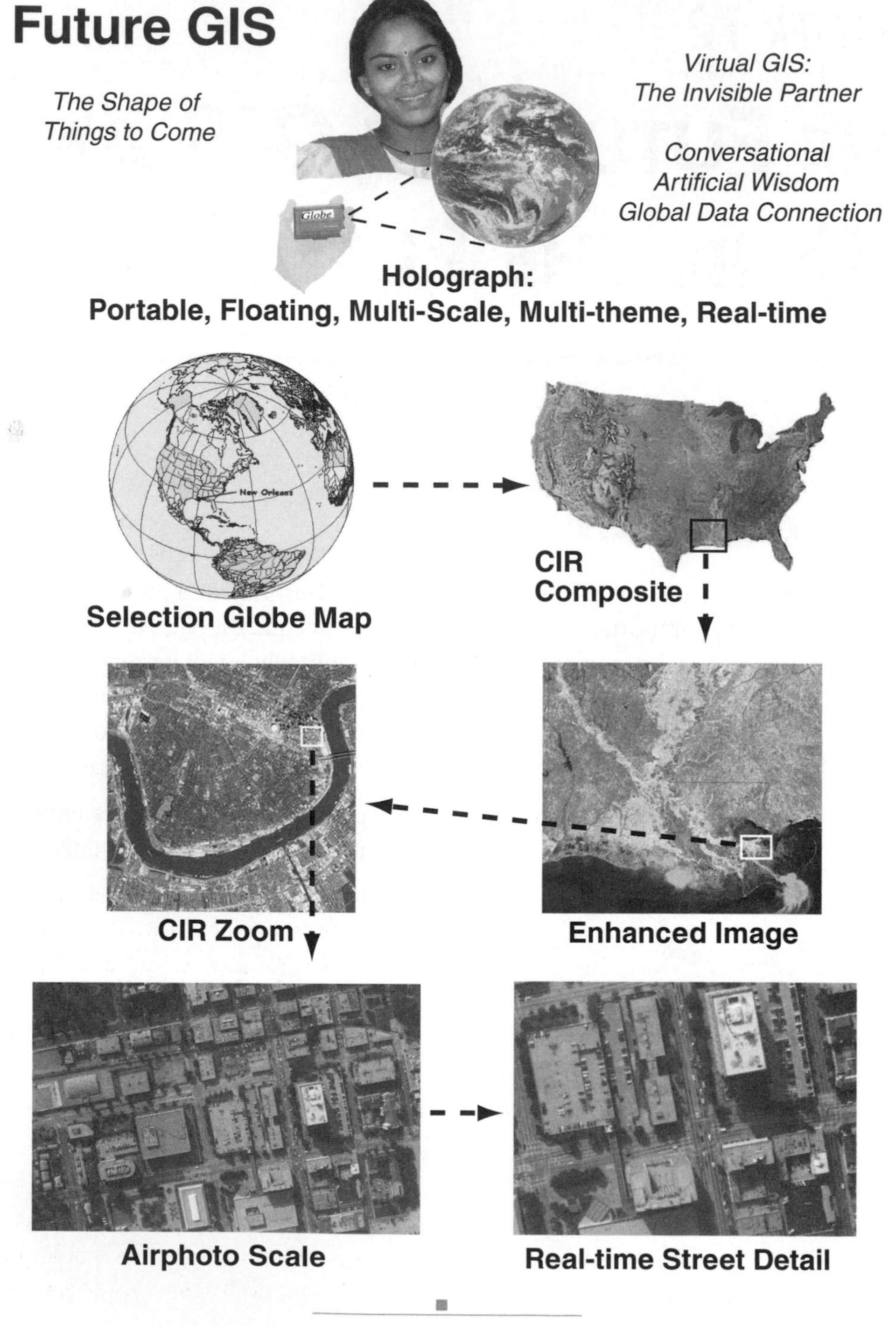

Fig. 12-1: The developing role of GIS.

GIS and the Future

The title of this chapter, "The Future of GIS and GIS in the Future," reflects the two-way view of GIS—its future and its role in the future. Technology development shapes GIS, and GIS shapes technology development. The evolution of computer-based technology is certainly one of the most obvious and exciting aspects of the Information Age. Trends in technology advancement are interesting and promising, and speculation about what is coming can be useful and fun.

The consequences of the many advances in technology are very important, particularly for an interdisciplinary and applied discipline such as GIS. Where are these trends taking GIS, and what are the real and social values that will result? What does it all mean? These are the fundamental questions faced by those who will live in the future, and that is all of us.

This chapter is only a brief and selective look at the future, but the main point is that GIS is being transformed from a specialized computer technology into a social tool. As we enter the twenty-first century, many changes are underway, not least of which is our way of thinking. Whereas the fundamentals of GIS are sound, a major evolution is definitely in progress.

Although speculation about future hardware, software, data, media, and other issues are covered in this chapter, perhaps it is useful to offer an illustration to help establish a benchmark or point of reference set in the future, so that we understand where GIS may be going. This helps to put the following pages in the context of a process, rather than simply being a description of things that may be developed. A future GIS quite unlike what is available today certainly is coming, only the time cannot be predicted.

Figure 12-1 shows a future user with her global GIS (a Global Information System?), a real-time Earth view holograph from a small portable "GIS player" (a pager-type device, the only equipment) with many scale options, connected global data sets, and themes upon voice command. This is her "virtual partner." GIS exploration is by natural language conversation (any language), with advanced artificial "wisdom" (insight, understanding). She asks (verbally, of course; keyboards are a relic of the past) first to view the real-time world, a satellite view of the globe at that moment. Then she asks to make a globe map for selection of a site or theme; in this case, the location of New Orleans, Louisiana, USA.

From that she wants a color infrared (CIR) composite view of the country and then a specialized zoom into the New Orleans region using enhanced digital processing for environmental study. She continues to zoom in, first with a CIR view of the city (the Mississippi River is the black U-shaped feature), and then to an air-photo scale view of buildings, followed by a street-level detail to monitor real-time landscape events. Feature selection and maps are only a command away.

Technologic Change

Image 1: Foundations for the 20th Century Machine Age

Electricity

Mechanical Mass Transportation

Internal Combustion Engine: Technology for the Individual

Mass Communications

Image 2: Foundations for the 21st Century Digital Age

Digital Revolution

Shrinking Technology

Visualizing the world as never before

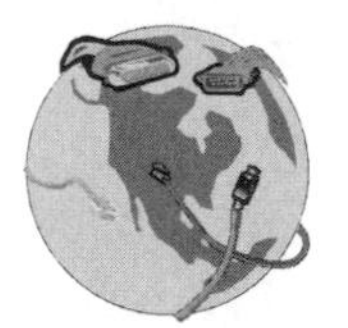

Global Linkage

Image 3: Human Resource Development

Empowerment

Humans are the heart of technology

Fig. 12-2: Changes in the nature of technology over time.

There are many other possibilities with this scenario, largely limited only by the imagination. It is introduced here to give an idea of where GIS may be heading and how these last pages are only peeks into a very exciting future.

Technologic Change

To understand what is happening today that is making tomorrow, it is useful to realize the nature of technology and the process of change. The transition into the twenty-first century represents much more than a mere change in the calendar; the new millennium is a new era because technology has taken such an important lead role in human development. Much of the world is experiencing a revolution that is largely technology driven and that is changing the way we live and think. (Whether that is good or bad is left to debate.)

Because of rapidly evolving computer-based technology, the world is undergoing a major paradigm shift in many fields, especially science, medicine, education, engineering, business, communications, and even international relations, among others. Although it is impossible to predict inventions and events, certainly the world will continue a period of tremendous change.

Actually, a similar evolution occurred in many parts of the world at the turn of the twentieth century, when electricity and the internal combustion engine were catalysts (initiating mechanisms) for technologic growth and social impact. Image 1 of figure 12-2 demonstrates a few vehicles of these changes. Think of how electricity has changed the world and human lives since the late 1800s. It is difficult to imagine society without it. Transportation certainly has changed, moving it from a very slow mode of movement for a select few to a very rapid mechanism for the many. A trip that once took months now takes only hours. The impact has been almost immeasurable.

The remarkable technologies of today, only the foreshadows of the twenty-first century, are transforming communications, visualization, and information systems of all types. As we begin the new millennium, what dreams are to become realities? Image 2 shows only a few of the technologies that mark the beginning of this astonishing era. In some ways, these are today's equivalents of the items in image 1. Surely, today's near-incredible advances may seem equally old fashioned in a few decades.

Technology for its own sake may be interesting and fun, but that is not adequate reason to support or engage in its continued evolution. Henry David Thoreau, a mid-1800s American writer, said, "We should not become the tools of our tools." What is important is the *human resource development* aspect that technology creates. Technology, particularly GIS, helps to *empower* people to do the things they need to do and want to do. GIS is a powerful technology for the Information Age because it offers the tools to accomplish a wide range of tasks, to help solve diverse problems, and to support a large number of applications. GIS is an empowering technology for human resource development (image 3).

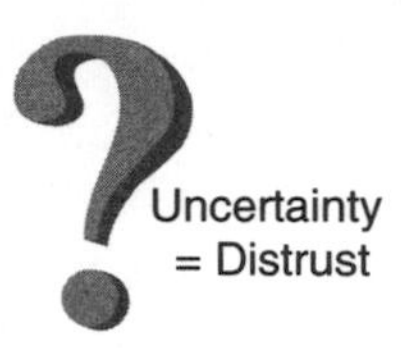

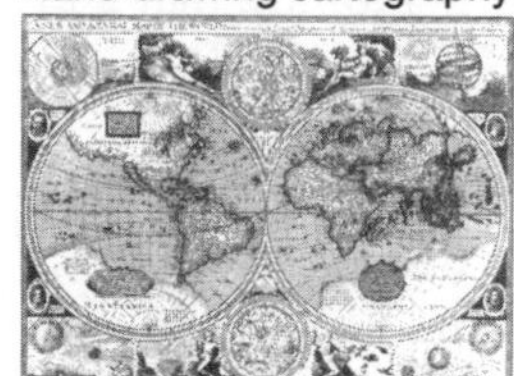

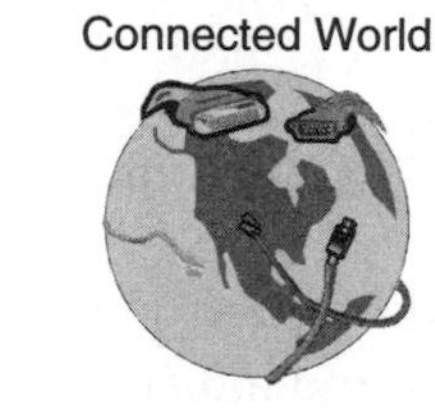

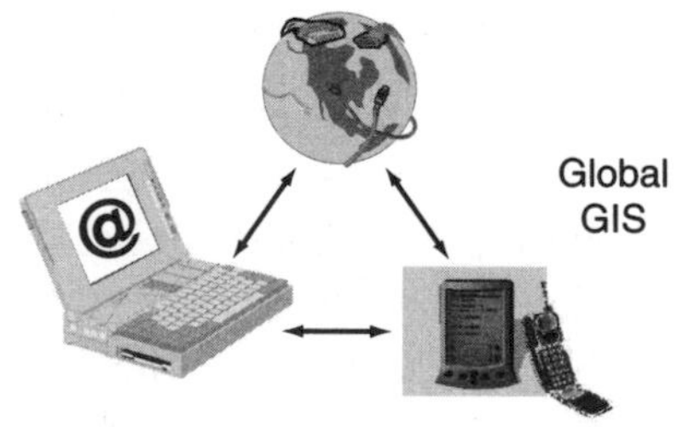

Fig. 12-3: Stages of technology acceptance.

Acceptance of New Technology

For the most part, humans are a bit suspicious of new ways of doing what they have always done successfully in established ways. Accepting new technology can be somewhat disturbing because it calls for change, and there can be distrust of unknowns. Most people and most institutions (businesses, schools, governments) go through four basic stages of technology acceptance.

These acceptance stages apply to most new developments, whether they are printed books that replace hand-written scrolls; the microscope, which opened new frontiers and new fears; the radio and newspaper, which brought information to the common people; or modern electronic equipment that introduces benefits and controversy (using calculators in school is still an issue in some places). These stages of acceptance are:

- *Reluctance to use* (image 1 of figure 12-3): There is fear of the unknown and comfort in the old ways. People prefer to stay with proven, traditional methods. The old ways are best and the new ways represent potentially discomforting change. File folders stuffed with paper are still around and the true "paperless office" probably will never exist.
- *Traditional use* (image 2): Once the initial fear of trying the new technology is overcome, the first uses are to reproduce traditional products or methods. In GIS, this included computer mapping (little more than automatic drawing). These are new ways of doing old things, and people are usually convinced of the benefits of the introduced technology. They are now willing to invest the time and effort to learn and adopt it.
- *Comprehensive use* (image 3): Once the technology is established and becomes a standard part of the infrastructure and knowledge base, users begin to appreciate the expanded possibilities and find they have the ability to do new things that were not possible before. In GIS, there are new modes of analysis and presentation, including easy overlays, making instant buffers, and site suitability analysis from multiple formats of data. The Internet helps to develop a connected world, which may bring an integrated world, at least in terms of global data sets and location-independent communications.
- *Innovative use* (image 4): The really exciting part of technology evolution is the possibility of accomplishing tasks and insights not previously known or considered. New doors are opened; new discoveries and new concepts are possible. Sometimes dreams are achieved. New methods mean new ways of doing new things. In effect, completely new paradigms are developed. GIS has been at the forefront of innovative technology for many fields and certainly it will continue to be as it continues to evolve. Illustrated in image 4 are two advanced products from Chapter 9, the 3D drape of land use on exaggerated elevation and the extruded features on the 3D Volcano Island. The concept of global GIS is becoming a new, exciting world paradigm.

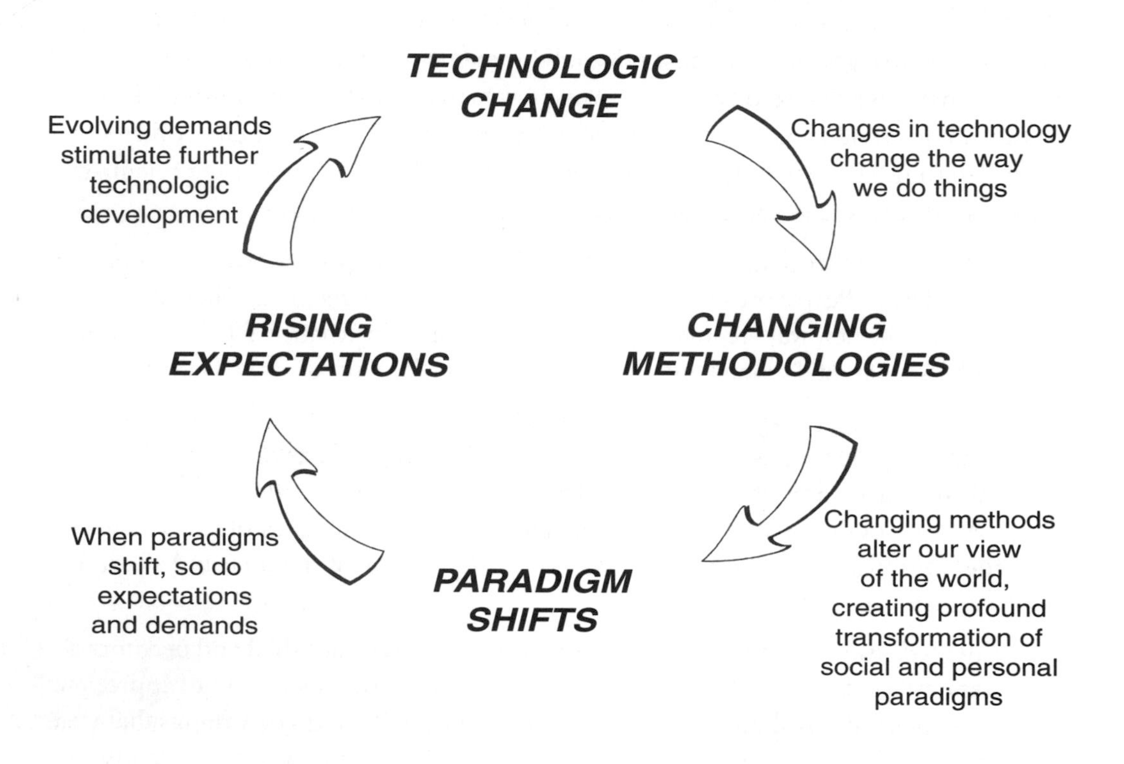

Fig. 12-4: The effects of technologic change.

Technology Impact Cycle

As technology evolves, it has impact beyond mere introduction of new equipment. The effects of technologic change can be expressed as a cycle or sequence of impacts. Figure 12-4 shows how technologic development affects society and its fundamental procedures and thinking.

- *Technologic change to changing methodologies:* When a new technologic development occurs, it changes our methodologies, the way we do things. In GIS, technology radically transformed geographic analysis and presentation (cartography) from the old manual procedures to digital methods. Technology has a powerful influence on science, applications, and society.
- *Changing methodologies to paradigm shifts:* Once methodologies are transformed, our way of thinking begins to shift. Our view of the world, the way we think about it, and how we work within it are transformed; operational paradigms change. The ways we perceive data and how to use it, for example, have been greatly altered by the computer. GIS itself is a major new geographic paradigm.
- *Paradigm shifts to rising expectations:* When our thinking is changed, our expectations are revolutionized. No longer are the old ways or old products suitable; only the new is acceptable. Demands evolve. The arrows rarely turn in the opposite direction; our path is always forward in this cycle.
- *Rising expectations to technologic change:* Because the methods, paradigms, and expectations are set on a course, they in turn stimulate additional technologic development. Change creates further change. It is a continuing cycle of evolution and reformation.

The technology impact cycle is a model of changes created by development. It shows that GIS, for example, has been much more than machinery and convenience; our view of the world has been profoundly restructured.

Future Technology Change

Image 1: Growth Trend

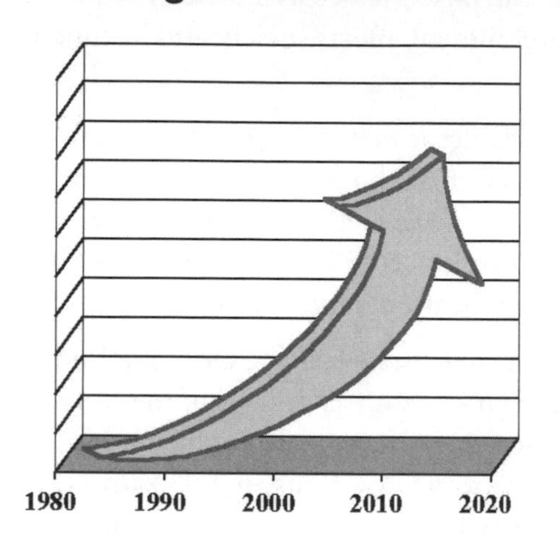

More change
will occur
in the next 10 years
than occurred
in the previous
50 years

Image 2: Types of Growth

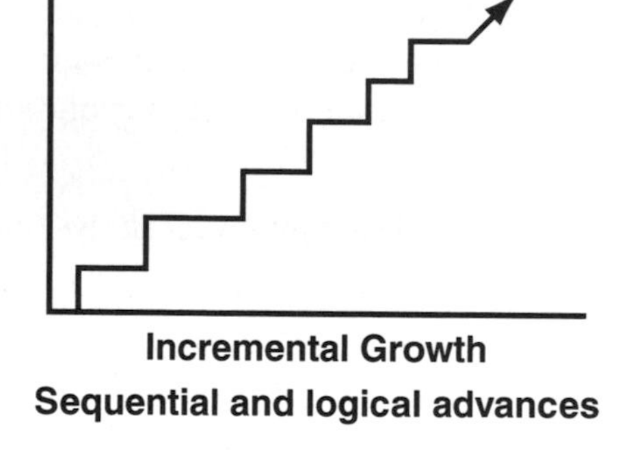

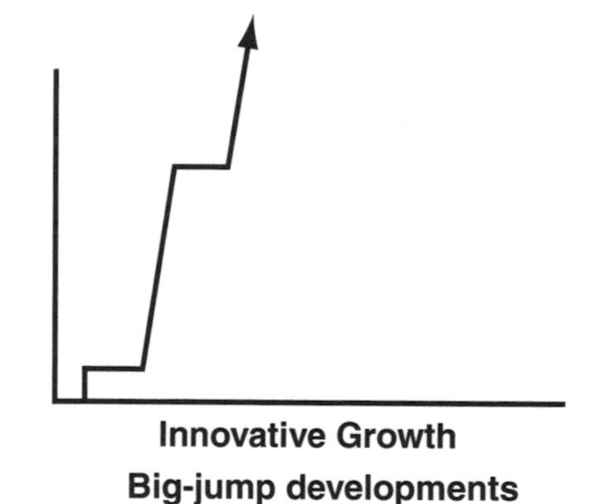

Image 3: Project Chronology of Changes

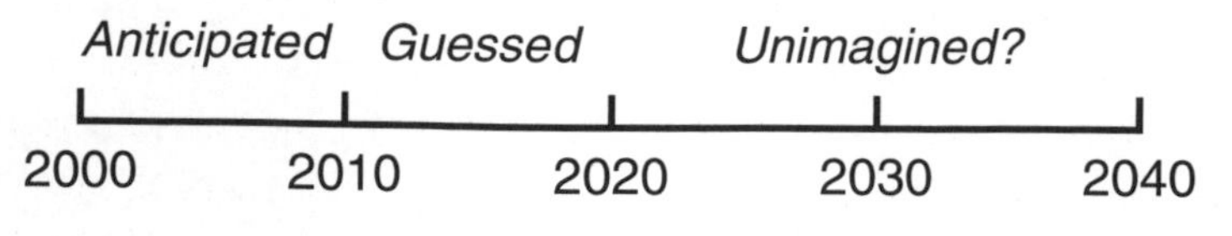

Fig. 12-5: Technology growth trends.

Future Change

Not only is the pace of change becoming more rapid, but the rate of change is also increasing. The term *rate of change* refers to the fact, for example, that the world will experience far more change in the next decade than occurred in the previous fifty years (image 1 of figure 12-5). If that is difficult to believe, modify the number to the previous twenty years; still, it is an astounding notion and the anticipation is both a bit frightening and very exciting. Even though there is debate about the current rate of technologic change compared to the first half of the twentieth century, there is no dispute that technology is having a profound impact on global societies and that we are in an extraordinarily dynamic world today.

Change occurs in two basic ways: incrementally and innovatively (image 2). Incremental linear changes are the small sequential advancements built upon existing systems, typically in the form of increased capabilities in current technology, such as the next-generation processing chip. Growth may be uneven, but is largely predictable.

Innovations are unique creations, the big jumps that are often a result of breakthroughs, discoveries, or inventions; completely new developments. Both types of change occur in the process of technologic development. The concept of non-silicon microscopic computer memory has been discussed for years, but it is such a potentially major advancement that it will be remarkably innovative when successfully developed and marketed.

Looking into the future of technology and GIS is not always accurate, despite trends and despite soundness of theory and current evolutions. Some developments in GIS have been anticipated for a long time but have yet to materialize significantly, such as natural language speech-controlled programs. Nonetheless, there are stable trends and the user community can count on some advances. Both types of future developments are discussed here: the sure ones and those that *probably* will (or may) occur, sooner or later.

What does the future hold? Here is a bold statement that may summarize technology development in the first part of the twenty-first century: developments of the first decade are likely anticipated at this time, and those of the following decade can be guessed, but those of the following years may very well be unimagined today. Changes in the near future, say the next ten years or so, will be astounding, though largely anticipated, or at least imagined (but there are sure to be surprises as well). Image 3 presents a speculative chronology.

What do these developments mean for GIS? As a computer technology, GIS will evolve along with its information counterparts. Potential advances in hardware, software, networks, data, GIS operations, and other issues are discussed in the following pages. These represent only a few of the pending changes in store. Focus is primarily on personal and office technology and applications, although there are greater advances awaiting in research and very large capacity computing environments. Still, today's personal computers were yesterday's mainframes, which themselves were beyond the expectations of society not long before that.

Hardware

Image 1: Evolution of Desktop Computers

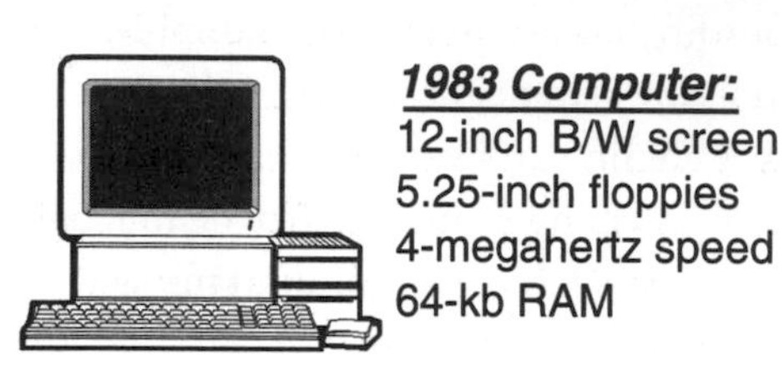

2000 Computer:
17-inch color flat-screen
CD writer, DVD drives
1.5-gigahertz speed
40-gigabyte hard drive
256-megabyte RAM

Image 2: Changing Media

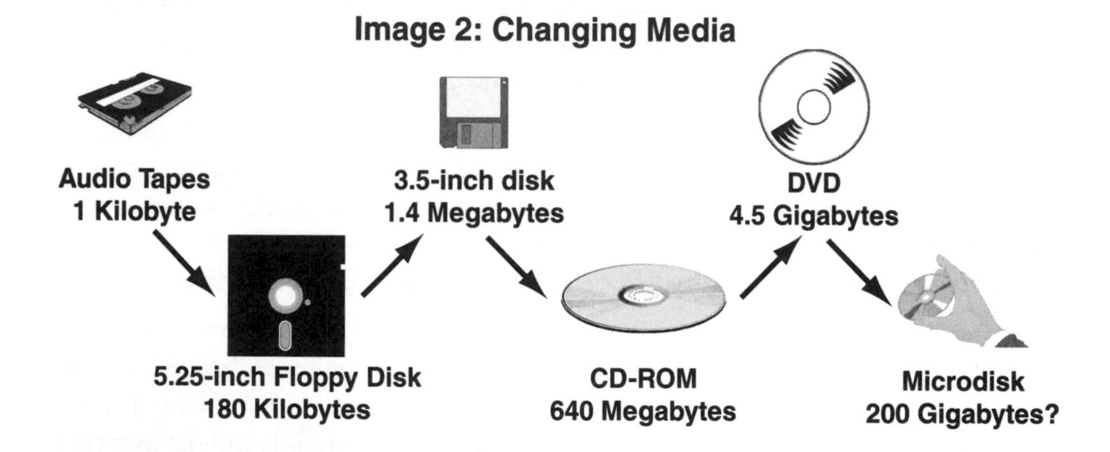

Image 3: Controls and Displays

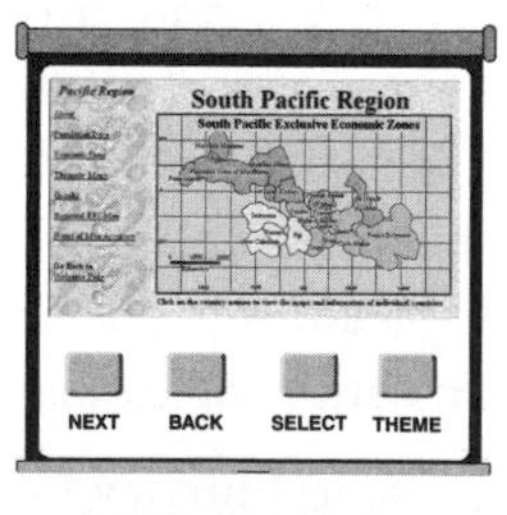

Touch Screen Soft Panels

Portable Wall Display "Smart Board"

Video Glasses

Image 4: Virtual Reality

"Virtually" Anywhere
Selected Scales

Fig. 12-6: Advances in hardware associated with technologic change.

Hardware

Advances in computer hardware are the most obvious developments. Computers are becoming smaller, faster, more powerful, and less expensive. Moore's law states that microchip power doubles about every 18 months (and therefore, the power of computers doubles); for example, the 1 gigahertz for standard commercial machines in 2000 may be at 2 gigahertz by 2002. A comparison of machines from the early 1980s and 2001 shows substantial differences (image 1 of figure 12-6). For GIS, this means impressive computing power, offering the ability to accomplish more sophisticated analysis in more efficient time. Perhaps the user will have better attention to the task at hand rather than focusing on the system.

Memory, both operating (RAM) and storage, is improving at remarkable speed as well. Hard disks on the personal computer now hold huge amounts that were inconceivable a few years ago; undoubtedly, today's capacity will be relatively weak next year and probably laughed at the year after. Possible innovations include shrinking the memory medium past magnetic and silicon minimums, down to bacterial and even protein and molecular levels.

The imagination is challenged to envision hardware below visibility and memory in the tera- or petabytes. *(Bytes*: Measures of data size in programming "words." *Megabyte* (MB) = 1,000,000 bytes; *gigabyte* (GB) = 1,000 MB or 1,000,000,000 bytes; *terabyte* (TB) = 1,000 GB, and *petabyte* (PB) = 1,000 TB.) Portable storage, or computer media, is shrinking, with today's DVDs (digital video disks) holding an equivalent of approximately seven CDs, or over 3,200 3.5-inch high-density "floppy" disks on each side. Even small and higher capacity disks are coming. Image 2 shows how media have progressed in the past two decades.

Perhaps a return to touch screens and "soft panel" controls will help GIS users (specialists or general public) in the operation of systems; that is, pointing, drawing, or button clicks by fingers on the monitor rather than by mouse manipulation. Large, flat displays (foldable and portable when needed), even wall-sized "smart" screens (with interactive controls) or video glasses with voice control, will remove the small window between the user and data (image 3). Stereo is used now, though usually for specific applications, such as topography and 3D views.

Virtual reality, one of the most exciting potential technologies for GIS, may open worlds previously unreachable, even unimaginable. "Field work" on a virtual Mars (or anywhere on Earth) will be possible, with choices for scale and location, such as several kilometers height to study major landforms or for touring deep in its crust for geologic exploration, or maybe a few centimeters height for soil investigation (image 4).

Scanners will become more "intelligent" for digitizing maps, perhaps to the point of true automatic data entry. With voice control and other software advances to be discussed, the data entry process will be unlike today's tedious and time-consuming methods. The ongoing advances in hardware will continue to astound, and today's states of the art will probably be jokes in a few years.

Software Developments

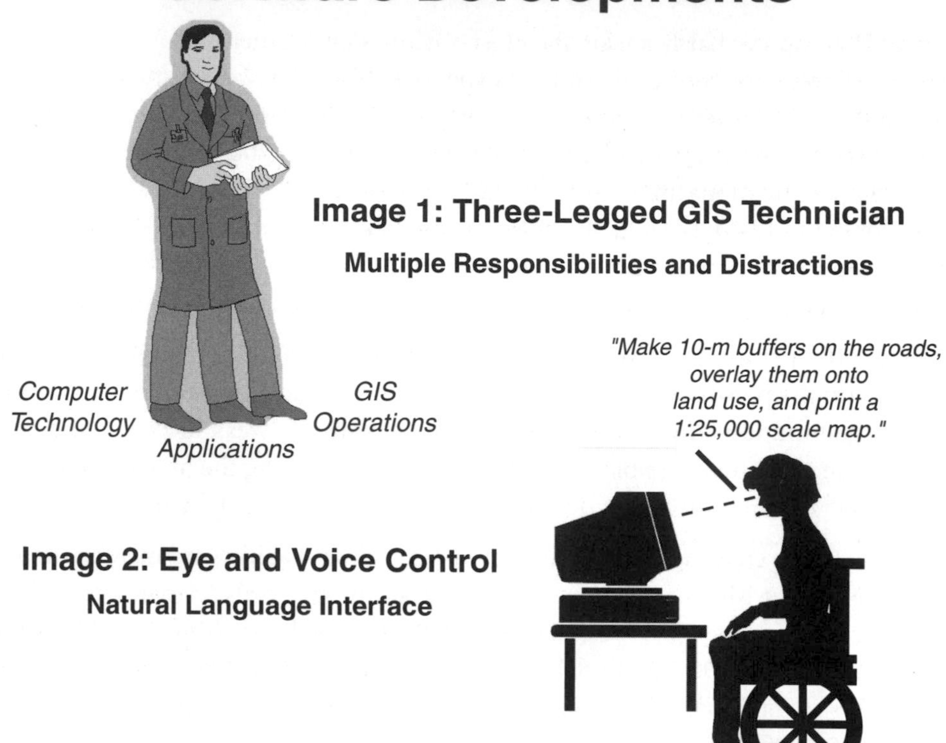

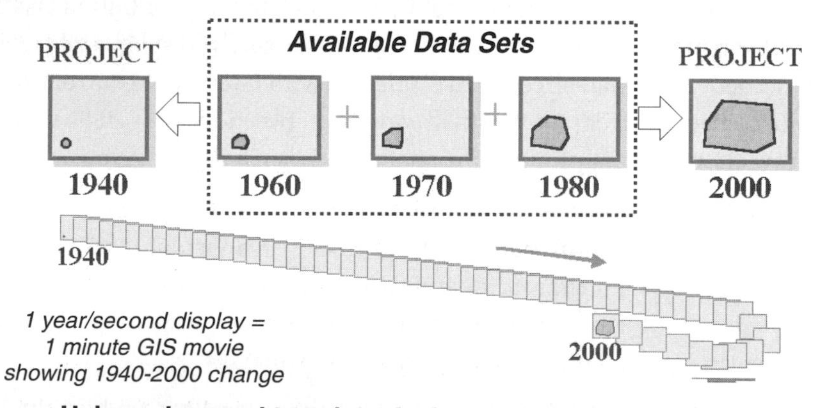

Fig. 12-7: Advances in software associated with technologic change.

Software I

Software evolution has been, and will continue to be, exciting and sometimes surprising. A few of the advances GIS can look forward to are presented in this and the next three sections, yet there are many others underway that are not addressed. GIS software is becoming much more powerful, easier to use, and more integrative of other programs and capabilities. It is no longer accurate to speak of a single GIS program, because vendors are incorporating additional specialized "modules" or programs that enhance the core software, such as image processing (for remote sensing data), GPS, and spatial statistics, to name a few.

In GIS, the operator is often distracted by the demands of software and computer systems; concentration on the application or task at hand is interrupted and important project continuity can be lost. Even though point-click icons, smart menus, and operation wizards are helpful assistants, the GIS operator must do a "three-legged dance": one foot in maintaining the computer system, one foot in GIS operations, and one foot in the applications (image 1). This is tricky, but fortunately there are software developments that have promise for assistance.

GIS programs seem to be easier to work than in the past, although the additional power does add challenge to learning (usually an acceptable trade-off for everyone except the casual low-end user). For example, software developers realized (after too many years) that most users do not want to rely on thick manuals or to develop new programming skills to make the GIS work as needed; better functionality and more friendly operations are becoming standard. These types of improvements are particularly valuable in GIS analysis, such as in What-if scenarios, where the user must be able to make small changes quickly and easily, and to concentrate on results rather than detailed operations.

Natural language speech recognition will be a significant advance for GIS. "Conversation" with the program may produce more intelligent instructions, such as a single command stating "Make 10-meter buffers on the roads and overlay them with land use, then print a 1:25,000 scale map." Eye and voice control may soon support physically handicapped GIS users, a wonderful new advantage for the GIS community (image 2).

GIS programs are becoming more multimedia, using pictures from external sources (e.g., ground photograph of the study area), animation, and sound (verbal description rather than a wordy text box on a map). Animation can present the fourth dimension, time, more appropriately than a sequence of static maps.

Image 3 shows three spatial data sets, from 1940 to 1980. Advanced spatial trend analysis software is applied to interpolate spatial growth between the years and then project 20 years backward and forward. The result is an animation of 60 images, shown at one per second, presenting a time series model in a one-minute GIS video of landscape change over 60 years. Projects will soon be presented as "GIS shows," with mixed media delivered on CD-ROM, DVD, and the Internet.

Software Development
Advanced Systems

Image 1: Expert Systems

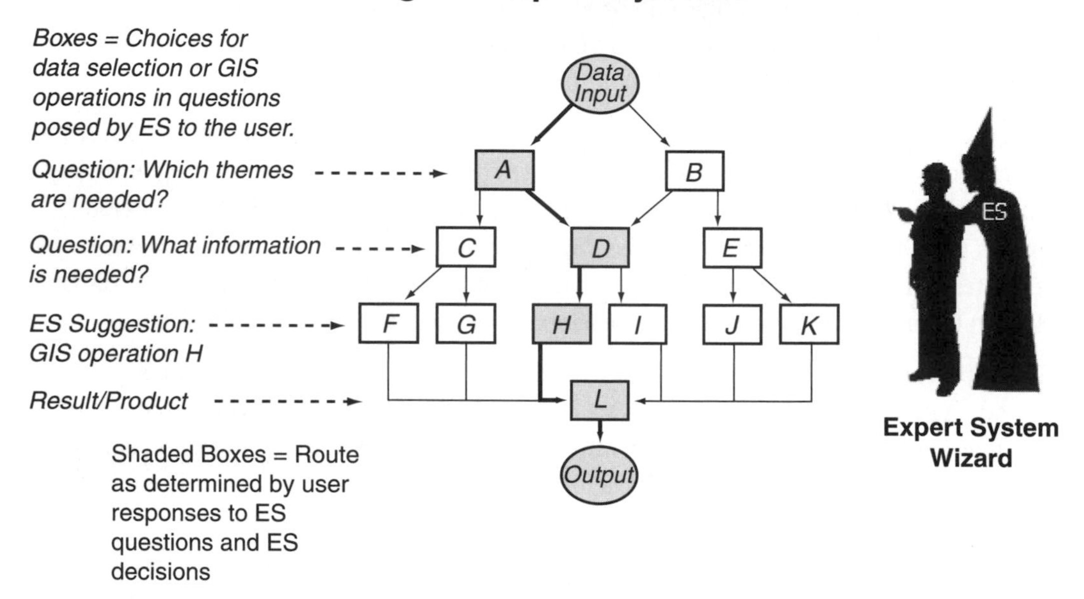

Image 2: Artificial Intelligence

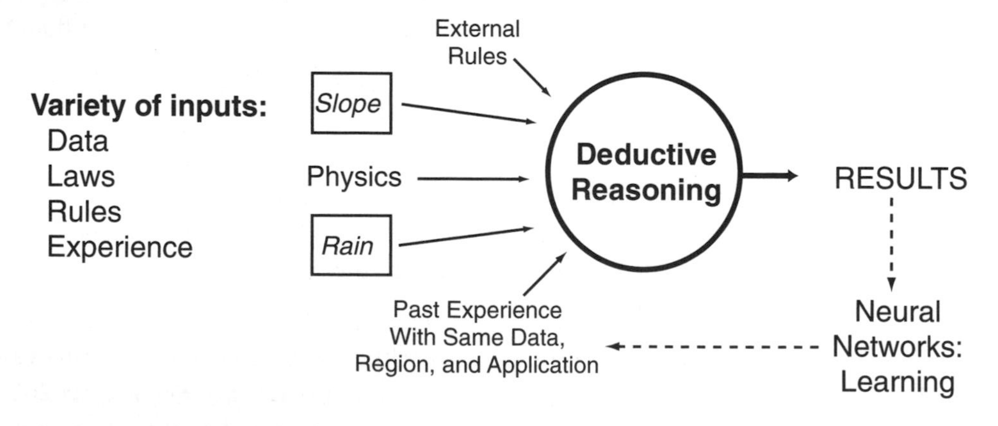

Fig. 12-8: Expert systems and artificial intelligence for GIS.

Software II

The integration of advanced computing developments into GIS will help revolutionize the field. Six concepts are briefly discussed in the next three sections.

Expert Systems (ES)

An expert, by definition, is highly skilled and knows how to accomplish tasks in an efficient and effective way. An expert system simulates those skills by guiding the user in problem solving and achieving needed results. Program wizards (the small pop-up step-by-step guides to specific operations) are elementary versions, but real expert systems lead users through complex decision pathways while automatically determining, from user inputs along the way, what is important and relevant to the ultimate goal.

The ES uses established rules, but it can also make inferences (deductions) from the data and user response to determine the next logical step in the decision path. Image 1 of figure 12-8 is a simplified pathway diagram through a complex series of possible choices. (Note the ES wizard directing the user.)

Artificial Intelligence (AI) and Neural Networks

AI is the ability of computer programs to simulate human thought processes, instead of relying strictly on static programmed directions. Expert systems use simple AI when they make decisions that have not been programmed. Full-functioning AI is not here yet, but an AI module in GIS modeling, for example, can process numerous possible combinations of themes, data, and rules to present the most logical model options.

Image 2 shows slope and rain input to predict runoff and erosion for a region. AI also uses previous experience with these themes, as well as physics (water must run downhill, not uphill; cannot violate universal laws) and other rules, such as that the rate of precipitation determines accumulation. The program uses deductive reasoning principles to calculate the potential erosion. This frees the GIS user from all tedious (mind-numbing) work, giving more time to concentrate on the final decisions that only humans can make.

Neural network programs are simulations of the brain's neural connections to help produce artificial learning. They assist in building AI for specific applications; for example, learning the effects of various amounts and rates of precipitation on a variety of slopes in a region. Then it also learns and incorporates differences between GIS in urban planning and in geological exploration. The AI improves as "experience" is gained.

Software Development

Advanced Analysis

Image 1: Fuzzy Logic

From binary data to transitional

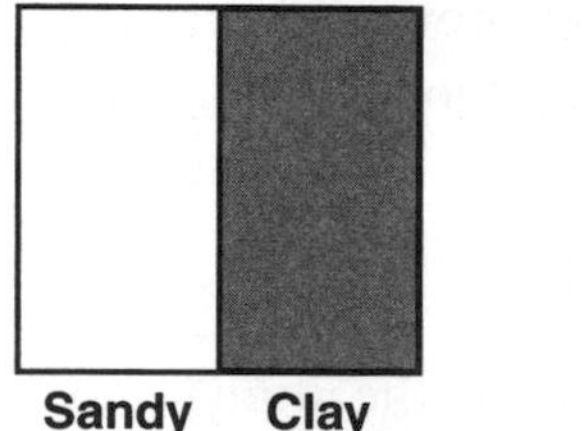

Image 2: Exploratory Data Analysis
Visual exploration of data relationships

Is there a relationship between X and Y?

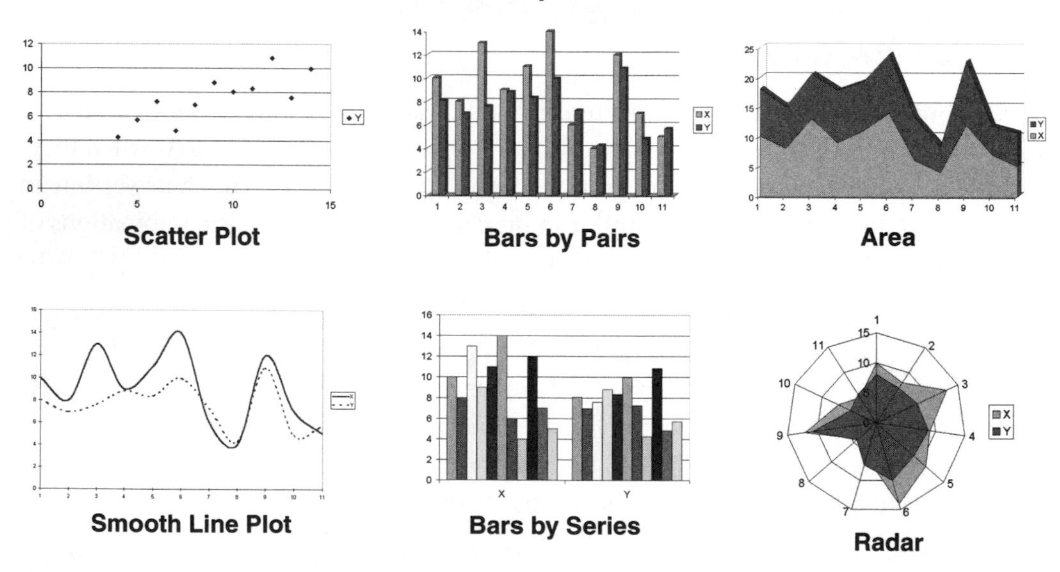

Fig. 12-9: Fuzzy logic and exploratory data analysis.

Fuzzy Logic

Digital data format has always been binary; which implies only two states of existence, such as 0/1, true/false, on/off, yes/no, black/white, and so on. Fuzzy logic introduces the ability to calculate partial values between absolute values; for example, no/somewhat (a little of no and yes)/yes, or black/shades of gray/white. The color of the sky no longer has to be blue/not-blue, but can be transitional shades between light and dark blue.

For GIS, this can mean interpretation and mapping of continuous data and transitional zones, from one soil type to another when there is a progression with no abrupt border (image 1 of figure 12-9), or the delineation of the gradual zones from a dense population to more sparse. Social behavior, such as voting, is often "fuzzy" in spatial context. Fuzzy logic in GIS may permit more realistic analysis in numerous applications.

Exploratory Data Analysis (EDA)

Many GIS users are not prepared to apply complicated statistics techniques to spatial data; assistance is welcomed. Exploratory data analysis (EDA) is a relatively new approach, integrating graphics with multiple analytical tools for searching relationships and suggesting analysis and modeling options. In effect, EDA searches for meaning using visual analysis, thereby relieving the user of predetermining which statistical methods will be needed for insight.

Image 2 shows various graphs from a set of data to begin exploration of relationships between various measures of X and Y for some phenomenon. This new paradigm is a bit controversial in that the software does the initial exploration, presenting the user with options for further work (some experts think there is no substitute for having advanced statistical knowledge). For GIS, EDA offers a powerful set of tools to help understand complex data. In the absence of statistical expertise, it can be an important aid in spatial analysis. The relationship of crater size and distribution can be investigated, for example, with EDA assisting with a variety of statistical tests.

Software Development
Fractals and Chaos Theory

Image 1: Fractals

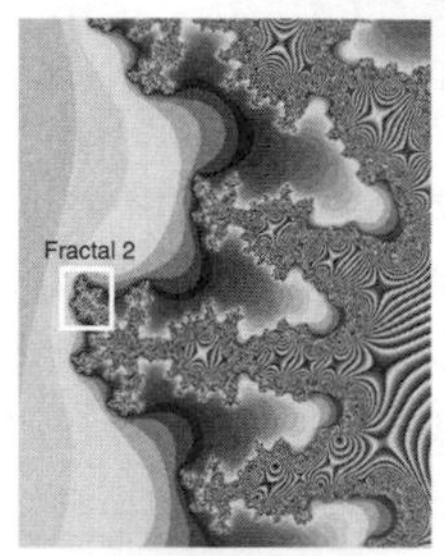

Fractal 1

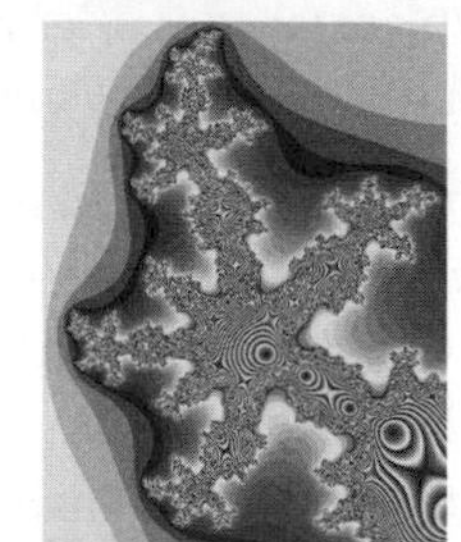

Fractal 2

Dendritic Drainage System

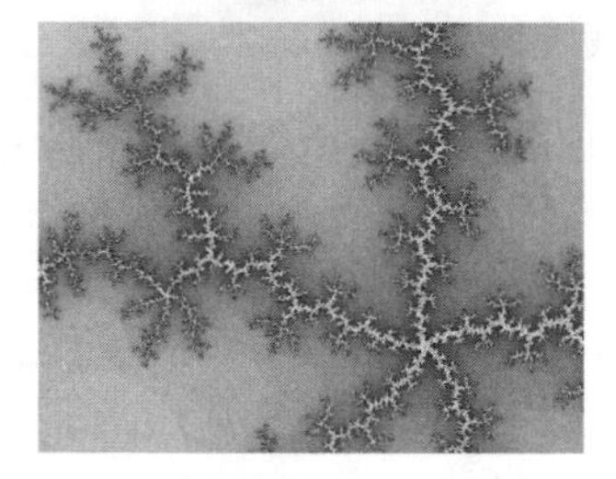

Mandelbrot Fractal

Does the dendritic system have fractal patterns?

Image 2: Chaos Features

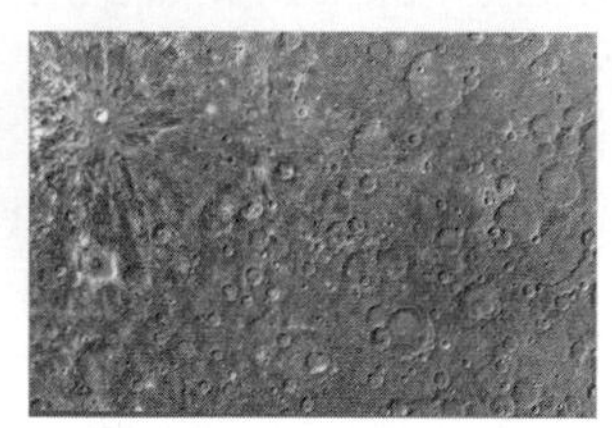

Planet Mercury Craters
Size and Distribution

Hurricane Dynamics
Andrew Sequence
Morphology, Path, Strength

Earth Weather Patterns
Known Models;
Unknown Specifics

Fig. 12-10: Fractals and chaos theory.

Fractals

Fractals are patterns that have some consistency, regardless of scale; they exhibit repeating characteristics of complicated shapes from one scale to another (termed *self-replicating patterns*). Consider a spiral and the mathematical formula that makes it. The spiral continues in a consistent pattern when magnified because the formula continues.

Image 1 of figure 12-10 shows a pair of fractal patterns (a Julia set), with increasing scale. When fractal 1 is magnified, each small part also presents the same type of pattern in fractal 2 (even when complicated), primarily because the structure is based on a set formula that continues regardless of the scale. The dendritic drainage system photography appears to have fractal pattern morphology (shape), somewhat like the Mandelbrot fractal image next to it. Does nature exhibit fractals? There may be many examples in nature.

At least two valuable applications to GIS are under development: image compression and landform mapping. By detecting complex patterns in an image that can be expressed by mathematical formulas, the image can be saved as a mathematical expression rather than as a large number of raster cells with individual values. The data size of the image or file can be greatly reduced. Similarly, many landforms have complex shapes, but if fractal patterns can be interpreted, the landscape can be presented as mathematical expressions, making analysis and modeling more efficient and effective.

Consider fractal 1 as a complex coastline and note how details continue in an organized way as the scale increases. Application of fractal analysis can reveal patterns that are then used to model the coast and permit comparison with other regions. Perhaps a look at satellite imagery at one scale can predict details at larger, unseen scales.

Chaos Theory

Chaos means disorder and randomness; chaos theory involves description, explanation, and understanding of why some things in nature *seem* to be chaotic, but are really organized in a complicated, unknown way. The path of a hurricane or the distribution of craters on the moon, for example, are not actually random or erratic; there are specific reasons and laws why they occur as they do, even if we do not understand them. Further, a small change in one component can change others and eventually lead to large changes. The introduction of a set of hills or an urban heat island in a weather system model can make changes in the amount of rain, for example.

The practical side of chaos theory is to assist in making landscapes and processes seem less chaotic (which is an "anti-chaos theory," if you will). It considers the nature and influence of every input and tries to sort out the complexities. (Exploratory data analysis may be useful in this type of investigation.) For GIS, an operational chaos theory program in the analytical logic can assist in understanding complex systems, both physical and cultural.

Data Developments
Global Infrastructure

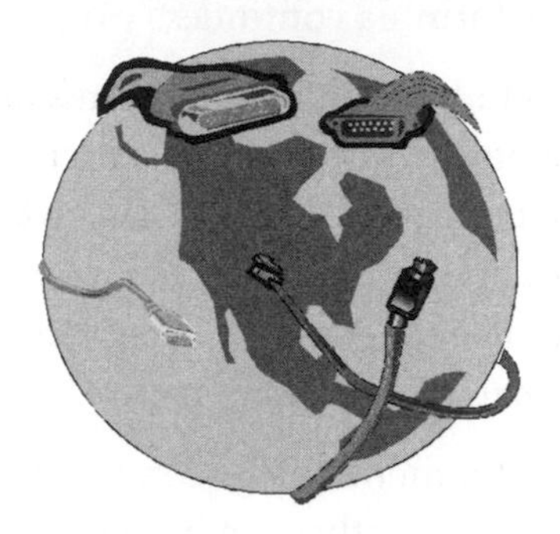

Image 1: Global Data Infrastructure

Data-Connected World:
Many sources.
Many formats.
Affordable.

Image 2: Global Remote Sensing Coverage

250-channel sensors.
1-meter resolution.
Many countries.
Data overload?

Australia: Entire nation in a single image

Fig. 12-11: Data development, global infrastructure.

Even though the basic models are known, the specifics are still mysterious for many phenomena, such as weather patterns on Earth, the size and distribution relationships of craters on Mercury, or the spatial dynamics of a hurricane (image 2 of figure 12-10).

Data

It is clear that more, better, and less expensive data will be available for GIS work in the near future. On-line archives are becoming common, methods of data collection are improving, and there is more dedication to building large databases for multiple uses, particularly by the web-based community. The day of complete open data for all users is still to come, but impressive progress is being made. Perhaps one day there will be a true global information infrastructure that is largely open to anyone, at affordable prices.

These systems will connect collected data from many sources and will comprise many types of data, from scanned and georeferenced maps to very-high-resolution imagery, all with a variety of scales, dates, and formats that can be imported into any GIS (image 1 of figure 12-11). With true open GIS (which includes a full range of functions and capabilities we now expect on desktop systems), the open data model will provide universal utility for a wide range of GIS users, from the novice to the technical professional.

Global databases are available now, with the amount and types of data increasing almost daily, particularly satellite imagery (image 2). Remote sensing technology provides multichannel spectrophotometers, instruments that can sense 250 different parts of the electromagnetic spectrum at once. Although the United States and the Soviet Union (now Russia) were the initial space powers, many nations are in the satellite data business (e.g., France, Japan, Canada, India, Sweden, and others). With commercial satellite imagery now down to 1-meter resolution, the real problem will be the amount of data rather than a shortage.

Data Developments
Data Use and Distribution

Image 1: GPS and Real-Time Distribution

1-meter resolution
Location independent
Immediate availability

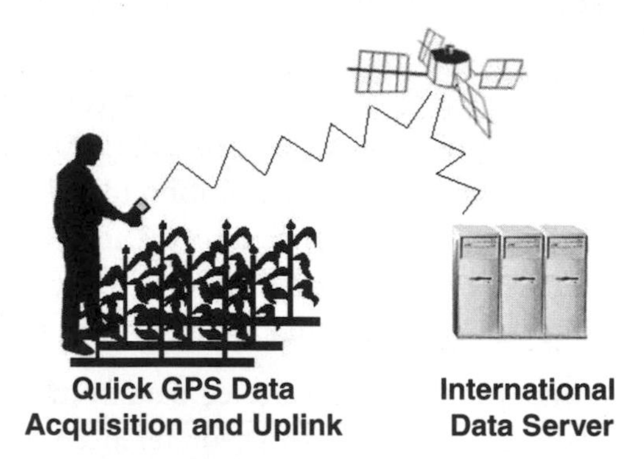

Image 2: Easy Data Exchange

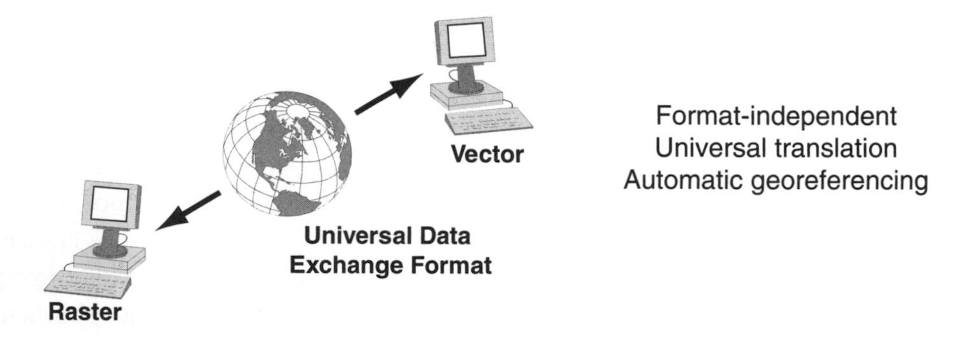

Format-independent
Universal translation
Automatic georeferencing

Image 3: Entire Nation on Disk

Inexpensive
Commercial and public
E-commerce

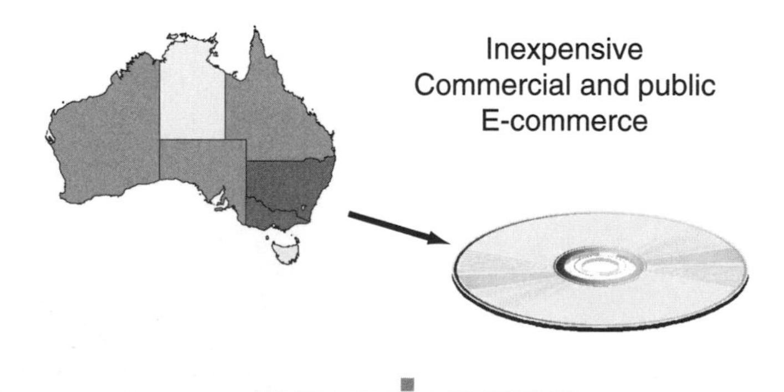

Fig. 12-12: Developments in GIS data use and distribution.

GPS is revolutionizing data collection, with high precision, near real-time distribution. It is now possible to acquire field data at 1-meter accuracy (or better if needed) almost anywhere in the world, and then uplink to a central server for ultimate distribution as global network data (image 1 if figure 12-12). Even today, fieldwork can be spatially precise and relayed "on the fly." Future developments are left to the imagination.

One of the promises already underway is easier exchange of data, regardless of format. The difference between vector and raster will no longer include the difficulties of importing or exporting; all GISs will accept both types seamlessly (image 2). Like today's easy point-and-click incorporation of the various graphics formats (JPEG, TIFF, GIF, and so on), perhaps automatic georeferencing will be available soon, which will be a tremendous benefit to GIS.

Commercial data is becoming more prevalent and less expensive. For example, a multiple CD-ROM set of topographic map data (multiple scales) of the entire United States can be purchased for less than U.S. $100 (image 3). Similar regional and national data sets are becoming available. The outlook for the data business is excellent. E-commerce on the Internet is creating new vendors and opening new markets. Data developments for GIS are growing very rapidly.

The Internet

Users from anywhere and everywhere

INTERNET

Data

Contacts

Assistance

Information

Discussion Groups

Distributed GIS

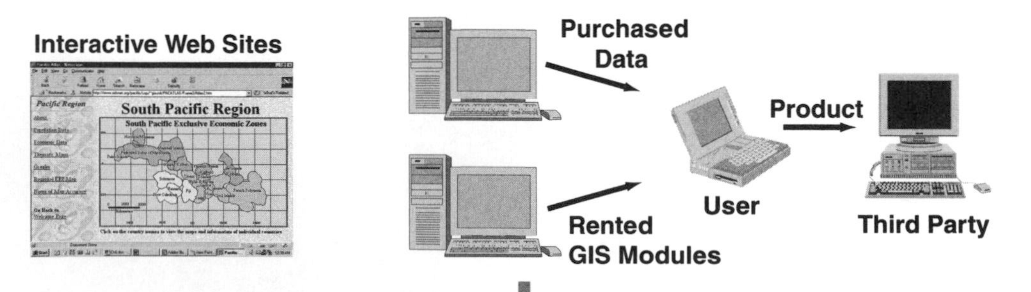

Fig. 12-13: Impact of the Internet on GIS.

The Internet

Until the early 1990s, networks were largely local or regional, usually connecting institutions for restricted communications and data exchange. With the growth and success of networks and the development of an easy communications format (World Wide Web), the Internet became the global medium for communications, ushering in the Information Age. The benefits have been very high for GIS in various ways, as depicted in figure 12-13.

Data acquisition and project delivery are two primary Internet applications for GIS. The Net is replacing tapes and disks as a means of delivering GIS data. The number of web-based data sources is growing rapidly, from government and university sites (typically free) to commercial (e-commerce) sites that sell data and services. The Internet is usually the first place to search for data today, and in many ways data is becoming more available because of relatively easy access. In fact, it has become necessary for all GIS-related businesses to have Internet "presence" in order to stay competitive.

The Internet also provides connections to professional contacts and expertise from around the world. Instant assistance and professional dialogs are available, which greatly enhance GIS productivity, particularly for users largely out of the world mainstream of activities, such as much of the Third World. User web sites (growing daily) offer a wealth of ideas, insight, and even potential collaboration. Discussion groups, connections to projects, and information sites abound. "Distance education" is discussed in material to follow.

Distributed GIS is becoming an important component. A basic version is the presentation of GIS work on web pages, either display of finished maps and graphics or using simple interactive capabilities, such as selection of themes and exhibiting tables of data. Interactive GIS is becoming a significant business. Distributed GIS may also mean distribution of GIS software, data, or operations; for example, software running from a server in one location using data from several other sites, and sending products to a third-party customer located elsewhere. Software and data rental services may be a wave of the future, freeing the occasional user from purchasing and maintaining expensive systems.

Other Issues

There are numerous issues emerging with GIS as it develops into a standard part of the computing and general applications world. A few of the more obvious are discussed in the following, namely privacy, open records, data piracy, and other legal issues. These represent only a small sampling of the existing and coming concerns.

- *Privacy*: Increased availability and quality of data are sure to create surprises regarding previously unknown information. For example, satellites now cover the world, revealing activities and landscapes that were once believed to be secret. The privacy of nations has been invaded to some degree. Personal data may be more exposed, purposely and accidentally, as large databases become more accessible. Already many towns have public property records on web-based GIS that give property and owner information. There are fears that such data will be used in unintended and damaging ways. Issues of personal rights and privacy are raised, leading to concerns regarding laws and legal procedures.
- *Open records laws and ownership*: Many citizens believe that public data should be open to the public, especially if it was collected or created by public organizations. The demand for *open records* laws has grown in the recent years, usually resulting in public access to information previously guarded by user agencies. Questions about who owns and controls data have begun. What should be private and what should be public are ongoing questions. Controversy continues and is not likely to be settled in the near future. The Internet is making this issue even more confusing.
- *Data piracy*: As data resources grow and availability increases (especially over the Internet), and as data becomes more important in everyday life, the danger of theft and misuse increases. Stealing and illegal reproduction of data are relatively new crimes, but their impacts can be significant and major. Hackers can break into databases over the Internet, either taking or corrupting data for fun or profit. Ownership questions raised previously make this an unclear issue at times. If someone gains unauthorized entry into a database of public information and copies selected parts, is that person guilty of theft or just illegal entry? In some countries, this may not be perceived as stealing in the sense of removing property, but the legal issues are complicated and evolving.
- *Liability and other legal issues*: The legalities involved in many aspects of data use raise new concerns. If GIS is used to decide boundary disputes, for example, there may be concerns of accuracy, precision, and responsibility. Who pays for the damage that poor-quality data may cause? Who is legally liable for maintaining data standards and access? These types of issues are leading to the new specialty of geoinformation law.

GIS Education

Training and education for GIS is a contemporary issue that will continue to grow in importance in the future. The field (and its associations, such as GPS, remote sensing, and others) is growing very rapidly; some estimates are as high as 30 percent a year. The need for trained professionals is much higher than the current supply.

A novelty in formal education only a few years ago, GIS has become a major part of the curriculum today. For example, the discipline of geography has offered cartography and air-photo interpretation for decades, but now GIS and digital remote sensing have become the standard courses. Other disciplines also use and teach GIS, such as geology, sociology, anthropology, computer science, business, criminal justice, and numerous others.

Although higher education offers the major professional curriculum in GIS, grade schools are now using it, including the elementary level. There are active K-12 programs around the world (kindergarten to 12th grade in the United States; equivalents in other nations). Certainly, GIS is much more than a mere computer novelty in education.

The Internet opened doors to many educational possibilities. Although long-distance education in GIS is just beginning, it can be a major component of the new Information Age. Virtual classrooms, on-line tutorials and workshops for both schools and industry, and even certificates and degrees are part of the GIS educational infrastructure today. A single web site can serve students anywhere in the world, an obvious benefit for the many professionals and trainees who do not have access to formal schools. On-line courses and workshops can be independent of time and place, able to be studied at any time from anywhere. The market is truly global.

The future will bring new opportunities for human resource development in GIS. As the technology grows, so too will the demands and challenges. On-line training and education, already in progress, will surely grow to tremendous proportions. How do the educational paradigms of today evolve to meet the needs of tomorrow? This is one of the major questions in education today. Where do we go from here?

GLOSSARY

THE RELATIVELY SIMPLE DEFINITIONS USED HERE are applied to GIS, even though many terms have other meanings and associations.

Algorithm: A computer program operation (also called a routine) used to solve a particular problem or carry out a set of specific steps. Usually a small set of codes in a larger program.

Accuracy: Data accuracy refers to truth or validity according to reality, within acceptable limits. Example: accurate names and classifications. See figure 6-1. See also *Spatial accuracy.*

Application: The practical, applied use of geographic information systems (GIS), usually considered the purpose or reason for GIS operations and data. Example: forestry planning.

Artificial intelligence (AI): Advanced computer programs that simulate human intelligence using logical rules and inferences (rational assumptions). Example: AI can be applied to complicated landscape analysis by helping to group features into categories. See also *Expert systems.*

Aspect: The direction a slope faces, usually expressed in cardinal directions (north, east, and so on). Important in determining sun exposure as it relates to considerations such as soil moisture and the effects on vegetation. GIS can generate and use aspect data in ecological resource evaluation. See figures 9-13 and 9-14.

Attribute: Data description, characteristic, or quality. Describes or explains the observations (records). Some GISs store attribute data as class numbers, whereas others use text. Attributes are usually the columns in a database. Example: Area B (the record) = pine, which is code 2.

Binary: Having only two possible values or states. For example, 0 = Off, 1 = On; or Yes/No. Computer instructions normally use binary data. See *Fuzzy logic* for alternate model.

Boolean operation: A multiple-condition query that represents relationships based on criteria specified by logical operators, such as AND, OR, NOT, and AND/OR. Example: "Show all forests AND brush areas." See figure 8-4.

Buffer: The zone (corridor or area) on each side of or surrounding a feature, as constructed by GIS operations. Example: a 2-km environmental protection buffer zone around a pollution site. Other names are used by various software programs, such as "spread." See figures 8-17 to 8-19.

CAD: Computer-aided drafting, or computer-aided design; typically used for engineering and design work, and now being adopted for mapping and integration with GIS. (Alternate: CADD, computer-aided drafting and design.)

Cartography: The science and art (both are important) of making maps. It is a science because there are objective standards, but also an art because aesthetic judgment is needed.

CD-ROM: Compact disk read-only memory, the physical medium for which is typically referred to as a "CD." Small compact storage disk about 12 cm (4.75 inches) wide, read by a laser device, holding approximately 640 megabytes of data on one side.

Cell: See *Grid cell.*

Chain: The connection between vertices and nodes. Single link or part of a line or polygon. A two-part line consists of two chains. A square has four chains. The word *line* is avoided as a synonym because of the confusion with the same-named GIS feature. Sometimes called an arc. See figure 3-9.

Continuous data: Data values that transition from one number to the next, with no definite boundary between the two. Example: temperature maps show definite zones for convenience, but in reality temperature typically transitions from one measure to the next, without a defining border. GIS mapping of continuous data is difficult. See figures 2-4 and 6-8.

Coordinate system: Location system using an X-Y (horizontal and vertical) grid to permit accurate Earth location of a feature. Example: Latitude-Longitude, a world coordinate system. A GIS coordinate system can also include depth or height (elevation) as the third dimension, termed the Z coordinate. See also *Z value.*

CPU: Central processing unit. The "brains" of a computer that computes and directs functions. Informally called the "box" to separate it from the monitor and other parts of the computer system.

Database: Collection of organized data in a manageable structure, usually resembling a table, with rows and columns. Rows are the *records* (observations) and columns have the *attributes* (descriptions). See figure 2-11. See also *Relational database* as a special type.

Database approach: Using the database for inventory or analytical operations when possible. Because the database has most of the data and numerous data handling options, many tasks can be performed with more efficiency and accuracy in the database rather than using graphics. See figure 2-14.

Database management system (DBMS): A computer program used to store and manage information. In GIS, DBMS software is usually included as part of the system.

Data entry: The process of loading data directly into a GIS, usually into a database, either manually or automatically. See also *Digitize* and *Digitizer.*

Data set: A collection of related data, sometimes referring to a single file but usually designates a set of related files. Provides data for the database.

Data standards: Specified criteria to be maintained for consistency of quality, definition, accuracy, precision, and other GIS properties that are important to users. Standards are valuable for data that is transferred or exchanged so that the recipient is assured of certain qualities. Although the details are controversial, each project or institution may include or impose standards within itself in order to give all users confidence in the data. See figure 6-13.

Data structure: The organization and format used to define data so that the software can use it. Example: a raster structure uses a grid cell format to define GIS data. See also *Raster* and *Vector* for examples.

Data visualization: Presentation of data in visual form to enhance understanding. A map is easier to interpret than a complicated database. See figure 1-5.

Derived data: Data made from other data. Two input data sets or themes can be combined to make a new third theme, which therefore becomes the output data derived from the input data. See figure 6-4.

Digital elevation model (DEM): Data set of gridded digital surface data, expressed in elevation measures, which can be shown in either 2D (flat view) or 3D (realistic perspective). A digital terrain model (DTM) is a special version that uses heights above mean sea level.

Digitize: To convert maps or images into digital data. Special instruments are used, such as a digitizer (see *Digitizer*). See figure 5-15.

Digitizer: An electronically sensitive table that uses a tracing device for copying paper map data into a computer for conversion to digital data. Manual digitizing: human tracing of map features with a small hand device, which relays the locations of features (points, lines, polygons) into the GIS. Automatic digitizing: scanning equipment that traces the features, which is faster and more accurate than manual digitizing but requires human guidance. See figure 5-10.

Discrete data: Distinct and noncontinuous data, with definite values that are not transitional from one number to the next; definite boundaries and identities. For example, a district consists of defined borders, with a distinctive name. Opposite of *continuous data.* See figure 2-4.

Distributed system: Set of computers and devices that may be widely separated by location but linked through a network. Example: a central database located in one city, with employees with computers in other cities linked to that database. The system is "distributed" over the region. Also termed *distributed GIS.* See figures 6-15 and 6-16.

Electromagnetic spectrum: The range of wavelengths or frequencies of electromagnetic radiation extending from very short gamma rays to the very long radio waves, with a narrow band of visible light in the middle. Each portion presents different physical properties that electronics can sense, thereby giving multiple views of the world outside the human visual range. See figure 5-4.

Expert system: An advanced computer program that uses rules to simulate a human expert's skills and reasoning in making specific decisions. Example: an expert system for site suitability analysis guides the user in making data choices and selecting GIS operations to achieve each step in the process. See figure 12-8.

Fractal: A feature that uses mathematics to define its shape (usually complicated) and that is "self-repeating"; that is, the formula continues the shape regardless of scale. Coastlines often demonstrate fractal patterns. See figure 12-10.

Fuzzy logic: Recognition of digital values between classical binary numbers 0 and 1. Rather than expressions of only yes/no or black/white, fuzzy logic uses intermediate states, such as maybe or sometimes between yes and no, or various gray tones between black and white. A new, more flexible computing logic. See figure 12-9.

Geocoding: Translating or transferring geographic coordinates into X-Y digital format for use in a GIS. Can also include assigning locations, such as street addresses, to features not based on a world coordinate system.

Geographic data: Data that connects or relates to location. Can include *spatial data* and associated *nonspatial data.* For example, the location of a political area is spatial, as is its area; the name and demographics are nonspatial. See figure 2-2.

Georeferenced: Properly aligned or registered to a fixed coordinate system. GIS data must be set in correct cartographic space so that other data can be associated or connected with it without distortion or error.

GIS: Geographic information system. A computer-based technology and methodology for collecting, managing, analyzing, modeling, and presenting geographic data for a wide range of applications. As a computer system, it combines databases and graphics operations to make a variety of products, from lists to maps. GIS also refers to a methodology that uses geographic information technologies. See figures 1-6 and 1-7.

GIS infrastructure: The components making a complete GIS, which includes organization and people (the most important part), applications, methodologies, data, software, and hardware. See figure 1-9.

GIS paradigm: The synergistic combination of GIS technology components (hardware, software, data, and products) with GIS principles and procedures plus GIS as a business and profession, to make a guiding philosophy and framework to understand and use GIS. See figures 1-13 and 6-20.

GPS: Global Positioning System. A system of satellites that transmits signals used by special receivers on the ground for precise determination of location, sometimes within meters. Receivers are small, hand-held units or large, highly accurate systems. Data can include elevation and speed. See figure 5-7.

Grid cell: One cell containing a single GIS (or remote sensing image) value or attribute. The cell is one part of a large grid that makes a scene. The grid is a raster format. The cell is also termed a *raster* or *pixel* (picture element). See figure 3-1.

Hard copy: A map or graphical depiction of a theme on paper. In effect, something that can be held. It is the opposite form of a digital file.

Hologram: A 3D image made by splitting a laser beam, presenting a view that appears to be realistic in depth. Holograms also store large amounts of data and represent a new frontier in data storage and display.

Icon: A small graphic symbol or picture on the GIS display that represents a computer operation or a data set. Clicking on the icon initiates the process or selects the data. This has become the preferred method of operating GIS, rather than typing commands.

Information: Data combined and integrated to indicate something; meaningful data. Note that data by itself is not necessarily information, but merely a fact or collection of facts. See figure 2-1.

Information Age: The "post-industrial" era in which technology and economies are based on information. High-tech communications, digital technology, skilled labor, data analysis, and other elements are essential components replacing Industrial Age society. See figure 1-1.

Information system: In modern terms, a system supporting or delivering information in one form or another. Usually high technology, typically computer based. GIS is a major information system.

Interactive: Marked by real-time communication between computer and operator. As the operator puts in instructions, the computer reacts, producing results as soon as input is completed.

Label: Data identity text for features on the GIS display. Example: the first digitized polygon is labeled Polygon 1, which may be changed to a name in the database, which is then used as the text identifier on the display. See figure 11-15.

Land information system (LIS): Also called a land records system (LRS). A database system containing physical, quantitative, legal, and other descriptions of land, such as elevation, land value, and ownership. Not necessarily a complete GIS.

Landsat: U.S. Earth resources satellite system that produces images of land in multiple electromagnetic spectrum bands. Its thematic mapper (TM) senses areas as small as 28 meters on a side (784 square meters). The program began in 1972 and has provided most of the satellite image data for the United States. See also *Spot*, the French land resource satellite. Other nations have satellite Earth imaging programs.

Layer: One GIS data file, usually devoted to a single theme. Generic name for a GIS file, also known as theme and coverage.

Line: A one-dimensional feature on a map having length and direction, but normally no width. A GIS line feature begins and ends with nodes. Example: a railroad in a GIS is depicted as a line, starting at a node and ending at a node. Any width or thickness is a cartographic indicator of an identity or magnitude, not the actual width. See figure 2-3.

Map algebra: Using numbers assigned to features, usually in the raster format, for analysis, typically in some type of overlay. Each cell or feature is assigned a number and algebraic operations combine two themes to make a third, new theme. Example: total rainfall for 6 months adds the numbers in each cell of a raster theme to make a total rainfall theme. See figure 8-11.

Map projection: See *Projection.*

Map scale: The depiction or calculation of map distance to its corresponding real-world distance. Scale is expressed as a ratio of map units to real-world units. The nature of the units does not matter, but metric or English measurements are normally used. Example: 1:50,000 (also 1/50,000, the mathematical ratio) means that 1 centimeter on a map equals 50,000 centimeters of real distance, or 1 map inch = 50,000 real-world inches. Units must be the same on each side of the equation. See figures 6-5 to 6-7 for other considerations of scale.

Metadata: Literally, "data about data," but for GIS, it is information about the data included with a data set, project, or map. Normally includes the essential information about the origin, scale, projection, dates, and other important factors that give the user confidence that the data being used is known and is appropriate for the intended application. See figure 6-14.

Model: In GIS, a model is (1) a representation of reality, (2) an attempt to duplicate or display a real-world process in a simplified manner, or (3) a set of procedures for accomplishing a specific task. Various types of models are discussed in association with figures 10-8 to 10-15.

Network:

1. For GIS data: A set of interconnected line features that represent paths of movement. Example: a road system. There are specialized GIS network spatial operations, such as finding the shortest and fastest routes between two locations. See figures 9-21 to 9-23.

2. For computer technology: A system linking computers (and other devices, such as printers) by communications lines. They may be local (within an office or building) or wide-area (regional). The Internet is a global network. See also *Distributed system.*

Node: The data structure point (tip) at each end of a chain (which includes intersections of chains). A node is not a physical point, but only a location indicator for the end of a chain. A node should not be confused with the GIS feature termed *point.* See figure 3-9. See also *Vertex.*

Open GIS: A complicated concept that basically refers to *distributed*, shared, holistic, and integrated GIS open to many users for various applications. Opposite of the standard private in-house project. Can include widely available data, shared GIS operations (distributed software), distributed personnel, and other components that have open access to the necessary resources. Controversial but a grand possibility for global GIS. See figure 6-19.

Output: Product constructed for presentation, either by display (on the monitor), printing (map), reporting (text), or electronic data (on a CD-ROM or over a network). See Chapter 11.

Overlay: The combining (or superimposing) of two or more themes or layers to produce new data or a new theme. Example: an overlay of vegetation, soils, and slopes to make a new erosion potential theme. See figure 8-3.

Parcel: A map feature designating land ownership and rights. A parcel map is devoted to legal property depiction and descriptions.

Peripherals: Hardware linked to computers to provide additional assistance. Printers and digitizers are standard GIS peripherals. See figure 1-10.

Perspective view: An oblique view, or something other than horizontal (as from the ground) or directly overhead (standard map). A slightly slanted view of raised Z data (such as elevation) gives an appearance of reality. Convenient for data visualization when standard maps are difficult to interpret. Can be used with any Z data, including population or economics. See figures 9-19 and 9-20.

Pixel: Modified abbreviation of *picture element*, the smallest portion of an image to contain a data value. A raster *grid cell* is a pixel.

Plotter: A graphics drawing device normally used to produce large maps. Typically higher quality than standard printers, a plotter produces continuous lines (as opposed to a printer), and may use colored pens, ink jet spray, or electromagnetic means of printing. A common GIS peripheral.

Point: A single location object or event (occurrence), having only one X-Y coordinate location and no length or width (no spatial dimensions). In GIS, a point locates something that is either too small to show realistically or has no real spatial measurement. It is the simplest GIS feature. Example: a house (too small to show in detail in the image area) or an accident site (an event having no measurement). See figure 2-3.

Polygon: An enclosed area on a map or theme, having length and width; a feature with three or more sides that completely surrounds an area. It has perimeter and area. Examples: a political district or a forest site. See figure 2-3.

Precision: See *Spatial precision.*

Projection: A special shape used to fit a portion of the globe onto a flat view; converting spherical data to 2D presentation. There are many projections, each with specific characteristics for particular uses. GIS themes usually need a projection in order to be properly georeferenced (or located accurately on Earth). See figure 5-13.

Proximity analysis: GIS procedures to investigate the relationship of features in terms of nearness, connectivity, or other properties of distance. Also referred to as *neighborhood analysis.* See figure 9-1.

Query: "Asking" a question or making a request to the database, typically in the form of a command for specific data. Example: "Show all cities with populations > 10,000." See figure 7-3. Chapter 7 discusses various types of queries.

Raster: A cell that contains a single GIS or remote sensing image value. The cell is part of a larger grid system (rows and columns) that makes an entire theme layer or image. A raster system is a data structure format using the gridded cell-value arrangement. See also *Grid cell* and *Pixel.* See figure 3-1.

Recode: Renumbering values or renaming text for easier interpretation and use; reclassification. Typically a process of reducing data to make a data set more simple or generalized. Example: reducing 45 soils to a group of 5 soils. See figures 7-12 and 8-1.

Relational database: A powerful and flexible type of database that allows multiple linkages of data so that each field can be related to all other fields. Typically, a relational database query consists of defining specific conditions from one or more fields to find records meeting the needed conditions. Example: "Find (1) all properties valued over $50,000, AND (2) properties that are used for farming, AND (3) properties that cover over 50 hectares." See figure 7-4.

Remote sensing: Gathering data some distance from the target. In GIS, this usually means imaging the land (or ocean) from above, such as by aircraft or satellite. Landsat and Spot satellites, among others, provide remote sensing images of Earth. Typically, various parts of the electromagnetic spectrum are sensed, giving different data and visual versions of the same view. See figures 5-3 and 5-5.

Site suitability analysis: A GIS process normally involving several operations (such as overlays and buffers) to identify the best or worst sites for a particular purpose. For ecological analysis, it is often termed *site sensitivity analysis.* It can also mean selection of the least suitable site, termed *site unsuitability.* Also termed *site selection.* See figure 10-1, and Chapter 10 text for detailed discussion.

Spatial accuracy: Refers to the correct size, shape, and distribution of a feature or features. It is the "spatial truth" presented, meaning that the map feature represents the real-world feature truthfully, within acceptable limits. A vector format typically presents features with more spatial accuracy than do raster formats. Not necessarily the same as *spatial precision.* See figures 3-8 and 6-1.

Spatial analysis: Investigation and interpretation of spatial data. GIS operations to analyze spatial features for specific applications. Example: using recode, overlay, and buffer operations to select the best sites for industrial location. See figure 8-20.

Spatial data: Data that occupies geographic (mappable) space and usually has specific location according to some coordinate system (such as Latitude-Longitude) or system of location (address). Spatial data has physical properties, such as size and shape. In GIS, the three primary types of spatial features are *points, lines,* and *polygons.* Better termed *geographic data.* See figure 2-2.

Spatial precision: Refers to the exact and true location of a feature. Although it may be spatially accurate (see *Spatial accuracy*), it still may be incorrectly located. See figures 3-8 and 6-2.

Spatial relationships: The variety of ways features relate to one another spatially. Map properties such as distance, distribution, density, and pattern among features are important GIS characteristics. See figure 2-6.

Spot: French remote sensing satellite similar to the U.S. Landsat. It can produce stereo images and can gather data at 10-meter resolution.

Subset: A selected part of a theme that serves as a separate theme. Example: selection of Brazil from a large theme of South America in order to work on a Brazil project. See figure 7-9.

Technologic impact cycle: The model of the effects of technologic change on science and society. Technologic change creates changes in methodologies (the way we operate), which in turn impacts our view of the world, thereby changing our overall paradigms. Changed paradigms create changes in expectations, which in turn stimulate further technologic change. See figure 12-4.

Terrain analysis: GIS operations and applications using or concentrating on the third spatial dimension, such as elevation, depth, or height (Z data). Example: incorporating X-Y-Z data to make 3D displays of a landscape for watershed delineation. See figure 9-10.

Thematic map: A map or data layer devoted to a single theme. Example: a geologic theme with specific data. The opposite is the general reference map, with a diversity of features and themes. See figure 11-3.

Topology: In GIS, a powerful mathematically derived data structure used to define spatial relationships between and among features. Special programming applied to GIS data so that the features have "intelligence" in terms of connections with other features. Used with vector systems. See Chapter 4.

Tracking GIS: A new form of GIS operations and applications pertaining to following features over time and space. Uses changes in measurement, speed, location, and other properties important to understanding spatially dynamic features or phenomena. Example: tracking migration of birds, accounting for distance, location, changes in the number of birds, and the geographic features encountered that may have influence. See figure 9-9.

UTM (Universal Transverse Mercator): High-resolution alternative coordinate system to Latitude-Longitude. Based on 60 global zones of 16 degrees longitude width.

Vector: In GIS, vector refers to a data structure that defines points, lines, and polygons by their true position and dimensions. Vector displays appear to be true line drawings, much like a standard map, but really consist only of the locations of the nodes and vertices. Example: a square is defined by the coordinates of its four corners rather than by four actual lines. This allows high accuracy at any scale and is more map-like than the gridded raster system. See figure 3-9.

Vertex: The directional turning point on a chain; a *shape point* that defines a change in direction of a line or polygon side. It is given a coordinate label for identification. Distinct from a node (start or end point). The plural is *vertices* or *vertexes.* See figure 3-9.

Viewshed: A GIS calculation and display that shows the surface area visible from a given location, considering the visual obstructions that may interfere, such as mountains. Viewshed themes are constructed to show paths and areas of visibility. See figure 9-16.

Visualization: See *Data visualization.*

X-Y coordinates (values). A coordinate system for precise location of a feature. Example: Longitude = X and Latitude = Y.

Z value: The third coordinate, after X and Y, which locates a point in space in terms of elevation (height or depth). Example: the Z value of a point on a contour line indicates its elevation above sea level.

INDEX

E

F

G

H

I

L

M